Lecture Notes in Computer Science 15745

Founding Editors

Gerhard Goos
Juris Hartmanis

Editorial Board Members

Editors
Yingqian Zhang
Eindhoven University of Technology
Eindhoven, The Netherlands

Milan Hladik
Charles University
Prague, Czech Republic

Hossein Moosaei
Jan Evangelista Purkyně University
Ústí nad Labem, Czech Republic

ISSN 0302-9743   ISSN 1611-3349 (electronic)
Lecture Notes in Computer Science
ISBN 978-3-032-09191-8   ISBN 978-3-032-09192-5 (eBook)
https://doi.org/10.1007/978-3-032-09192-5

Preface

This volume contains the peer-reviewed papers presented at the 19th Learning and Intelligent Optimization Conference (LION 19), https://intelligent-optimization.org/lion19/, held in Prague, Czech Republic, from June 15 to 19, 2025.

LION 19 continued the successful series of internationally recognized LION events (LION1: Andalo, Italy, 2007; LION2 and LION3: Trento, Italy, 2008 and 2009; LION4: Venice, Italy, 2010; LION5: Rome, Italy, 2011; LION6: Paris, France, 2012; LION7: Catania, Italy, 2013; LION8: Gainesville, USA, 2014; LION9: Lille, France, 2015; LION10: Ischia, Italy, 2016; LION11: Nizhny Novgorod, Russia, 2017; LION12: Kalamata, Greece, 2018; LION13: Chania, Greece, 2019; LION14 and LION15: Online, 2020 and 2021; LION16: Milos Island, Greece, 2022; LION17: Nice, France, 2023; and LION18: Ischia, Italy, 2024).

LION 19 focused on exploring the intersections of Artificial Intelligence, Machine Learning, and Operations Research, as reflected in the diversity of contributions and keynote presentations delivered by four distinguished researchers:

Keynote 1: **Carola Doerr** (CNRS, Sorbonne University, France), *"Automated selection of black-box optimization algorithms"*
Keynote 2: **Maximilian Schiffer** (Technical University of Munich, Germany), *"Optimization-augmented machine learning pipelines"*
Keynote 3: **Bissan Ghaddar** (Western University, Canada), *"Learning for Nonlinear Optimization Problems"*
Keynote 4: **Jakub Marecek** (Czech Technical University, Czech Republic), *"Fairness in repeated uses of AI systems"*

The conference received 70 full paper submissions, of which 40 were accepted for presentation following a single-blind peer review process in which submissions received on average three reviews each. We would like to take this opportunity to thank all the reviewers for their valuable time and contributions to the review process. Additionally, 36 abstracts were submitted, and 26 were accepted for poster presentations at LION 19. This volume includes the 40 accepted full papers.

We also wish to acknowledge and thank the organizers of the following special sessions:

- *(Deep) Reinforcement Learning in OR Optimization*, organized by Lin Xie (Brandenburg University of Technology), Jinkyoo Park (Korea Advanced Institute of Science and Technology), and Yaoxin Wu (Eindhoven University of Technology)
- *Enhancing Exact Combinatorial Optimization Solvers with Machine Learning*, organized by Quentin Cappart (Polytechnique Montréal)
- *Sustainability in Surrogate Models, Bayesian Optimization, and Parameter Tuning*, organized by Antonio Candelieri (University of Milano-Bicocca) and Laurens Bliek (Eindhoven University of Technology)

– *Learning and Intelligent Optimization for Physical Systems*, organized by Konstantinos Chatzilygeroudis (University of Patras) and Michael Vrahatis (University of Patras)

More than 80 participants from around the world gathered in Prague to engage in a rich and dynamic program.

The LION 19 Best Paper Awards recognized outstanding contributions based on originality, technical quality, and potential impact. The awards, sponsored by Airbnb, included a certificate and a monetary prize.

- **Best Paper:**

  "Multi-Agent Soft Actor-Critic with Coordinated Loss for Autonomous Mobility-on-Demand Fleet Control", by Zeno Woywood, Jasper I. Wiltfang, Julius Luy, Tobias Enders, and Maximilian Schiffer

- **Runners-Up:**

  "Do LLMs Understand Constraint Programming? Zero-Shot Constraint Programming Model Generation Using LLMs", by Yuliang Song and Eldan Cohen

  "Learning to Solve the Min-Max Mixed-Shelves Picker-Routing Problem via Hierarchical and Parallel Decoding", by Laurin Luttmann and Lin Xie

We are deeply grateful to the local organizers from Charles University, Jan Evangelista Purkyně University, and Unicorn University, and to the sponsors: RSJ Investments Group, Airbnb, Inc., and EWG EU/ME – the EURO Working Group on Metaheuristics.

Finally, we thank the local organization team for their dedication and support in making the conference a success.

August 2025

Yingqian Zhang
Milan Hladik
Hossein Moosaei

Organization

General Chairs

Milan Hladik	Charles University, Czech Republic
Hossein Moosaei	Jan Evangelista Purkyně University, Czech Republic

Technical Program Committee Chair

Yingqian Zhang	Eindhoven University of Technology, Netherlands

Steering Committee

Roberto Battiti (Head of the Steering Committee)	University of Trento, Italy
Francesco Archetti	Consorzio Milano Ricerche, Italy
Christian Blum	Spanish National Research Council (CSIC), Spain
Mauro Brunato	University of Trento, Italy
Carlos A. Coello-Coello	CINVESTAV-IPN, Mexico
Clarisse Dhaenens	University of Lille, France
Paola Festa	University of Naples, Italy
Martin Charles Golumbic	University of Haifa, Israel
Youssef Hamadi	Tempero Tech, France
Laetitia Jourdan	University of Lille, France
Nikolaos Matsatsinis	Technical University of Crete, Greece
Panos Pardalos	University of Florida, USA
Mauricio Resende	University of Washington, USA
Meinolf Sellmann	InsideOpt, USA
Yaroslav Sergeyev	University of Calabria, Italy
Dimitris Simos	SBA Research, Austria
Thomas Stützle	University of Brussels, Belgium
Kevin Tierney	Bielefeld University, Germany

Program Committee

Ahmed Kheiri	University of Manchester, UK
Andoni Irazusta Garmendia	Universidad del País Vasco, Spain
André Hottung	Bielefeld University, Germany
Andrea Schaerf	University of Udine, Italy
Antonio Candelieri	University of Milan – Bicocca, Italy
Brian Plancher	Barnard College, USA
Ciriaco D'Ambrosio	University of Salerno, Italy
Constantin Waubert de Puiseau	Bergische Universität Wuppertal, Germany
Dachuan Xu	Beijing University of Technology, China
Dario Landa-Silva	University of Nottingham, UK
David Hartman	Charles University, Czechia
Dinesh Krishnamoorthy	Eindhoven University of Technology, Netherlands
Dionis Totsila	Inria, France
Dmitri E. Kvasov	University of Calabria, Italy
Fabio Amadio	Inria, France
Federico Berto	Korea Advanced Institute of Science and Technology, South Korea
Francesca Guerriero	University of Calabria, Italy
Francesco Morri	Inria, France
Frédéric Saubion	Université d'Angers, France
Giusy Macrina	University of Calabria, Italy
Vladimir Grishagin	Lobachevsky State University of Nizhni Novgorod, Russia
Guillaume Derval	University of Liège, Belgium
Hossein Moosaei	University of Jan Evangelista Purkyně, Czechia
Ilias S. Kotsireas	Wilfrid Laurier University, Canada
Inneke Van Nieuwenhuyse	Hasselt University, Belgium
Ioannis Tsikelis	Inria, France
Jan Kronqvist	KTH Royal Institute of Technology, Sweden
Jayanta Mandi	Katholieke Universiteit Leuven, Belgium
Kevin Tierney	Bielefeld University, Germany
Konstantinos Chatzilygeroudis	University of Patras, Greece
Konstantinos Parsopoulos	University of Ioannina, Greece
Lampros G. Printzios	University of Patras, Greece
Laura Genga	Eindhoven University of Technology, Netherlands
Laurens Bliek	Eindhoven University of Technology, Netherlands
Laurent Moalic	Université de Haute-Alsace, France
Laurin Luttmann	Leuphana Universität Lüneburg, Germany
Léo Boisvert	École Polytechnique de Montréal, Université de Montréal, Canada

Lin Xie	Brandenburgische Technische Universität, Germany
Luca Di Gaspero	University of Udine, Italy
Luca Grillotti	Imperial College London, UK
Marc Sevaux	Université de Bretagne Sud, France
Marie-Eléonore Kessaci	University of Lille, France
Mauro Brunato	University of Trento, Italy
Michael Khachay	Ivannikov Institute for System Programming of the Russian Academy of Sciences, Russia
Michael Römer	Universität Bielefeld, Germany
Milan Hladík	Charles University, Czechia
Miroslav Rada	Prague University of Economics and Business, Czechia
Neil Yorke-Smith	Delft University of Technology, Netherlands
Ornella Pisacane	Università Politecnica delle Marche, Italy
Paola Festa	University of Naples Federico II, Italy
Quentin Cappart	École Polytechnique de Montréal, Université de Montréal, Canada
Renato De Leone	University of Camerino, Italy
Robbert Reijnen	Eindhoven University of Technology, Netherlands
Roberto Battiti	University of Trento, Italy
Serdar Kadioglu	Fidelity Investments, USA
Sicco Verwer	Delft University of Technology, Netherlands
Sigrid Passano Hellan	NORCE Norwegian Research Centre, Norway
Sonia Cafieri	École Nationale de l'Aviation Civile, France
Tatiana Tchemisova	Universidade de Aveiro, Portugal
Thomas Stützle	Université Libre de Bruxelles, Belgium
Tomáš Kroupa	Czech Technical University in Prague, Czechia
Vincenzo Piuri	University of Milan, Italy
Vittorio Maniezzo	University of Bologna, Italy
Yaoxin Wu	Eindhoven University of Technology, Netherlands
Yingjie Fan	Leiden University, Netherlands
Yury Kochetov	Institute of Mathematics of the Siberian Branch of the Russian Academy of Sciences, Russia
Zaharah Bukhsh	Eindhoven University of Technology, Netherlands

Local Organizing Committee

Milan Hladik	Charles University, Czech Republic
Hossein Moosaei	Jan Evangelista Purkyně University, Czech Republic

David Hartman   Charles University, Czech Republic
Martin Černý   Charles University, Czech Republic
Elif Garajová   Charles University, Czech Republic
Matyáš Lorenc   Charles University, Czech Republic
Jaroslav Horáček   Charles University, Czech Republic
Petra Příhodová   Charles University, Czech Republic
Tung Anh Vu   Charles University, Czech Republic
Ludmila Petkovová   Unicorn University, Czech Republic

Sponsors

Contents

Autoregressive RL Approach
for Mixed-Integer Linear Programs

Paul Mingzheng Tang, Moses Hong De Lee, and Hoong Chuin Lau[✉]

School of Computing and Information Systems, Singapore Management University,
80 Stamford Rd, Singapore 178902, Singapore
`hclau@smu.edu.sg`

Abstract. Mixed-Integer Linear Programming (MILP) is a widely used method for modeling combinatorial optimization problems. Due to the NP-hard nature of many of these problems, efforts using machine-learning (ML) have been proposed to generate heuristics to speed up solvers, while maintaining optimality. While there is prior work on using graph neural networks (GNNs) to produce high-quality partial solutions, the methods used are non-auto-regressive which model the prediction of variables as conditionally independent due to concerns with solver runtimes. In this paper, we propose a novel auto-regressive reinforcement learning (RL) framework using GNNs which directly optimize for optimality and solver runtimes. Experimental results show our RL method outperforms the benchmark Predict-And-Search (PNS) method on harder real-world problems (55.7% speedup) with time limits and matches performance on easier problems.

Keywords: Graph Neural Networks · ML for OR · Reinforcement Learning

1   Introduction

Mixed-Integer Linear Programming (MILP) has widespread applications in various industries, from logistics to finance. These problems are combinatorial optimization problems that can be formulated using linear constraints and objective functions. Commercial solvers have been developed to tackle MILPs in general; however, due to scalability issues, there is ample opportunity for data-driven machine learning (ML) methods to learn domain-specific heuristics and solve MILPs more efficiently.

[1,3] is a primal heuristic that obtains a feasible solution for MILPs by assigning all or a subset of variables. Prior ML-based diving approaches often assign a subset of variables in a non-autoregressive manner (i.e., predicting all variable assignments in parallel). These methods frame the task as learning a probability distribution over variable assignments, which is trained to match the distribution of assignments sampled from a pool of reference solutions. A key assumption in

Y. Zhang et al. (Eds.): LION 2025, LNCS 15745, pp. 1–17, 2026.
https://doi.org/10.1007/978-3-032-09192-5_1

these approaches is that variable assignments are conditionally independent of one another.

In contrast, this paper proposes a novel autoregressive constructive heuristic for generating partial solutions to MILPs. By framing the task as learning a policy that maximizes a cumulative reward. Incorporating both the optimality gap and a time limit, we train a reinforcement learning (RL) model that directly optimizes the optimality gap within a specified time budget, closely aligning with real-world requirements.

Additionally, our method leverages the additive nature of partial-solution construction by interleaving a variable propagation step into the process. This reduces the number of variables that need to be predicted, distinguishing our approach from existing methods.

The contributions of this paper are as follows: i) a novel constructive heuristic that directly optimizes for a desired optimality gap within a specified time limit; ii) a variable assignment propagation procedure, interleaved into the RL environment, that reduces the number of assigned variables while maintaining solution feasibility.

2   Related Work

2.1   Graph Neural Networks(GNNs) for MILPs

Before selecting an ML method to solve a MILP, the problem must be represented in a form that downstream models can ingest. MILPs are flexible: we can add or remove decision variables and constraints, and both variables and constraints are permutation invariant. Following [4], we represent a MILP as a bipartite graph. This graph can be processed by GNNs to produce predictions at the graph level, as well as at the variable-node and constraint-node levels. A common architecture [4,6,9,11] is the bipartite Graph Convolutional Network (GCN), which uses separate learnable weights for message passing from constraint to variable nodes and from variable to constraint nodes. Our work extends the bipartite graph input representation and utilizes the bipartite GCN as the underlying architecture.

2.2   Reinforcement Learning (RL) for Combinatorial Optimization

There is extensive literature on applying RL to combinatorial optimization, but most approaches target specific problems–such as the Traveling Salesman Problem (TSP) or the Maximum Independent Set (MIS)–and often rely on hand-engineered decoding techniques [2].

RL-MILP [10] is an RL framework that trains a GNN model to learn a local-search heuristic for mixed-integer programming. In contrast, our approach formulates the task as learning a constructive heuristic that directly optimizes both solution quality and solver runtimes.

3 Preliminaries

3.1 MILP Formulation and Terminology

Formally, MILPs can be formulated as shown in Eq. 1.

$$\arg\min_{x}\{c^{\top}x \,|\, x \in \mathbb{Z}^{q} \times \mathbb{R}^{n-q}, Ax \leq b, l \leq x \leq u\}, \tag{1}$$

where

- x is a solution vector comprising q integer and $(n-q)$ real components,
- $c \in \mathbb{R}^{n}$ is the objective-function coefficient vector,
- $A \in \mathbb{R}^{m \times n}$ is the constraint-coefficient matrix,
- $b \in \mathbb{R}^{m}$ is the right-hand-side (RHS) constant vector,
- $l, u \in \mathbb{R}^{n}$ are the component-wise lower and upper bounds on x.

We adopt the following terminology throughout the paper:

- **LHS**: the left-hand-side linear expression in each constraint (row of Ax).
- **RHS**: the right-hand-side constant term in each constraint (b).
- **Sense**: the inequality symbol ($\leq$, $\geq$, $=$). Although one can standardize a MILP to a single sense (e.g., all constraints as "$\leq$"), we retain each original sense, since different inequalities may require distinct feature-engineering treatments.
- **LB / UB**: the lower and upper bounds on the LHS value given a partial assignment x. We denote these bounds by LB_{x} and UB_{x}, respectively.

3.2 Bipartite Graph Representation

A key development in applying ML techniques to MILP problems has been the introduction of graph-based representations. Gasse et al. [4] proposed a lossless representation of MILPs as bipartite graphs. This formulation captures the structure of a MILP in a way that can be efficiently processed by graph neural networks (GNNs).

In this representation, one set of nodes corresponds to the MILP's variables, while the other represents its constraints. Edges connect each variable to the constraints in which it appears, and the edge weights encode the corresponding coefficients from the constraint matrix. This approach enables the encoding of both the problem's structural information and its numerical data.

The bipartite graph representation offers several advantages:

1. It naturally captures variableâĂŞconstraint interactions.
2. It enables the use of graph-based ML models (e.g., GNNs) that can effectively learn from graph-structured data.
3. It scales to large MILP instances, since the size of the graph grows linearly with the number of variables and constraints.

This graph-based approach has been used in various ML frameworks for MILP solving, including those that predict branching decisions in branch-and-bound algorithms and those that generate initial solutions or guide the search process [4,9,11].

3.3 Reinforcement Learning

The fundamental components of RL include the environment, states, actions, and rewards. Let $\mathcal{S}$ denote the set of possible states, $\mathcal{A}$ the set of possible actions, and $\mathcal{R}$ the set of possible rewards. At each time step t, the agent observes the current state $s_t \in \mathcal{S}$, takes an action $a_t \in \mathcal{A}$, and receives a reward $r_t \in \mathcal{R}$ from the environment. The environment then transitions to a new state s_{t+1} according to a transition probability function $P(s_{t+1}|s_t, a_t)$. The agent's behavior is governed by a policy $\pi : \mathcal{S} \to \mathcal{A}$, which maps states to actions. The agent's goal is to learn a policy π that maximizes the expected cumulative reward.

3.4 MDP Formulation

Previous work commonly predicts all decision variables in a non-autoregressive manner, primarily due to concerns over long training times associated with autoregressive models. As a result, these methods explicitly model the probability distribution of variable assignments under the assumption of conditional independence, as shown in Eq. 2.

$$p_\theta(x|M) = \prod_{i=1}^{n} p_\theta(x_i|M) \tag{2}$$

As noted in previous work, a key limitation of non-autoregressive prediction is its inability to capture multi-modal distributions. In contrast, our approach employs an autoregressive model that avoids the strong assumption that optimal solutions to an MILP are unimodal. We empirically evaluate the benefits of our method on MILP instances known to exhibit multi-modality.

To formulate an autoregressive model, we build upon an idea proposed in the autoregressive ablation study from [11]. The study introduced a fixed number and order of variables to be predicted at each step and explored three different variable ordering strategies: input order, objective coefficient order, and fractionality order. The formulation is presented in Eq. 3.

$$p_\theta(x|M) = \prod_{d=1}^{D} p_\theta(x_d|x_{d-1}, x_{d-2}, ..., x_1; M) \tag{3}$$

Our work proposes to allow the GNN to vary the number and order of variables predicted per step, as well as allow the agent to terminate the partial solution construction process to allow the solver to complete the remaining assignments.

Our approach allows the GNN to vary both the number and order of variables predicted at each step, and enables the agent to terminate partial solution construction early, deferring the remaining assignments to the solver (Fig. 1).

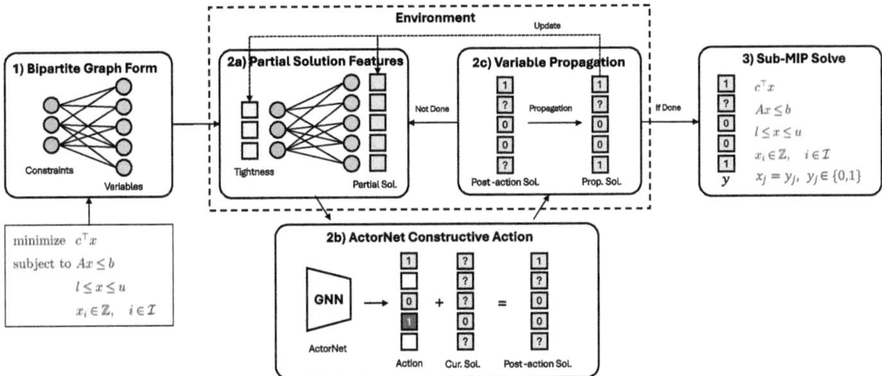

Fig. 1. Overall framework: our approach employs a reinforcement learningâĂŞbased constructive heuristic that predicts a subset of variable assignments at each step. Once a terminal state is reached, the MILP is solved with the partially assigned variables.

4 Methodology

We propose a novel reinforcement learning framework to solve MILPs.

1. **Bipartite Graph Construction.** We begin by converting the MILP into a bipartite graph, with one node set for variables and another for constraints. Edge weights encode the corresponding entries of the constraint matrix A.
2. **Reinforcement Learning Environment.** The core of our approach unfolds as an episodic interaction between an RL agent and the environment, structured into three sub-steps:

   2a) *Partial-Solution Features.* At each time step, the environment extends the bipartite graph representation with the following: (i) a *tightness metric* for each constraint (i.e., how close the constraint is to binding due to the current partial assignment), and (ii) a *partial solution assignment*, initially all "unknown."

   2b) *ActorNet Constructive Action.* The GNN-based *ActorNet* ingests the current bipartite graph with updated features and outputs a selection and assignment of a subset of variables, producing a *post-action partial solution*.

   2c) *Variable Propagation.* Given the newly assigned variables, our propagation algorithm infers additional assignments implied by the post-action partial solution, yielding a *propagated solution*. This solution–if feasible–is used to update both the tightness metric and the partial solution assignment for the next step.

   The agent repeats steps 2a–2c until a termination criterion is met (maximum number of steps reached or infeasible).
3. **Sub-MIP Solve and Reward Generation.** Once the episode terminates, we are able to fix the assignments according to the partial solution generated using one of two methods:

- *Direct Fixing:* apply the post-processing technique of [11] to complete and fix the assignments;
- *Trust-Region:* invoke the trust-region method of [6] to adjust assignments within a local neighbourhood.

Finally, we solve a small "Sub-MIP" to obtain a feasible full solution and compute the episode reward based on the resulting optimality gap and total solve time.

By iteratively selecting variable subsets and propagating implications, the GNN-based ActorNet learns to capture rich conditional dependencies among variables, directly optimizing for both solution quality and solver runtime.

Proximal Policy Optimization (PPO) In our approach, we used the state-of-the-art RL PPO training method, requiring an Actor Network that is used to select actions in the environment, and a Critic Network, used to estimate the value of the state-action pair taken at that specific state. The model architecture is elaborated further in 4.2.

4.1   Environment Design

State Space. We extend the variable and constraint node feature-engineering used in Predict-And-Search [11] to include the additional information introduced by the partial solution condition. This modified bipartite graph representation defines the observed state space for our RL agent.

To add the partial solution condition into the bipartite graph representation $(\mathcal{G}, \mathcal{C}, \mathcal{V})$, we concatenate the additional feature column to $\mathcal{V}$, where 0 and 1 represents a fixed assignment and 2 represents an unknown assignment.

With the inclusion of a partial solution, we enhance the features for the constraints. In this section, we introduce a new feature, *tightness*, which measures the range of possible values the LHS of the constraint can take.

We define *tightness* as shown in Eq. 4

$$\alpha = \frac{\text{RHS} - \text{LB}_x}{\text{UB}_x - \text{LB}_x},$$

$$T_{c,x} = \begin{cases} \alpha, & \text{if c.Sense is } \geq \\ 1 - \alpha, & \text{if c.Sense is } \leq \\ max(\alpha, 1 - \alpha), & \text{if c.Sense is } = \end{cases} \tag{4}$$

The *tightness* feature has the following properties:

- When $T_{c,x} = 1$, it implies that the constraint c is *binding*. This allows us to infer the assignment of these variables given a partial solution, and is used in our variable propagation method.
- When $T_{c,x} \leq 0$, it implies that the constraint c is *redundant* and can be relaxed. We also make use of this in our variable propagation method.
- when $T_{c,x} > 1$, there is no feasible assignment of the remaining unfixed variables for this constraint c.

Action Space. The action space for assigning values to n binary variables is defined as:

$$\mathcal{A} = \{0, 1, 2\}^n,$$

where each action vector $\mathbf{a} = (a_1, \ldots, a_n)$ corresponds to one of three possible assignments for variable i:

0/1: Fix variable i to $a_i = 0$ or $a_i = 1$
  2: Defer assignment (leaving a_i unresolved)

State Transition. In our RL environment, executing action a_t in state s_t yields the next state s_{t+1} and reward r_{t+1}. The transition proceeds as follows:

1. **Action Filtering.** Mask a_t by the set of variables already fixed in s_t, producing a filtered action a'_t. Represent each new assignment in a'_t as a tuple (v, x) where variable v is assigned a value x for each assignment.
2. **Variable Propagation.** By iteratively assigning each v to the value x, we compute the updated bounds of the constraints' LHS which allow us to find implied assignments to other variables and obtain a larger propagated partial solution. The variable propagation routine is described in (Algorithm 1), which
   - uses the *Propagate-Binding* subroutine (Algorithm 2) on binding constraints, and
   - uses the *Propagate-Redundant* subroutine (Algorithm 3) on redundant constraints,
   to infer any additional assignments or detect infeasibility.

Termination Criterion. To prevent the risk of over-assigning variables, which could lead to infeasible solutions, the environment terminates when the number of assigned variables reaches a predetermined proportion. This strategy not only caps excessive assignments but also maintains sufficient flexibility during solve, reducing the likelihood of infeasible outcomes.

Upon triggering the termination condition, the chosen variable assignments are configured and solved using Gurobi. The output from this phase yields metrics that are used in formulating a reward function designed to balance solution quality with solver runtimes (Fig. 2).

Reward Computation. The purpose of the reward is to guide the actor to produce actions that match our desired outcomes. There are 2 main objectives we want to optimize for: 1) optimality and feasibility, 2) solver runtimes, which is affected by solver time.

Additionally, to use Proximal Policy Optimization (PPO) efficiently, the rewards produced by the environment should not be sparse. We introduce an intermediate reward (action reward) to avoid a sparse reward landscape.

The reward function is computed in a lexicographic manner, prioritizing in the following order:

Algorithm 1: Propagate Assignment Algorithm

Input   : Problem P, Current Solution S, Assignments(Dict) A
Output: isFeasible

1 **while** A *is not empty and isFeasible* **do**
2   Variable v, Assignment $x = A$.pop()
3   **if** $S[v] = x$ **then**
4   continue
5   **if** $S[v] = !x$ **then**
6   **return** isFeasible $=$ False
7      $S[v] \leftarrow x$
8   **for** *Constraint C, weight W in v.Constraints* **do**
9   Update $C.lb$ and $C.ub$ given v, w, x
10   isFeasible $=$ PropagateBinding(S, A, C)
11   PropagateRedundant(S, A, C)
12   **if** *not isFeasible* **then**
13   **return** isFeasible $=$ False
14   `#PropagateBinding (PB) and PropagateRedundant (PR) details in Algorithms 2 and 3`
15 **return** isFeasible $=$ True

Algorithm 2: PB Algo

Input   : CurrentSol S,
          Assignments A,
          Constraint C
Output: isFeasible

1 **if** $C.sense$ *is* $\leq$ *and* $C.lb(S) = C.rhs$ **then**
2   Minimize assignments of $v \in C.vars$ if $S[v]$ unassigned.
3 **if** $C.sense$ *is* $\geq$ *and* $C.ub(S) = C.rhs$ **then**
4   Maximize assignments of $v \in C.vars$ if $S[v]$ unassigned.
5 **if** *new assignments conflict with A or S* **then**
6   **return** isFeasible $=$ False
7 Push new assignments to A

Algorithm 3: PR Algo

Input   : CurrentSol S,
          Assignments A,
          Constraint C

1 **if** *($C.sense$ is $\leq$ and $C.ub(S) \leq C.rhs$) or*
2 *($C.sense$ is $\geq$ and $C.lb(S) \geq C.rhs$)* **then**
3   **for** $v \in C.variables$ **do**
4          $v.Degree \leftarrow v.Degree - 1$
5   **if** $v.Degree = 0$ **then**
6   Assign v which minimizes Objective

Fig. 2. Propagate-Binding (PB) and Propagate-Redundant (PR) algorithms.

1. Feasibility (r_f)
2. Optimality (r_o)
3. Speed (r_t)

An exploration reward (r_a) and a step penalty (stepP) are given regardless of the 3-stage reward results above. This is formulated as in Eqs. 5 to 9.

$$r = r_a - \text{stepP} + \begin{cases} r_f, & \text{if infeasible} \\ r_o, & \text{if feasible} \\ r_o + r_t, & \text{if near-optimal} \end{cases} \tag{5}$$

$$r_f = -\frac{\text{numViols}}{\text{numConstrs}} \times \phi_1 \tag{6}$$

$$r_o = (1 - \text{optGap}) \times \phi_2 + \phi_3 \tag{7}$$

$$r_t = \frac{\text{BST} - \text{CST}}{\text{BST}} \times \phi_4 \tag{8}$$

$$r_a = \underbrace{\phi_5 \times \mathbf{1}[\text{num1s} > 0]}_{\text{num 1's reward}} + \phi_6 \times \underbrace{\frac{max(\text{num2s} - \text{solLen}/3, 0)}{2/3 \times \text{solLen}}}_{\text{num 2's reward}} \tag{9}$$

The reward components are as follows:

- r is the **Total Reward**.
- r_a is the **Action Reward**. The action reward functions as an intermediate reward, which provides a positive reward for 2 properties: i) assigning at least one 1; ii) assigning a large number of 2's in the output action. The reason for actions having at least one assignment of 1 is to encourage each action have significant information as problem instance solutions tend to have a sparse number of 1's which are crucial to the optimality of the solution. Additionally, encouraging small actions by providing a positive reward when the number of 2's assigned (deferring the variable assignment) will allow the variable propagation method to mask further actions and provide more information in the next step. Large actions can easily reach infeasible solutions. The choice of normalizing by solLen/3 is to allow the model to learn better solutions from a random actor output (having $\frac{1}{3}$ probability of assigning a 2).
- r_f is the **Feasibility Reward**. The feasibility reward assigns a negative reward when the solution or partial solution is infeasible. It computes the percentage of violated constraints by solving a maximum satisfiable version of the problem instance.
- r_o is the **Optimality Reward**. The optimality reward assigns a positive reward for reducing the optimality gap optGap $= \frac{\text{obj} - \text{BKS}}{|\text{BKS}|}$, scaled by ϕ_2. This includes a fixed reward for obtaining a feasible solution ϕ_3.

- r_t is the **Time reward**. The time reward assigns a positive reward which corresponds to the percentage speed-up obtained from solving the current sub-MIP, which has a current solve runtime CST, compared to the base solve time BST.

The choice of ϕ scales the various components, which we use $[20, 20, 10, 10, 2, 10]$ for ϕ_1 to ϕ_6 respectively. The step penalty stepP is set to $\frac{1}{\text{BST}}$. Lastly, we clip the reward to ensure more stable training for large negative values to -30. Formally, $r = max(r, -30)$.

4.2   Model Training

We employ two primary networks in our framework: the Actor Network and the Critic Network. Additionally, an Observer network is used to enrich the state features that are passed to both the Actor and Critic Networks.

Network Components.

- **Observer Network** builds on the GCN architecture from Predict and Search [6], it is pre-trained via supervised learning for one-step variable solution prediction. As shown in Fig. 3, the Observer (OBS GNN) is integrated into both Actor and Critic networks.
- **Actor Network** is a Probabilistic Actor that uses the enriched state features to select actions for the current step. An additional skip connection carries the current solution directly to the layer immediately preceding the output. This connection helps retain critical information from the previous state's partial solution.
- **Critic Network** In parallel, the Critic network aggregates the state features before processing them through its MLP to predict the Q-value of the state-action pair.

Decoding Method We investigate two distinct decoding strategies for configuring the solver:

- **Neural Diving:** This approach fixes variables directly and selects the optimal solution from k candidate samples.
- **Trust Region Method:** Here, the solver is constrained to search within a predefined radius to balance exploration and exploitation [6].

Additionally, we explore the impact of excluding variables assigned to 0 in the predicted solutions. Potentially, predicting only variables to fix as 1 (while omitting 0s) may suffice and even improve solver performance. By restricting assignments to a subset of promising variables, the solver retains flexibility to explore complementary assignments, provided the selected variables (fixed as 1) are high-quality candidates.

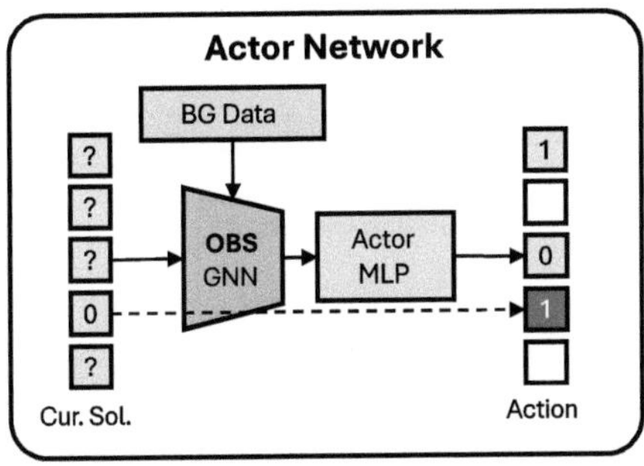

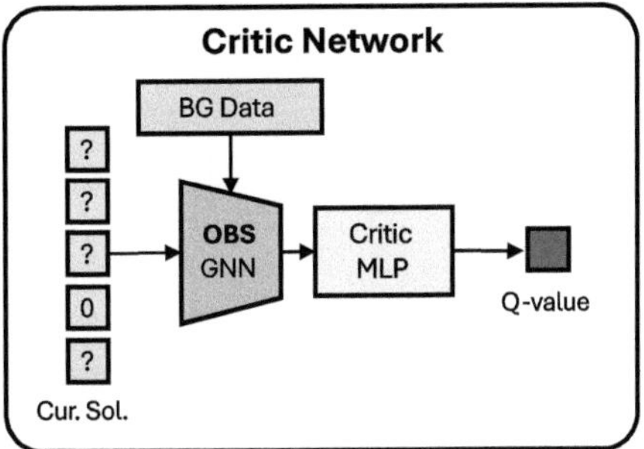

Fig. 3. Actor-Critic Network Architecture: Both the Actor and Critic Network utilize the pre-trained Observer Graph-Neural-Network (OBS GNN) to produce latent graph embeddings of the state features. The Actor's primary goal is to select actions while the critic's role is to evaluate the value of the actions taken within a given state, which are modeled using Multi-Layer Perceptrons (MLP).

Similar to the imitation learning scheme to learn the strong branching rule for a Branch and Bound GNN model in [4], our work uses behavior cloning to warm-start the actor net using an expert heuristic as shown in Fig. 4. The expert heuristic samples actions containing a small fraction of 0's and 1's from a reference solution, and terminates if the size of the partial solution is sufficiently large. Details for the implementation of the expert heuristic is shown in Algorithm 2 in Appendix A.

5   Experiments

5.1   Datasets

We use the following Distributional MIPLIB [8] datasets to evaluate our method, taking 50 instances each (30 for training, 10 for validation, 10 for testing):

1. Set Covering Problem (SC)
2. Combinatorial Auction (CA)
3. Generalized Independent Set Problem (GISP)

We benchmark our RL method against the Predict-and-Search (PNS) approach [6], which trains a one-step, non-autoregressive prediction model using the same trust region decoding strategy.

Preliminaries. To reflect real-world scenarios requiring quick solutions, each dataset was solved as a MILP with a 300 s time limit, from which the top 100 solutions were extracted. Optimal solutions were found for SC, while CA included some suboptimal ones, and GISP had many.

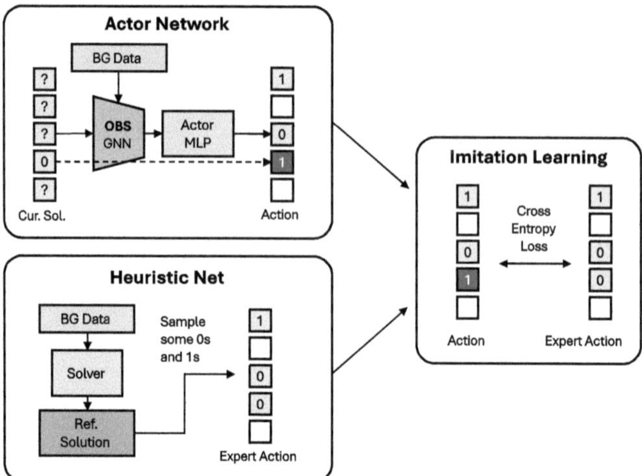

Fig. 4. Expert Heuristic. The expert heuristic iteratively samples a small percentage of 0's and 1's to assign to the current partial solution from a reference solution found in the base solve. Trajectories simulated by this expert heuristic is used to train the actor net using behavior cloning.

Set Covering Problem. We use the Set Covering Problem as an easy benchmark, focusing on medium-difficulty instances from Distributional MIPLIB. Each instance contains 1000 variables and 1000 constraints, with average solve times under 10 s. Although the PNS paper suggests these instances are too easy for the solver, we include them to assess performance on simpler problems.

Combinatorial Auction. For comparison with PNS, we evaluate on the CA Medium dataset using the same trust region parameters, applied to our RL method. The key difference is that our approach uses propagation to predict a much smaller subset of variables.

Generalized Independent Set Problem. To evaluate on a problem closer to real-world instances–where Gurobi is unable to reach an optimal solution within a 300-second time limit–we use the GISP hard instances. GISP can model forest harvesting optimization problems [7] and presents a challenging task even on relatively small Erdős–Rényi graphs with 150 nodes and an edge probability of 0.3. After 300 s, Gurobi reports an optimality gap (between the best bound and best-known solution) ranging from 5% to 20%, providing an opportunity for our method to surpass the BKS.

5.2   Metrics

We report the average primal gap and percentage speedup as primary evaluation metrics, following the PNS paper.

Average Primal Gap. To compare our RL model with the PNS baseline, we use the average primal gap metric, computed against the best known objective value z^* from the 300 s base solve–even if better solutions are later found. To ensure consistency across hardware, solver times are calibrated using a scaling factor, following the Neural Diving paper [11].

Speedup. Speedup offers a more intuitive measure, reporting the percentage of time saved by a method relative to the base solve in reaching a solution with objective value at least as good as the base method's best-known solution (BKS).

Evaluation Configurations. All evaluations are performed with the same configurations. The hardware we used has a one Intel(R) Core(TM) i7-14700HX @ 2.1Ghz with 20 Cores and 28 logical processors, 64GB of RAM and and one NVIDIA GeForce RTX 4070 GPU. For software, we used Gurobi 12.0.0 Gurobi Optimization, LLC (2025) [5] and pyTorch version 2.5.1. [12]

5.3   Results

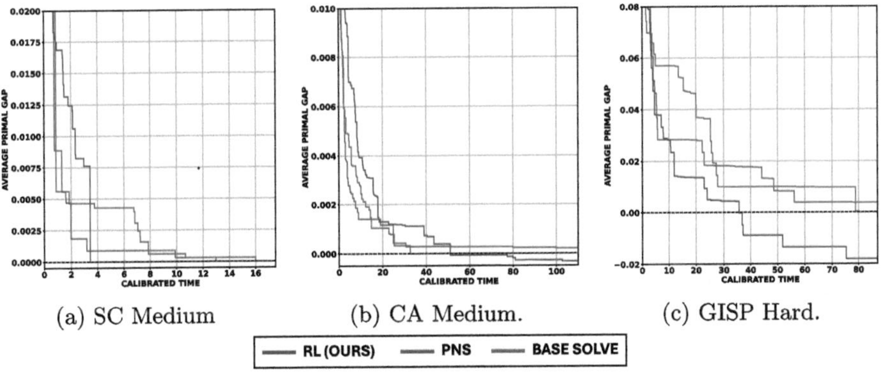

(a) SC Medium   (b) CA Medium.   (c) GISP Hard.

Fig. 5. Average primal gap over calibrated time for SC, CA and GISP datasets. Note that the average primal gap can go below 0 because the BKS objective value used is according to the base solve with a 300 s time limit. A negative gap implies a better solution found compared to the base solve. The graphs displayed are of the best decoding methods - see decoding methods ablations in Appendix B.

As shown in Fig. 5, our RL-based method outperforms the benchmark PNS approach for all three data sets.

SC Medium Dataset. Both methods achieved a 0% optimality gap within the time limit. However, our RL method converged much faster, achieving a 90.5% speedup over the base solver, while PNS showed a more modest 6.5% improvement.

CA Medium Dataset. Our method initially took 22.2% longer to match the base solver's best-known value but later found better solutions, yielding a negative average primal gap of 0.04% within 15.34 calibrated seconds. In contrast, PNS reached within 0.02% of the base value but made no further progress.

GISP Hard Dataset. Our RL method outperformed both benchmarks, reaching the base solver's best-known value 55.7% faster and achieving a better solution with a negative average primal gap of 1.8% in 75.2 calibrated seconds–faster than the base solver's 80.2 calibrated seconds. PNS failed to reach the base solver's best-known value.

Methodological Considerations. The observed under-performance of PNS relative to our base solver, particularly on the combinatorial auction dataset, warrants discussion. Two key factors may explain this discrepancy:

1. **Dataset Scale**: Although we matched the original variable and constraint structure, we trained the PNS model on fewer instances due to computational limitations.
2. **Solver Configuration**: Our experiments used Gurobi's default multi-threaded setting to reflect practical usage, whereas the original PNS results were obtained using single-threaded execution.

Table 1. Average number of predicted ones and zeros and propagated assignments (For RL only) based on the best results shown in Fig. 5.

Problem Instance	PNS		RL			
	Predicted		Predicted		Propagated	
	0's	1's	0's	1's	0's	1's
SC_Medium	500	0	19.5	1.6	0	0
CA_Medium	300	0	15.7	1.2	178.2	0.2
GISP_Hard	600	5	143	3.2	706.9	0

Variable Propagation Analysis. The figures shown in Table 1 present the average number of predicted ones and zeros, as well as propagated assignments, for a terminal partial solution obtained using our RL method. Compared to PNS, the results demonstrate how our method reduces the number of predicted variables needed to construct a high-quality partial solution that directly optimizes for solver speedup.

Note that there are no propagated variables for **SC Medium**. The set covering constraints take the form $\sum_{v \in S_i} v \geq 1$, and each variable typically appears in many constraints. Given this, for `PropagateBinding` to be triggered, all but one variable in the constraint must be assigned a value of 0. For `PropagateRedundant` to be triggered, only one variable in the constraint needs to be set to 1. However, this propagation only relaxes the constraint—it does not lead to direct variable assignment unless the degree of the variable is zero. In contrast, the **CA** and **GISP** constraints follow different forms: $\sum_{v \in S_i} v \leq 1$ for CA, and $x_i + x_j \leq 1$ for GISP.

6 Conclusions and Future Work

In conclusion, this paper presents a novel autoregressive RL framework for a constructive MILP partial solution heuristic, which directly optimizes to obtain solutions within an expected optimality gap and specified time limit.

Although our method outperforms the benchmark PNS method and the base Gurobi solver, it should be noted that there are limitations to our framework. A common issue with running RL algorithms is *catastrophic forgetting*, which we currently avoid by terminating model training at its onset. A possible future direction could be to explore other RL techniques to improve the stability of the training.

Furthermore, further investigation of the use of more sophisticated GNN models, such as Graph Transformers [10], could be used to further improve current performance. This paper limits itself to simple MLPs for downstream networks with concerns regarding inference speed and memory usage.

Lastly, there exist more sophisticated propagation methods (consider the presolve methods utilized in commercial solvers that simplify MILPs) that have the potential to propagate more assignments given the same partial solution. Future work could also consider more complex actions, such as repairing infeasible partial solutions, as described in [13].

Acknowledgement. This research/project is supported by the National Research Foundation, Singapore under its AI Singapore Programme (AISG Award No: AISG2-100E-2023-118).

A   Heuristic Expert Algorithm

Algorithm 2: HeuristicNet Forward Pass

Input   : Partial Solution S, Reference solution R
Output: action
1 coverage $\leftarrow \frac{\sum (S \neq 2)}{|S|}$
2 **if** *coverage* > *coverageThreshold* **then**
3 $\quad \lfloor$ **return** Terminate action

4 $k1 \leftarrow \max(1, \lfloor 0.05 \times |S.ones| \rfloor)$
5 $k0 \leftarrow \max(1, \lfloor 0.005 \times |S.zeros| \rfloor)$

6 sampledOnes $\leftarrow$ RandomSample($S.ones$, $k1$)
7 sampledZeros $\leftarrow$ RandomSample($S.zeros$, $k0$)
8 #Filter sample 0's and 1's by variables already assigned in S
9 action $\leftarrow$ sampledOnes + sampledZeros - S
10 **return** action

B   Decoding Methods Ablation

We perform an ablation study exploring the effectiveness of the PNS's trust region decoding against ND's fixed variable decoding, when applied to our RL method. Specifically, the sub-MIP solve uses the trust region constraint $\sum_{v \in X_0} v + \sum_{v' \in X_1} (1 - v') \leq \delta$ instead of the fixed variable constraints $v =$

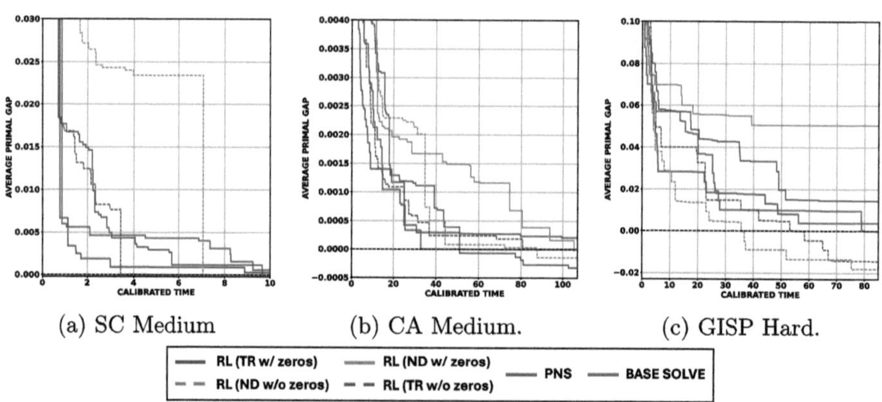

Fig. 6. Average primal gap over calibrated time for set covering, combinatorial auction and generalized independent set datasets, including the various decoding method configurations. ND and TR refer to Neural Diving decoding and Trust Region decoding respectively. w/ zeros and w/o zeros refer to whether the zeros in the terminal partial solution are kept / removed prior to solving.

$0, \forall v \in X_0$ and $v' = 1, \forall v' \in X_1$, where X_0 and X_1 are the assigned 0's and 1's in the partial solution respectively.

We can see from the results in Fig. 6 that for SC and CA datasets the PNS method outperforms the ND method. ND for GISP however can perform well when configured to decode without zeros. Across all 3 datasets, ND with zeros (missing in SC diagram because it performs too poorly) does not perform well, which is expected given the lack of flexibility during prediction.

References

1. Berthold, T.: Primal heuristics for mixed integer programs. Ph.D. Thesis, Zuse Institute Berlin (ZIB) (2006)
2. Berto, F., et al.: Rl4co: an extensive reinforcement learning for combinatorial optimization benchmark. arXiv preprint arXiv:2306.17100 (2023)
3. Eckstein, J., Nediak, M.: Pivot, cut, and dive: a heuristic for 0–1 mixed integer programming. J. Heuristics **13**, 471–503 (2007)
4. Gasse, M., Chételat, D., Ferroni, N., Charlin, L., Lodi, A.: Exact combinatorial optimization with graph convolutional neural networks. Adv. Neural Info. Process. Syst. **32** (2019)
5. Gurobi Optimization, LLC: Gurobi Optimizer Reference Manual (2025). https://www.gurobi.com
6. Han, Q., et al.: A GNN-guided predict-and-search framework for mixed-integer linear programming. In: The Eleventh International Conference on Learning Representations (2023)
7. Hochbaum, D.S., Pathria, A.: Forest harvesting and minimum cuts: a new approach to handling spatial constraints. Forest Sci. **43**(4), 544–554 (1997)
8. Huang, W., Huang, T., Ferber, A.M., Dilkina, B.: Distributional MIPLIB: a multi-domain library for advancing ml-guided MILP methods. arXiv preprint arXiv:2406.06954 (2024)
9. Khalil, E.B., Morris, C., Lodi, A.: MIP-GNN: a data-driven framework for guiding combinatorial solvers. In: Proceedings of the AAAI Conference on Artificial Intelligence, vol. 36, pp. 10219–10227 (2022)
10. Lee, T.H., Kim, M.S.: RL-MILP Solver: a reinforcement learning approach for solving mixed-integer linear programs with graph neural networks. arXiv preprint arXiv:2411.19517 (2024)
11. Nair, V., et al.: Solving mixed integer programs using neural networks. arXiv preprint arXiv:2012.13349 (2020)
12. Paszke, A.E.A.: PyTorch: an imperative style, high-performance deep learning library. In: Wallach, H., Larochelle, H., Beygelzimer, A., d'Alché-Buc, F., Fox, E., Garnett, R. (eds.) Advances in Neural Information Processing Systems, vol. 32. Curran Associates, Inc. (2019)
13. Salvagnin, D., Roberti, R., Fischetti, M.: A fix-propagate-repair heuristic for mixed integer programming. Mathematical Programming Computation (2024). https://api.semanticscholar.org/CorpusID:259922376

Algorithm Configuration in the Unified Planning Framework

Dimitri Weiß[1]([✉]), Andrea Micheli[2], and Kevin Tierney[1]

[1] Decision and Operation Technologies, Bielefeld University, Universitaetsstrasse 25, 33615 Bielefeld, Germany
{dimitri.weiss,kevin.tierney}@uni-bielefeld.de
[2] Digital Industry Center, Fondazione Bruno Kessler, Via Sommarive 18, 38123 Trento, Italy
amicheli@fbk.eu

Abstract. The Unified Planning Framework (UPF) provides convenient access to automated planning technology. It allows for problem formulation independent of a planning engine and the utilization of planners available on the system. However, choosing a suitable parameter configuration of the planning engine for a given problem constitutes a significant challenge. Manually finding a high-quality configuration requires domain knowledge and a considerable time investment, contradicting the intended ease-of-use of the UPF. This issue is addressed by Algorithm Configuration (AC) techniques, which aim to automatically find high-quality configurations. Algorithm runtime as well as quality of solutions found by the parameterized algorithm have been shown to be improved by AC methods in wide-ranging problem settings, which includes planning. We integrate three state-of-the-art AC methods into the UPF and perform AC runs with planning engines which are integrated in the UPF. To this end, we perform AC in runtime, solution quality and anytime planning scenarios on problem instance sets from several International Planning Competitions (IPC). We demonstrate that AC methods provide performance improvements for the IPC.

Keywords: Automatic algorithm configuration · unified-planning framework · planning

1 Introduction

The Unified Planning Framework (UPF) [15] opens planning to a wider group of users by simplifying the use of planning technology in several ways. It offers the possibility to formulate planning problems in a simple, planner-independent way, by using Python data structures and provides a standardization of the possible interactions with planning engines. Furthermore, it allows reading in planning problems from files and converting them to its native format. Alongside the possibility to transform the problem in several ways, e.g. grounding, it is

Y. Zhang et al. (Eds.): LION 2025, LNCS 15745, pp. 18–30, 2026.
https://doi.org/10.1007/978-3-032-09192-5_2

possible to let the UPF choose which planning engine to use with a given planning problem.

If the planning engine chosen by the UPF is not up to task for the problem at hand, e.g., due to the planner not solving problems fast enough, the next step is to adjust it by setting its parameters. Finding a suitable parameter configuration, however, requires domain knowledge and can require significant time cost and effort. Since the intention of the UPF is to make it easier for non-experts to use planning technology, this poses a contradiction to the goals of the UPF. However, we can overcome this limitation by means of algorithm configuration (AC), thus automating the task. AC methods including CALIBRA [1], ParamILS [12], GGA/OAT [3,16], Irace [14], SMAC [11], ReACT/ReACTR [6,7] and CPPL [4] have been shown to effectively improve the performance of algorithms in operations research. Additionally, AC has already been applied successfully in planning. The performance of the planner Fast-Downward [10] has been shown to be significantly improved on IPC domains from 2011 through applying AC [19]. FD-Autotune [5] successfully integrates ParamILS with Fast-Downward and places its performance in the upper half of the IPC 2011 contenders. Performance optimization through AC can target different performance metrics, e.g., algorithm runtime or solution quality. The primary contribution of this paper is the integration of three of the state-of-the-art AC methods in the UPF to allow for automated AC in several planning scenarios as well as a demonstration of their application.

This paper is organized as follows. Section 2 defines AC and outlines related work in planning. In Sect. 3 we describe which AC methods are integrated, which planning engines are targeted and how it is achieved. Section 4 describes an experimental setup and its results, which demonstrate the impact of AC in the UPF. Finally, we conclude and discuss possible future work with the AC integration in Sect. 5.

2   AC in Planning

AC offers the possibility to improve performance of parameterized algorithms. Planners are parameterized in many cases and can be adjusted to improve their performance by configuring their parameter settings. Note that AC approaches and approaches adjacent to AC were investigated in planning literature and partially implemented in planning applications. However, this branch of research was not pursued further since. In this section, we first define the underlying AC problem and then outline related work in the planning community.

2.1   Automated Planning

Automated planning is the problem of synthesizing a course of actions to achieve a desired objective (called a "goal"), given a formal description of a system to be controlled [9]. Over the years different flavors of planning have been proposed, depending on the assumptions on the system specification and the kind of plan being synthesized. For the sake of this paper, we focus on classical planning problems, where the system is described using a finite set of Boolean variables (called predicates) and the actions are deterministic.

2.2 AC Problem

The AC problem underlying the methods integrated in the UPF is in the offline setting. To define the problem, we follow the formal notation introduced in [18]. Given a parameterized algorithm $\mathcal{A}$ and a set of problem instances $i \in \mathcal{I}$, the objective is to find a single high-quality configuration. $\mathcal{A}$ has configurable parameters $p_1, \ldots, p_k$, with parameters p_i having a domain Θ_i, such that $\Theta \subseteq \Theta_1 \times \ldots \times \Theta_k$ defines the search space of feasible configurations. $\mathcal{I}$ represents the space of problem instances over which the probability distribution $\mathcal{P}$ is defined. Furthermore, the problem instances i can optionally be described by features $f_{i,1}, \ldots f_{i,d}$ in vectors $\boldsymbol{f}_i \in \mathbb{R}^d$. A cost function from the cost function space $\mathcal{C}$, denoted as $c : \mathcal{I} \times \Theta \to \mathbb{R}$, represents the performance metric. The goal in offline AC is to find $\boldsymbol{\theta}^* \in \Theta$ which minimizes the cost function

$$\boldsymbol{\theta}^* \in \operatorname*{argmin}_{\boldsymbol{\theta} \in \Theta} \int_{\mathcal{I}} c(i, \boldsymbol{\theta}) d\mathcal{P}(i).$$

Since the distribution $\mathcal{P}$ over $\mathcal{I}$ is unknown, an alternative problem is solved in practical applications. Using a set of training instances $\mathcal{I}_{\text{train}} \subseteq \mathcal{I}$ the aggregation function $m : \mathcal{C} \times 2^{\mathcal{I}} \times \Theta \to \mathbb{R}$ is minimized to approximate $\boldsymbol{\theta}^*$. Typically, m is calculated as the arithmetic mean or a comparable function by evaluating $\hat{\boldsymbol{\theta}}$ with $\mathcal{I}_{\text{train}}$. The objective in practical applications becomes

$$\hat{\boldsymbol{\theta}} \in \operatorname*{argmin}_{\boldsymbol{\theta} \in \Theta} m(c, \mathcal{I}_{\text{train}}, \boldsymbol{\theta}).$$

The cost function c, approximated by m, in offline AC can represent different performance metrics of the algorithm $\mathcal{A}$. In this work, we target the runtime, plan quality and the plan quality of the first plan found during anytime planning.

2.3 Configuration of Planners

Configuration of planners and the resulting performance improvement has been investigated in the planning community. The planner LPG [8], which is included in the UPF, is configured to improve performance on specific planning domains in [22]. In this work, the AC method used is ParamILS, and it significantly improves the performance of the planner on IPC 2011 domains. Another planner included in the UPF, Fast-Downward [10], is shown to improve performance on IPC 2011 domains using the AC method ParamILS in [19]. Fast-Downward and ParamILS are unified into FD-Autotune as described in [5]. This planner system participates numerous times in the IPC. In later work, Fast-Downward is configured by SMAC in [20]. Both, LPG and Fast-Downward are included in the Algorithm Configuration Library (AClib) [13]. AClib is a benchmark library commonly used for testing and development purposes in AC. We extend the idea of configuring planners by applying several state-of-the-art AC methods to a wider range of planning engines and integrating it in the UPF.

3 AC in the UPF

We enable the use of AC in the UPF by a generic AC interface. Based on this interface, four AC methods are integrated and target six of the planning engines available in the UPF. Our approach also allows integrating further AC methods and planning engines in the UPF[1]. In the following, we describe the AC interface, the included AC methods and the planning engines that can be configured.

3.1 Integration in the UPF

The generic interface manages information flows between planning engines and AC methods, as depicted in Fig. 1. The AC methods are initialized with an AC scenario, which includes planning engine name and minimization metric, parameter space definitions, problem instance sets, maximum allowed runtime for an evaluation, penalty costs for not solving a problem instance, AC runtime, number of available CPUs and tournament configuration (if applicable). We implement pre-defined parameter spaces for the planners included in the experiments, but a custom definition can be loaded from files in the PCS format [13], a standard format for describing algorithm parameters, or be formulated in code by the user.

During the AC process, AC methods suggest parameter settings which need to be evaluated by running the configured planner on problem instances. The AC method also chooses which problem instances from the sets the configuration needs to be evaluated on. The interface converts the configurations to a format suitable for the planning engine and calls the planning engine with the problem instance. If the planning engine produces a plan within the given time limit, its output is converted to a format suitable for the AC method in use and passed to it. The AC interface terminates planning engines should the time limit be reached while the planning engine still runs.

3.2 Integrated AC Methods

We integrate three state-of-the-art methods into the UPF. The first method, Irace [14], employs the racing principle to effectively utilize resources. In runtime optimization, configuration runs can be capped before the time limit is reached. In this way, configurations exhibiting high runtime are not evaluated for a longer time period than necessary. Furthermore, Irace includes forbidden and conditional parameter value constraints into its parameter space search, which prevents the evaluation of non-functional configurations. This is also the case for the next AC method, SMAC [11], which is a Bayesian Optimization approach to AC. SMAC includes forbidden and conditional parameter value constraints and is additionally capable to include problem instance features in the AC process to assist it. We integrate the Optano Algorithm Tuner (OAT)

---

[1] Our approach is available via pip at: https://github.com/DimitriWeiss/up-ac, an installation guide and documentation at: https://up-ac.readthedocs.io/en/latest/.

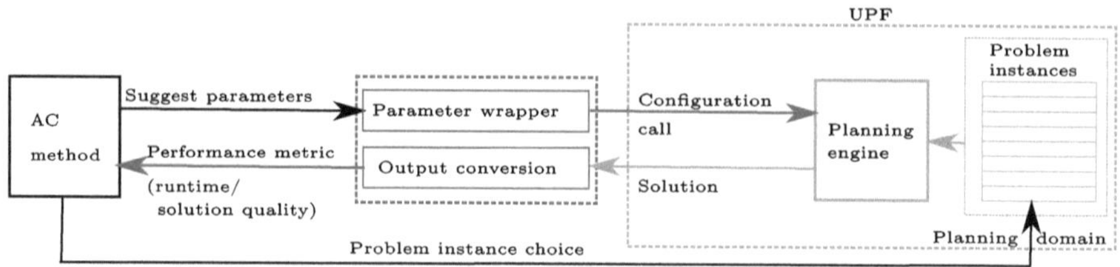

Fig. 1. Information flow between AC method, interface and planning engine.

[16] into the UPF. This AC method provides multiple AC approaches combined. OAT is based on the GGA [3] configurator, which uses a genetic algorithm variant that exploits dependencies between parameters. Finally, we integrate the ensemble-based algorithm configurator Selector, described in [23] which is currently under revision. The ensemble includes functionalities and models derived from GGA, SMAC and the realtime algorithm configurator CPPL. The Python iplementation of this method is available under the name "selector-ac" on PyPI.

3.3   Targeted Planners

The choice of planning engines to include is based on whether they are integrated in the UPF and whether they can be configured. The version of the UPF used in this work includes eight planning engines, two of which (Aries and FMAP) are not integrated with the AC interface. The planning engine Aries does not have parameters, which excludes it from AC. The multi-agent planner FMAP is not integrated because not enough multi-agent problem instances could be found for testing during the implementation.

We integrate the remaining six planning engines with the interface. The planner LPG [8] is based on local search and planning graphs. The parameter space of this planning engine includes numerous parameters only applying to specific planning domains, hence, we use a reduced parameter space definition both in the experiments in Sect. 4 and in the implementation. Fast-Downward [10] is a classical planning engine. It is based on heuristic search and has a unique and complex parameter space. The parameters are not set by assigning values to individual parameters but are rather nested functions passed as a single string. Due to this fact, we also use a fraction of the possible parameter space with Fast-Downward. The planning engine SymK is a classical optimal and top-k planner based on symbolic search, which extends Fast-Downward. Since it has the same parameter space, we handle it as described for Fast-Downward. The remaining three planning engines all have a parameter space of two parameters. ENHSP [17] is an expressive numeric heuristic planner. It is capable of processing a range of domains wider than any other planning engine integrated. The planning engine Tamer [21] is based on heuristic search and a satisfiability modulo theory solver. Pyperplan [2] is based on search heuristics. While it is intended to be used as a teaching or prototyping tool and does not offer state-of-the-art performance, we nonetheless integrate it with the AC interface.

Table 1. Planning domains from the IPC 2014, 2018, and 2023 used in the experiments.

IPC	2014	2018	2023
Domain	Barman	agricola	folding
	CaveDiving	caldera	labyrinth
	Childsnack	caldera-split	quantum-layout
	CityCar	data-network	recharging-robots
	Floortile	flashfill	ricochet-robots
	GED	nurikabe	rubiks-cube
	Hiking	organic-synthesis	slitherlink
	Maintenance	organic-synthesis-split	
	Openstacks	settlers	
	Parking	snake	
	Tetris	spider	
	Thoughtful	termes	
	Transport		
	Visitall		

4   Experimental Results

We conduct several experiments to demonstrate the possible performance gain of
the planning engines in the UPF through AC. The experiments include runtime,
plan quality and anytime planning optimization scenarios. The runtime or the
quality metric of a problem domain, either for the final or an intermediate plan,
are minimized for each planning engine. In the anytime optimization scenario,
the plan quality is also optimized. However, it is the quality the first intermediate plan that is returned by the planning engine within the time limit that is
optimized. For this, we use a custom function to compute the plan quality metric
that is minimized as per the domain file, or the plan length if no minimization
metric is specified. We use this function to compute plan quality for Pyperplan
and Tamer, since both planning engines do not output plan quality metrics. All
planning engines are configured in all scenarios, except for Pyperplan and Tamer
in the anytime scenario, since no anytime planning is implemented in the unified
planning framework for these planners.

4.1   Problem Domains

For the experiments, we use problem instances derived from the IPC to obtain
a challenging and diverse set of instances. The problem instances stem from the
classical satisfiability tracks of the years 2014, 2018 and 2023. We note that the
planners ENHSP and Tamer are numeric and temporal planners, respectively,
thus they are strictly more expressive than the other planners, and we expect
that they will underperform the other planners on these sets.

Every domain listed in Table 1 entails 20 problem instances, which amounts to 660 problem instances in total. We choose these instances because the planning problem underlying the instances can be used in runtime, plan quality and anytime planning optimization scenarios alike. Besides this, problem instances from the IPC are designed to be challenging, which fits the use case for AC.

Table 2. Numbers of instances available after filtering per planner.

Planner	Fast-Downward	SymK	LPG	ENHSP	tamer	pyperplan
Train instances	260	280	240	380	100	80
Test instances	140	120	120	180	60	40
$\sum$ Instances	400	400	360	560	160	120

4.2 Filtering Problem Instances

The planning algorithms contending in the IPC aim to solve as many of the problem instances of the track that they participate as quickly or to as good a solution quality as possible. However, it is not given that the planning engines are designed to perform best on exactly the selection of domains we use in the experiments. The UPF automatically evaluates whether the characteristics of a domain match the capabilities of a planning engine or not. Additionally, the UPF performs a sanity check of a problem definition when loading problem instances and domains. Both procedures exclude problem instances from use with a planner, if necessary.

We determine the problem instance sets to be used in the experiments as follows. First, the domains in Table 1 are filtered through the sanity check of the UPF. This excludes five domains, which results in 560 problem instances left. Next, the domains are filtered by the planning engine's capabilities, which reduces the set for all planning engines, except ENHSP, even further. For the AC process, we split the problem instances into training and test sets by domains. We roughly use a third of the domains for the test sets. The resulting numbers can be seen in Table 2.

4.3 Performance Gains on the Test Sets

All AC runs are performed on compute nodes with AMD Milan 7763 processors with 128 cores running at 2.45 GHz. The allocated AC runtime is 24 h. We choose this runtime due to the small parameter space of only two parameters some of the planners and to avoid overfitting on some of the training sets that are small compared to typical AC settings. We evaluate the performances of the planning engines in all scenarios with the default configurations and the configurations found by the AC methods three times each. Evaluations in the runtime scenario reaching the time limit are recorded as running for 300 s, while evaluations not finding results within the time limit in the quality and anytime planning scenarios are recorded as values of 10,000. The test and training instance sets used in the three scenarios are the same. The results on the test sets are listed as averages in Tables 3, 4 and 5 for comparison.

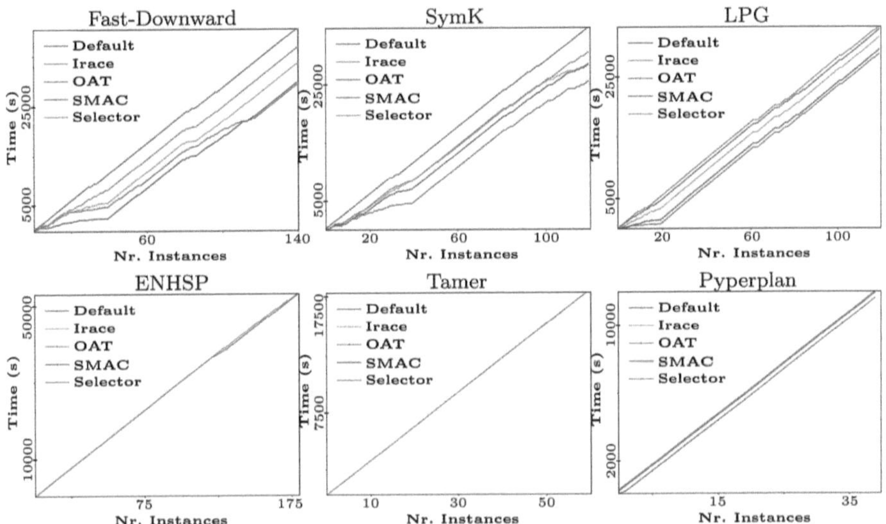

Fig. 2. Instance runtimes.

Table 3. Average performances in runtime configuration scenario.

Planner	Performance				
	Default	Irace	OAT	SMAC	Selector
Fast-Downward	288.77	238.55	208.50	211.37	261.09
SymK	288.73	236.90	211.55	235.58	252.95
LPG	181.89	174.89	159.11	163.30	184.42
ENHSP	293.34	294.41	293.42	293.34	293.41
Tamer	300.00	300.00	300.00	300.00	299.80
Pyperplan	290.28	300.00	298.07	299.87	300.00

Overall, the AC methods provided configurations partly reducing the runtimes of the planning engines, with few exceptions, as illustrated in Table 3. The potential time saving of using configured planners is depicted in Fig. 2. It is noteworthy that Selector provided a configuration performing worse than the

Table 4. Average performances in quality configuration scenario.

	Performance				
Planner	Default	Irace	OAT	SMAC	Selector
Fast-Downward	9530.28	6494.64	6510.64	6366.52	6511.64
SymK	8084.60	5296.35	4956.90	5265.79	5183.90
LPG	5619.19	6666.66	6432.89	6486.30	6006.66
ENHSP	9668.21	10000.00	9668.21	9723.34	9668.21
Tamer	10000.00	9837.45	10000.00	10000.00	9891.54
Pyperplan	9504.18	10000.00	10000.00	9504.18	10000.00

Table 5. Average performances in anytime planning configuration scenario.

	Performance				
Planner	Default	Irace	OAT	SMAC	Selector
Fast-Downward	9530.28	7387.12	6541.05	6562.22	6662.98
SymK	9319.27	9371.10	7181.87	9400.87	6418.85
LPG	8421.63	8437.77	8867.95	8443.15	8380.64
ENHSP	9723.52	9723.52	10000.00	9889.27	9723.52

default configuration. Due to it being an ensemble method and being comprised of multiple models, we assume that the low AC runtime allocation leads to too few evaluation cycles for Selector to perform well with a decent sized parameter space. On the other hand, Selector provided a configuration for Tamer that at least solved one problem instance within the time limit, contrasted to all other configurations for Tamer in this scenario. However, it can not be excluded that the solution of this problem instance was not due to the heuristic nature of Tamer and more of a coincidence. Similarly, the performance of ENHSP could not be significantly improved. Note that all AC methods, but Irace, concluded with the default configuration. The runtime of Pyperplan, however, was degraded in this experiment. Due to Pyperplan having a parameter space of only two parameters and the experiment running on the smallest instance set size, we conclude that this is a case of overfitting to the training set.

The results in the quality scenario mirror the results of the runtime scenario, as illustrated in Table 4. In this scenario, however, the performance of LPG could not be improved, likely due to the AC runtime being too short.

The results of the anytime scenario are shown in Table 5. While the performance of Fast-Downward could be improved by all four methods, finding a configuration of the other planners which yields initial plans with lower quality metrics is a challenge. The success of AC, however, is higher in this scenario. This can be explained by a higher amount of evaluations in this scenario, due to intermediate plans being found quicker than final plans. For ENHSP, Irace and

Selector provided the default configuration, while the configurations from OAT and SMAC resulted in worse performance.

Table 6. Average performances in runtime configuration scenario on training sets.

	Performance				
Planner	Default	Irace	OAT	SMAC	Selector
Fast-Downward	279.86	215.42	220.08	220.49	219.63
SymK	280.68	235.22	206.83	235.95	245.55
LPG	150.02	139.42	149.94	142.66	158.27
ENHSP	247.65	275.03	247.85	248.32	248.07
Tamer	246.02	242.79	231.17	232.65	231.47
Pyperplan	286.36	285.36	285.79	285.48	285.35

4.4 Performance Gains on the Training Sets

Performance gains on training sets provide further clarification about the results. The results illustrated in Table 6 confirm overfitting of small parameter space planners, e.g. very clearly for Tamer. The performance on the training set is improved, while there was no improvement on the test set. The rest of the results is in accordance to the results of the test sets.

On the training set of the quality scenario, illustrated in Table 7, The effect of the separation of domains between training and test set for small instance sets and small parameter spaces is illustrated with Pyperplan. The performance of different configurations is barely distinguishable to the default configuration on the training set but differs significantly on the test set, which amounts to unusable feedback for the AC methods. This also affects the other planning engines with only two parameters in an attenuated form.

Table 7. Average performances in quality configuration scenario on training sets.

	Performance				
Planner	Default	Irace	OAT	SMAC	Selector
Fast-Downward	9335.46	7435.61	6239.51	6730.42	6603.71
SymK	9889.02	7608.19	7044.91	7634.79	7570.87
LPG	5375.75	5957.98	5575.07	5088.04	5714.65
ENHSP	8080.96	9660.22	8107.26	8349.58	8107.27
Tamer	8108.93	7412.46	10000.00	8009.35	7412.46
Pyperplan	9376.98	9376.85	9376.85	9376.98	9376.98

The results in Table 8 mirror the results on the test sets. This can be explained by the higher amount of evaluations in the anytime optimization scenario, since an intermediate plan is generated quicker than a final plan, as is the criterion for solving an instance in the quality as well as the runtime scenario.

Table 8. Average performances in anytime planning configuration scenario on training sets.

Planner	Performance				
	Default	Irace	OAT	SMAC	Selector
Fast-Downward	9284.41	6922.65	6908.66	7022.21	7010.48
SymK	9359.49	9359.45	7527.87	9383.23	6842.50
LPG	8364.21	8337.44	9462.92	8336.96	8334.08
ENHSP	8107.27	8107.27	9340.31	8806.83	8107.26

5   Conclusion and Future Work

The results of the three AC scenarios show that AC can improve performance of planning algorithms. The impact of AC, however, is dependent on an adequate amount of resources allocated to the AC process. Planning engines with small parameter spaces can lead to overfitting, while planning engines with a bigger parameter space need to be configured for a reasonable amount of time. The computational effort of applying AC clearly outweighs the savings in runtime on the instance sets used in the experiments. The value, however, lies in improved performance of an algorithm in a competition or in the reoccuring runtime savings of a configured algorithm applied to further problem instances in practice. The same is true for quality and anytime quality. The results with Fast-Downward and SymK demonstrate the utility of AC in adequate settings. Given that the parameter spaces of Fast-Downard and SymK were reduced due to their vastness, it cannot be ruled out that parameters with high impact on the performance of the planning engines were not considered in the experiments. It is possible that higher performance gains can be achieved through AC with these two planning engines. We note that the AC runtime certainly needs to be increased accordingly since a larger parameter space would need to be searched. The experiments with ENHSP included the biggest instance set and demonstrated that the default configuration already generalizes over the diverse set of domains. Taking into account that the problem instance sets are very challenging and different domains were evaluated in testing than were used in training, the results are adequate.

Future work can be done with the integration of further AC methods' functionalities. For example, OAT includes a gray-box functionality that is not yet integrated in the UPF. Applying gray-box AC allows for capturing and processing intermediate algorithm output to learn about configurations and prematurely terminate their runs. This allows for more evaluations in the same allocated AC runtime and proved to further improve the AC process [24].

Acknowledgments. This work was partially supported by the AIplan4eu Grant Agreement number: 101016442 - AIPlan4EU-H2020-ICT-2018-20 / H2020-ICT-020-2, in Track A Agreement Number 1961741. The authors gratefully acknowledge the computing time provided to them on the high-performance computer Noctua 1 at the NHR Center PC2. These are funded by the Federal Ministry of Education and Research and the state governments participating on the basis of the resolutions of the GWK for the national high-performance computing at universities (www.nhr-verein.de/unsere-partner). Andrea Micheli was supported by the STEP-RL project funded by the European Research Council under GA n. 101115870.

Disclosure of Interests. The authors have no competing interests.

References

1. Adenso-Díaz, B., Laguna, M.: Fine-tuning of algorithms using fractional experimental designs and local search. Oper. Res. **54**, 99–114 (2006)
2. Alkhazraji, Y., et al: Pyperplan (2020). https://doi.org/10.5281/zenodo.3701399
3. Ansótegui, C., Sellmann, M., Tierney, K.: A gender-based genetic algorithm for the automatic configuration of algorithms. In: Principles and Practice of Constraint Programming, pp. 142–157 (2009). https://doi.org/10.1007/978-3-642-04244-7_14
4. El Mesaoudi-Paul, A., Weiß, D., Bengs, V., Hüllermeier, E., Tierney, K.: Pool-based realtime algorithm configuration: a preselection bandit approach. In: Lecture Notes in Computer Science, vol. 12096, pp. 216–232. Springer International Publishing (2020). https://doi.org/10.1007/978-3-030-53552-0_22
5. Fawcett, C., Helmert, M., Hoos, H., Karpas, E., Röger, G., Seipp, J.: FD-Autotune: domain-specific configuration using fast downward. In: ICAPS 2011 Workshop on Planning and Learning, pp. 13–17 (2011)
6. Fitzgerald, T., Malitsky, Y., O'Sullivan, B.: ReACTR: realtime algorithm configuration through tournament rankings. In: International Joint Conferences on Artificial Intelligence Organization (IJCAI), pp. 304–310 (2015)
7. Fitzgerald, T., Malitsky, Y., O'Sullivan, B.J., Tierney, K.: ReACT: real-time algorithm configuration through tournaments. In: Annual Symposium on Combinatorial Search (SoCS) (2014)
8. Gerevini, A., Serina, I.: LPG: a planner based on local search for planning graphs with action costs. In: Conference: Proceedings of the Sixth International Conference on Artificial Intelligence Planning Systems, pp. 13–22 (01 2002)
9. Ghallab, M., Nau, D.S., Traverso, P.: Automated planning - theory and practice. Elsevier (2004)
10. Helmert, M.: The fast downward planning system. J. Artif. Int. Res. **26**(1), 191–246 (2006)
11. Hutter, F., Hoos, H.H., Leyton-Brown, K.: Sequential model-based optimization for general algorithm configuration. In: Learning and Intelligent Optimization (LION), pp. 507–523 (2011)

12. Hutter, F., Hoos, H.H., Leyton-Brown, K., Stützle, T.: ParamILS: an automatic algorithm configuration framework. J. Artif. Intell. Res. (JAIR), 267–306 (2009)

13. Hutter, F., et al.: ACLIB: a benchmark library for algorithm configuration. In: International Conference on Learning and Intelligent Optimization (LION), pp. 36–40 (2014). https://doi.org/10.1007/978-3-319-09584-4_4

14. López-Ibáñez, M., Dubois-Lacoste, J., Stützle, T., Birattari, M.: The irace package: iterated racing for automatic algorithm configuration. Operat. Res. Persp. **3**, 43–58 (2016). https://doi.org/10.1016/j.orp.2016.09.002

15. Micheli, A., et al.: Unified planning: modeling, manipulating and solving AI planning problems in Python. SoftwareX **29**, 102012 (2025). https://doi.org/10.1016/j.softx.2024.102012, https://www.sciencedirect.com/science/article/pii/S2352711024003820

16. OPTANO: OPTANO Algorithm Tuner documentation (2022). https://docs.optano.com/algorithm.tuner/current/

17. Scala, E., Haslum, P., Thiebaux, S., Ramirez, M.: Interval-based relaxation for general numeric planning. In: Proceedings of the Twenty-Second European Conference on Artificial Intelligence, pp. 655–663. ECAI'16, IOS Press, NLD (2016). https://doi.org/10.3233/978-1-61499-672-9-655

18. Schede, E., Brandt, J., Tornede, A., Wever, M., Bengs, V., Hüllermeier, E., Tierney, K.: A survey of methods for automated algorithm configuration. Journal of Artificial Intelligence Research **75**, 425–487 (10 2022).https://doi.org/10.1613/jair.1.13676

19. Seipp, J., Braun, M., Garimort, J., Helmert, M.: Learning portfolios of automatically tuned planners. In: Proceedings of the International Conference on Automated Planning and Scheduling, vol. 22, no. 1, 368–372 (2012). https://doi.org/10.1609/icaps.v22i1.13538, https://ojs.aaai.org/index.php/ICAPS/article/view/13538

20. Seipp, J., Sievers, S., Hutter, F.: Fast downward smac. In: IPC 2014 (2014). https://api.semanticscholar.org/CorpusID:8317657

21. Valentini, A., Micheli, A., Cimatti, A.: Temporal planning with intermediate conditions and effects. Proc. AAAI Conf. Artif. Intell. **34**, 9975–9982 (2020). https://doi.org/10.1609/aaai.v34i06.6553

22. Vallati, M., Fawcett, C., Gerevini, A., Hoos, H., Saetti, A.: Generating fast domain-specific planners by automatically configuring a generic parameterised planner. In: ICAPS 2011 (2011)

23. Weiss, D., Schede, E., Tierney, K.: Selector: Ensemble-based automated algorithm configuration (2025). under revision

24. Weiss, D., Tierney, K.: Realtime gray-box algorithm configuration using cost-sensitive classification. In: Annals of Mathematics and Artificial Intelligence, pp. 1–22 (2023). https://doi.org/10.1007/s10472-023-09890-x

Learning to Repair Infeasible* Problems with Deep Reinforcement Learning on Graphs

M. Zouitine[1,2]($\boxtimes$), A. Berjaoui[1], A. Lagnoux[2,3], C. Pellegrini[2,3], and E. Rachelson[4]

[1] IRT Saint Exupéry, Toulouse, France
{mehdi.zouitine,ahmad.berjaoui}@irt-saintexupery.com
[2] Institut de Mathématiques de Toulouse, UMR5219, Université de Toulouse, CNRS, UT2J, F-31058 Toulouse, France
{agnes.lagnoux,clement.pellegrini}@math.univ-toulouse.fr
[3] UT3, F-31062 Toulouse, France
[4] ISAE-SUPAERO, Toulouse, France
emmanuel.rachelson@isae-supaero.fr

Abstract. In the last few years, deep learning has demonstrated significant potential in Operations Research across various tasks. In this work, we tackle the problem of repairing infeasible constraint satisfaction problems by introducing a novel deep reinforcement learning approach. Our method leverages graph deep learning to represent infeasible problems, utilizing a graph representation of Constraint Satisfaction Problems. By employing bipartite graph neural networks to encode the constraints of these problems, we train a deep learning agent to identify and extract a subset of constraints that restores feasibility solely from the reward signal, requiring no labeled data. We evaluate our approach using several bipartite graph neural network architectures and demonstrate its effectiveness in two domains: maximizing feasibility in Linear Programs and maximizing satisfiability in Boolean satisfiability problems. Our results show that the agent is competitive with existing heuristics in both solution quality and computational efficiency across these domains. An open source implementation of our methods is available at https://github.com/MehdiZouitine/Learning_to_repair_infeasible_problems_with_DRL_and_GNN.

Keywords: Infeasibility Analysis · Graph Neural Networks · Learning Based Heuristics · Linear Feasibility · Boolean Satisfiability · Deep Reinforcement Learning

1 Introduction

Constraints are fundamental building blocks of many Operation Research (OR) problems, as they model real-life scenarios. Constraint Satisfaction Problem

---

Throughout this paper, we use the terms "feasibility" and "satisfiability" interchangeably to refer to the same concept.

© The Author(s), under exclusive license to Springer Nature Switzerland AG 2026
Y. Zhang et al. (Eds.): LION 2025, LNCS 15745, pp. 31–47, 2026.
https://doi.org/10.1007/978-3-032-09192-5_3

(CSP) solving algorithms aim to find a solution that satisfies the constraints. However, in some cases, a problem may contain inconsistent or incompatible constraints, rendering it infeasible. In such situations, it becomes crucial to analyze which constraints are responsible for this infeasibility. More specifically, an important question to address is "What is the maximum subset of constraints that can be retained to make the problem feasible?" or its complementary question, "What is the minimum subset of constraints that can be removed to achieve feasibility?" This subset is called the maximum feasible subset of constraints and the discovery of such a subset is known to be NP-hard [2]. The answer to these questions can help users understand the main causes of infeasibility and enable efficient repair of the problem. Recently, Machine Learning (ML) has emerged as a powerful tool for discovering new heuristics to tackle NP-hard problems. Approaching these challenges from a statistical learning perspective allows for the rapid development of effective heuristics, reducing the need for extensive domain expertise, and improving the overall efficiency.

Although most ML approaches focus on learning solutions for specific OR problems, we explore how to repair infeasible CSPs and illustrate our method on two families of CSPs. Linear Feasibility Problems (LF) and Boolean Satisfiability Problems (SAT), using graph deep reinforcement learning (RL). More precisely, we present an approach to infeasibility repair using deep RL, expanding the scope of ML for OR. Our method encodes infeasible problems using bipartite Graph Neural Networks (GNNs). This encoding demonstrates robustness and flexibility in varying the size of the problem. By designing a deep RL policy, we eliminate the need for labeled data, enabling the model to learn efficient heuristics. We apply this policy to find the maximum feasible subsets in both LF and SAT problems, achieving superior generalization and efficiency compared to existing heuristics.

The paper is structured as follows. Initially, we review the pertinent literature in operations research, reinforcement learning, and deep learning. We then detail our methodology, explaining how we model CSPs as graphs, outline the Markov Decision Process (MDP), and describe the neural architectures designed to learn heuristics. Lastly, we provide a comprehensive evaluation of our approach in the LF and SAT domains, focusing on its efficiency (in terms of solution quality and computational time), generalization capabilities, and limitations.

2 Background and Related Work

Constraints Satisfaction Problem (CSP). CSPs [21] are a powerful framework for modeling and solving a wide range of complex problems in artificial intelligence and OR. At their core, CSPs involve finding a set of values for variables that satisfy a given set of constraints. Two notable examples of CSPs are LF and SAT problems. In LF, constraints are expressed in terms of linear combinations of variables. For example, a constraint might take the form $2v_1 + 3v_2 \leq 10$ and the goal is to find values of v_1 and v_2 that satisfy this and other similar constraints. This formulation is particularly useful for problems involving resource allocation, scheduling, and optimization in various industrial settings. In SAT problems, we deal with Boolean variables, and the constraints are typically expressed in Conjunctive Normal Form (CNF). In CNF, a constraint is a conjunction (AND) of clauses, where each clause is a disjunction (OR) of literals (variables or their negation). For example, a SAT constraint might look like $(v_1 \vee \neg v_2) \wedge (\neg v_1 \vee v_3)$, where v_1, v_2, and v_3 are boolean variables. The goal in SAT problems is to find an assignment of true/false values to these variables that satisfies all clauses simultaneously. SAT solvers are widely used in circuit design, software verification, and AI planning. Both the LF and SAT problems illustrate how CSPs can model diverse real-world scenarios, each with a unique constraint representation, showcasing the versatility and power of the CSP framework in problem solving across various domains.

Infeasibility and Unsatisfiability Analysis. Investigating infeasibility or unsatisfiability in CSPs often involves finding relevant subsets of constraints or clauses. These include Unsatisfiable Cores [25], Minimal Unsatisfiable Subsets [23], Irreducible Infeasible Subsets [11], and the focus of this paper: the Maximum Feasible Subset (MAXFS) in Linear Programming (LP) and the Maximum Satisfiability Problem (MAXSAT). Being NP-hard problems [1], exact solutions for MAXFS are achievable only for small instances. However, its importance spans numerous real-world applications [2,6,26], which requires efficient heuristics that provide high-quality solutions. Although there are many exact formulations [11,27,28], their practical application is limited due to the NP-hard nature of the problem [11]. Among heuristic approaches, Chinneck's LP based heuristics [9,10] are widely recognized and utilized, having demonstrated superior performance compared to other methods in experimental settings. In the SAT domain, the exact algorithms for MAXSAT are well-established, [7,36]. As in LF, there also exist efficient heuristic solvers like RC2 [17] for MAXSAT, represent the state of the art in this field.

Reinforcement Learning. RL [31] considers the problem of learning a decision-making policy for an agent interacting over multiple time steps with a dynamic environment. At each time step, the agent and the environment are described through a state $s \in \mathcal{S}$ and an action $a \in \mathcal{A}$ is performed; then the system transitions to a new state s' according to probability $p(s'|s, a)$, while receiving a reward $r(s, a)$. The tuple $M = (\mathcal{S}, \mathcal{A}, p, r)$ is called a Markov Decision Process (MDP) [29], which is often complemented by the knowledge of an initial state distribution $p_0(s)$. A policy π_θ parameterized by θ is a function $(s, a) \mapsto \pi_\theta(a|s)$ that maps states to distributions conditioned by actions. Training an RL agent consists of finding the policy that maximizes the discounted expected return $J(\pi_\theta) = \mathbb{E}[\sum_{t=0}^{\infty} \gamma^t r(s_t, a_t)]$, where $\gamma \in [0, 1)$ is a discount factor.

Graph Neural Networks. In recent years, GNNs have become powerful tools for various tasks defined in graph structures, using their ability to process graph data through message passing mechanisms [19]. The core idea behind GNNs is to exchange and aggregate information from the neighborhood of nodes and edges, allowing the network to capture complex dependencies inherent in graph-structured data. A particularly relevant type of graph in this context is the bipartite graph, which has been extensively studied in GNN research. Several architectures have been developed specifically to address problems such as assignment tasks [14,38] or to make predictions related to LP problems, including Mixed Integer Linear Programming (MILP) [8,13]. One of the key strengths of GNNs lies in their flexibility and generality. They can be trained on one set of graphs and applied effectively to different graph structures, demonstrating their robustness across various types of graphs [18].

Learning Based Heuristics. In recent years, a new set of methods known as learning based heuristics has emerged as a promising paradigm for solving optimization problems from a statistical learning point of view. In a foundational work, [34] introduced the pointer network, a novel architecture designed to solve permutation-based problems such as convex-hull and travelling salesman problems. In this view, [5] used RL to learn heuristics solely from rewards, tackling both traveling salesman and knapsack problems. RL is used to learn constructive heuristics functions that incrementally build solutions step by step, optimizing a cumulative reward based on partial solutions. The efficiency of such constructive heuristics is significantly based on the chosen neural architecture. Finding the right architecture is crucial for the heuristic to effectively navigate the solution space and make smart incremental decisions. [20] proposed an attention model rooted in transformer architecture [33], demonstrating impressive results on a variety of routing challenges. Recently, several advanced architectures have been proposed, each meticulously designed to address distinct OR challenges. Notable among these are solutions to the job shop scheduling problem using directed acyclic GNNs [37] and to assignment problems using bipartite GNNs [14,38]. The works most closely related to ours include research on predicting the feasibility of LP [8] and attempt to solve MAXSAT problems [24]. Both approaches are based on supervised learning techniques, in which the generation of labels for each instance is required to solve the problems to optimality.

3   Methodology

Problem Formulation. Given an infeasible problem, defined by a set of constraints $C = \{c_1, c_2, \ldots, c_m\}$, the problem cannot be solved because the constraints do not allow a feasible solution. In such cases, a natural question arises: How can we make the problem feasible? To address this, the goal is to find the minimum subset of constraints called the coverset $\bar{C} \subseteq C$, that must be removed so that the remaining set $C \setminus \bar{C}$ allows the problem to be feasible (illustrated in Fig. 1).

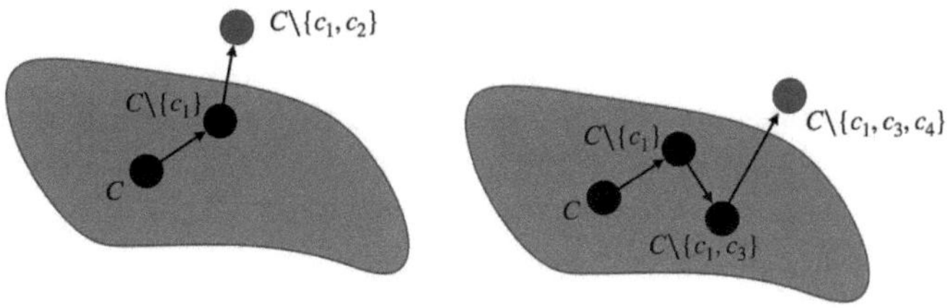

Fig. 1. Finding a good set $\bar{C}$ can be framed as a shortest path problem. The top solution removes only two constraints, while the bottom one leads to a longer path, requiring the removal of three.

Graph Representation. It is known that many constraint optimization problems [13,14,24,38] can be represented as a bipartite graph $\mathcal{G} = (\mathbf{C}, \mathbf{E}, \mathbf{V})$, where $\mathbf{C}$ denotes the set of constraint features, $\mathbf{V}$ the variable features, and $\mathbf{E}$ the edges (illustrated in Fig. 2).

In the LF domain, each node feature i in the vector $\mathbf{C}$ is constituted of the right-hand side weight b_i, a categorical variable specifying whether the constraint is an inequality or equality. The edges $\mathbf{E} = (e_{i,j})_{i,j}$ of the bipartite graph correspond to the weights of the variables within the constraints. The variable features $\mathbf{V}$ are set to zero, as our work focuses solely on feasibility, not on the coefficients of the objective function. In the SAT domain, the node feature vector $\mathbf{C}$ is set to 0. Each edge is set to 0 if the variable does not appear in the clause, 1 if the variable is included in the clause, and -1 if the variable appears in its negated form. Similarly to the LF domain, the variable features $\mathbf{V}$ are also set to 0, focusing solely on assessing the satisfiability of clauses without the influence of variable weights.

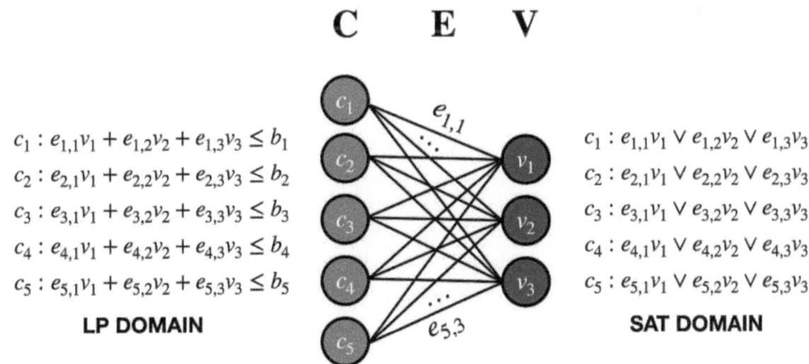

Fig. 2. Bipartite graph representation of LF constraints and SAT clauses.

Markov Decision Process. To learn the constraint selection policy, we define the underlying MDP as outlined below and illustrated in Fig. 3.

The **state** s_t represents the current state of the environment at time step t. Specifically, the agent takes the graph features $s_t := (\mathcal{G}_t, \bar{C}_t)$ that encapsulate the current problem instance and the set of constraints $\bar{C}_t$ that have been removed so far.

The **action** a_t refers to the index of the next constraint to be removed and added to the subset $\bar{C}_t$.

The **transition** function $p(s_{t+1} \mid s_t, a_t)$ is deterministic. The next state s_{t+1} is obtained by removing the constraint indexed by a_t and updating the bipartite graph accordingly. The state s_{t+1} can be terminal if the problem becomes feasible after this action.

The **reward** function r is designed to minimize the number of constraints removed to achieve feasibility. The reward in each step is defined as $r(s_t, a_t) = -1$ if the problem remains infeasible after removing the constraint indexed by a_t. The reward encourages the agent to remove the smallest possible set of constraints, as the constant -1 reward frames the MDP as a stochastic shortest path problem.

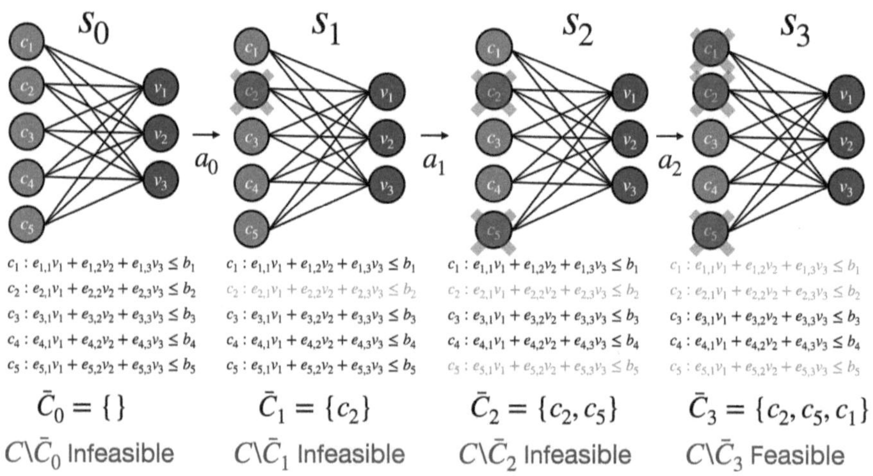

$$\bar{C}_0 = \{\} \qquad \bar{C}_1 = \{c_2\} \qquad \bar{C}_2 = \{c_2, c_5\} \qquad \bar{C}_3 = \{c_2, c_5, c_1\}$$

$C\backslash\bar{C}_0$ Infeasible $C\backslash\bar{C}_1$ Infeasible $C\backslash\bar{C}_2$ Infeasible $C\backslash\bar{C}_3$ Feasible

Fig. 3. MDP on a the LF domain. The agent's goal is to construct the smallest subset of constraints that restore the feasibility of the problem. At each step t, the agent selects a constraint to remove and increments the subset $\bar{C}_t$. The agent continues until the problem becomes feasible.

Bipartite GNN and Policy. Our approach focuses on learning rich representations of constraints, which can then be used by an RL policy (see Fig. 4).

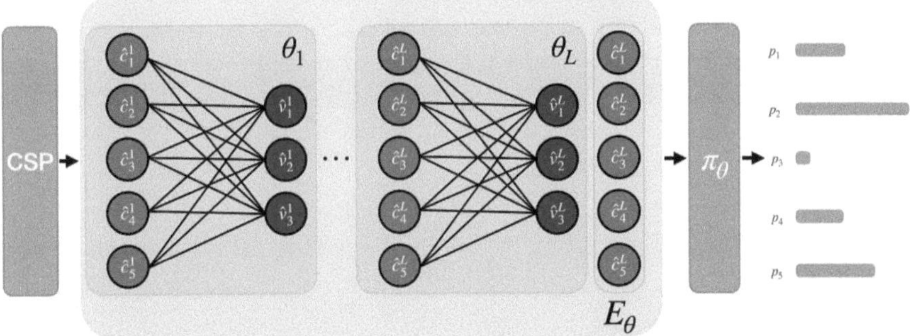

Fig. 4. The general framework of applying GNNs policy to solve MAXFS and MAXSAT problems involves converting the problem into a bipartite graph. Each pair of node and edge in the graph is assigned an initial embedding, which is iteratively updated through a message passing process. Finally, the policy outputs a score for each constraint.

In this work, we propose two encoder architectures for learning constraint embeddings. Both architectures take a bipartite graph $\mathcal{G} = (\mathbf{C}, \mathbf{E}, \mathbf{V})$ as input and output the embeddings for constraints and variables, denoted as $\hat{\mathbf{C}}$ and $\hat{\mathbf{V}}$, after L layers of processing. The first encoder E_θ^{GCNN} is based on the Graph Convolutional Neural Network (GCNN) model introduced by [13], which uses simple graph convolutions to construct constraint embeddings. This process involves message passing between nodes in the bipartite graph, with updates occurring in two stages: from variables to constraints and then from constraints to variables. The second E_θ^{DKA}, based on the Deep K-partite Assignment (DKA) model proposed by [38], employs a self-attention mechanism to learn constraint embeddings. Here, the embeddings are derived by averaging the edge embeddings, which are updated using attention layers within the DKA framework. The specific details of the message-passing mechanisms for both architectures are provided in [13,38].

Building on top of the constraint embeddings generated by the GNN, we propose a policy network that uses a simple Multi-Layer Perceptron (MLP) to score each constraint. This policy network, denoted as π_θ, is responsible for guiding the sequential selection of constraints to be removed to achieve feasibility. Mathematically, policy $\pi_\theta(\hat{\mathbf{C}})$ accepts constraint embeddings $\hat{\mathbf{C}}$ as input and outputs a score p_i for each constraint, representing the probability of removing that constraint at the current time step. The value network takes as input $\bar{g}$ defined as: $\bar{g} = \frac{1}{|C|} \sum_{\hat{c} \in \hat{C}} \hat{c} + \frac{1}{|V|} \sum_{\hat{v} \in \hat{V}} \hat{v}$. Here, $\bar{g}$ represents a global graph pooling of the graph embedding, where $\hat{C}$ and $\hat{V}$ are the sets of constraint and variable embeddings, respectively. The policy network is trained using Proximal Policy Optimization (PPO) [30] to maximize the expected cumulative reward and, equivalently, to minimize the number of constraints removed to achieve feasibility. Like the policy described in [5], our approach also incorporates a masking mechanism to ensure that the constraints are not removed more than once. Overall, this actor-critic architecture leverages the GNN-based constraint embeddings to effectively navigate the space of possible constraint removals. Following [5], we scale the node scores (before masking) within $[-K, K]$ using as an activation function tanh (see Eq. 1). Then we compute the final output probability p_i using a softmax:

$$u_i = \begin{cases} K \cdot \tanh\left(\pi_\theta(\hat{c}_i)\right) & \text{if } c_i \notin \bar{C} \\ -\infty & \text{otherwise} \end{cases} \qquad p_i = \frac{e^{u_i}}{\sum_{j=1}^{|C|} e^{u_j}}. \tag{1}$$

4 Experiments

Result on Linear Feasibility and Sat. To validate our approach, we generate an extensive benchmark consisting of various infeasible problems, including LF and SAT problems of varying sizes, defined by the number of variables (**v**) and constraints (**c**). Given the lack of large datasets of infeasible instances in both domains, we generated our own instances of different sizes to ensure a comprehensive evaluation. We train our agent in these diverse instances and test its performance on unseen new instances.

Algorithm 1. Applying the learned heuristic at inference

Require: Infeasible set of constraints C, trained encoder E_θ, trained policy π_θ,
$\quad \bar{C} \leftarrow \emptyset$
$\quad$ **while** `infeasible`$(C \setminus \bar{C})$ **do**
$\quad\quad \mathcal{G} \leftarrow$ `construct_graph`$(C, \bar{C})$ {Construct the graph}
$\quad\quad \hat{\mathbf{C}} \leftarrow E_\theta(\mathcal{G})$ {Compute constraint embeddings}
$\quad\quad P \leftarrow \pi_\theta(\hat{\mathbf{C}}, \bar{C})$ {Compute scores for each constraint}
$\quad\quad c \leftarrow \text{argmax}_{c' \in C \setminus \bar{C}} P(c')$ {Select constraint with highest score}
$\quad\quad \bar{C} \leftarrow \bar{C} \cup \{c\}$ {Add selected constraint to coverset}
$\quad$ **end while**
$\quad$ **return** $\bar{C}$

For the LF domain, the performance of our method is compared with several baselines, including a simple deletion heuristic described in [4]. We also compare our approach to the state-of-the-art heuristics, Chinneck's heuristics, which include two variants: Chinneck [11] and Fast-Chinneck [10]. Both heuristics are based on iteratively solving elastic LP, where the Chinneck heuristic aims to provide higher-quality solutions, while the Fast-Chinneck heuristic seeks a trade-off some solution quality for improved time efficiency. For smaller instances, we further compare our approach with exact solutions to assess its optimality, using the Big-M MILP formulation [11]. The deletion metric will be applied exclusively to very large instances, providing a reliable baseline in expansive settings (see Table 3). Due to the extensive execution time, the classical Chinneck heuristics will not be included in Table 3.

In the SAT domain (see Table 2), we compare our method to a state-of-the-art solver, RC2 [17], implemented in the PySAT library [16]. Additionally, we implement a random heuristic that iteratively selects a constraint at random (uniformly) for removal until satisfiability is achieved, providing an upper bound on the number of constraints removed. To further evaluate our method, we consider two types of settings: one in which each constraint is assigned equal weights, denoted as LF and SAT, and another in which weights are sampled from a uniform distribution, denoted as wLF (weighted LF) and wSAT (weighted SAT).

In each table, the cost of a solution refers to the number of constraints removed; in wLF and wSAT, it corresponds to the weighted number of constraints removed. In Table 1, the first column is not bolded as it represents the reference optimal solution cost.

Table 1. Average solution cost for various heuristics across 300 instances of MAXFS and weighted MAXFS; lower values indicate better performance. Bold text indicates the best overall heuristics, while gray cells denote the second best. Black cells signify that the method was not applied due to intractability.

	Big-M (Opt) [27]	Chinneck [11]	Chinneck-F [10]	Ours + GCNN [13]	Ours + DKA [38]
LF(c10v2)	2.83 ± 1.24	3.07 ± 1.48	3.16 ±1.48	2.88 ± 1.23	**2.84 ± 1.22**
LF(c20v5)	4.29 ± 1.54	4.81 ± 1.9	4.98 ±1.92	4.58 ± 1.71	**4.38 ± 1.55**
LF(c50v10)	11.09 ± 2.11	12.98 ± 2.79	13.24 ± 2.80	12.55 ± 4.17	**11.33 ± 2.27**
LF(c100v20)		25.30 ± 3.95	25.54 ± 3.99	26.06 ± 3.5	**21.84 ± 2.80**
LF(c150v30)		36.60 ± 4.78	36.97 ± 4.92	36.30 ± 3.51	**31.42 ± 3.50**
wLF(c10v2)	1.30 ± 0.74	1.46 ± 0.86	1.48 ± 0.87	1.37 ± 0.78	**1.32 ± 0.74**
wLF(c20v5)	1.80 ± 0.82	2.11 ± 1.03	2.18 ± 1.08	2.00 ± 0.83	**1.85 ± 0.89**
wLF(c50v10)	4.80 ± 1.27	5.98 ± 1.79	6.09 ± 1.84	6.00 ± 1.49	**5.04 ± 1.31**
wLF(c100v20)		11.61 ± 2.44	11.93 ± 2.45	11.99 ± 2.11	**9.38 ± 1.68**
wLF(c150v30)		16.73 ± 2.80	17.24 ± 2.92	17.09 ± 2.5	**15.99 ± 2.06**

Table 2. Average solution cost for various heuristics across 300 test instances of MAXSAT and weighted MAXSAT; lower values indicate better performance.

	RC2 [17]	Rnd	Ours + GCNN [13]	Ours + DKA [38]
SAT(c20v2)	**5.04 ± 1.23**	15.95 ± 2.32	7.53 ± 2.16	**5.04 ± 1.23**
SAT(c40v4)	**5.46 ± 1.46**	29.06 ± 5.42	5.65 ± 1.50	5.51 ± 1.61
SAT(c60v3)	**12.87 ± 2.11**	52.74 ± 3.71	17.37 ± 3.58	12.96 ± 2.32
SAT(c100v5)	**12.78 ± 2.22**	85.63 ± 6.92	19.75 ± 3.89	12.91 ± 2.33
SAT(c200v10)	**10.20 ± 2.38**	153.2 ± 22.3	19.85 ± 4.09	10.27 ± 2.38
wSAT(c20v2)	**2.32 ± 0.75**	7.90 ± 1.67	2.59 ± 0.91	**2.32 ± 0.75**
wSAT(c40v2)	**2.42 ± 0.82**	17.71 ± 2.16	3.40 ± 1.28	2.43 ± 0.82
wSAT(c60v3)	**6.04 ± 1.22**	26.22 ± 2.80	7.11 ± 1.54	6.08 ± 1.27
wSAT(c100v5)	**5.86 ± 1.35**	42.67 ± 4.32	9.56 ± 2.43	5.94 ± 1.39
wSAT(c200v10)	**4.57 ± 1.35**	76.11 ± 11.46	5.73 ± 1.69	4.81 ± 1.40

Better Solution Quality. In both domains, our RL approach exhibits very good results compared to other heuristics, achieving near-optimal solutions for small instances and outperforming Chinneck's heuristics and RC2 solver. The DKA-based agent demonstrates better overall performance compared to GCNN, as the attention mechanism can capture more sophisticated relationships, leading to better embedding of constraints.

Generalization to Unseen Larger Graph. In Table 3, we also demonstrate in the LF domain that our agent can generalize in a zero-shot manner to larger instances than those encountered during training. This capability underscores the flexibility and robustness of the model in handling increasing complexities of problems.

Table 3. Zero-shot generalization performance of a model trained on LF(150,30).

	Deletion [4]	Chinneck-F [10]	Ours + GCNN [13]	Ours +DKA [38]
LF(**c**200**v**40)	97.86 ± 18.26	48.76 ± 5.62	45.06 ± 12.83	**37.07 ± 9.35**
LF(**c**300**v**60)	159.6 ± 26.6	72.28 ± 7.72	62.98 ± 28.86	**28.71 ± 10.67**
LF(**c**600**v**60)	345.26 ± 69.2	192.88 ± 11.1	166.59 ± 58.81	**69.59 ± 26.3**
LF(**c**1000**v**80)	463.4 ± 145.3	342.5 ± 28.9	87.19 ± 89.67	**27.3 ± 20.70**

Better Time Efficiency. For medium and large instances, our model also demonstrates far better time efficiency than Chinneck's heuristics and is quite close to the Chinneck-Fast heuristic (see Fig. 5). For the largest instances, both the DKA and GCNN agents are significantly more efficient. This shows that our approaches scale well in terms of both solution quality and time efficiency compared to the already known heuristics.

Graph Convolutional Neural Network vs Attention. Both architectures show very good results: the GCNN model exhibits lower memory requirements, lower training, and inference time (see Fig. 5) while the DKA model leads to better solution quality, generalization, stability, and sample efficiency (see Fig. 6).

Additional Results on Time Efficiency. Most heuristics applied to the benchmarks (excluding the big-M formulation) rely on multiple calls to a solver to verify the feasibility of a given set of constraints. As demonstrated in Fig. 5, our RL approach is significantly more time efficient. This efficiency comes from the fact that the RL agent requires at most $|C|$ calls to the solver. In contrast, other approaches may invoke the solver far more frequently than the number of constraints. The Fig. 7 compares the number of solver calls across LF domains, illustrating that RL approaches are considerably more effective in reducing solver interactions.

Furthermore, as shown in Fig. 6 for the LF context, the use of attention-based encoders (DKA) is also highly effective in the SAT domains (see Fig. 8). The

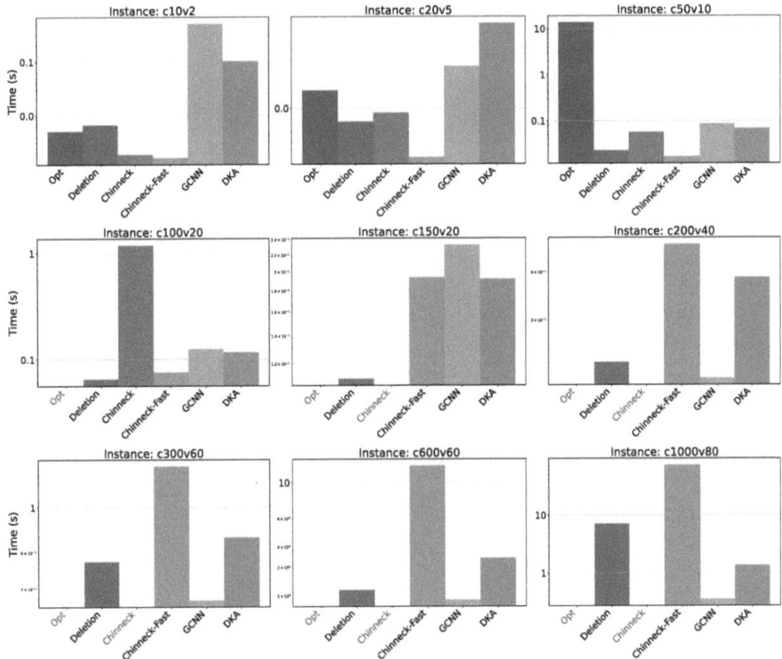

Fig. 5. Time comparison (log scale) of different heuristics across various infeasible LF instances. Red labels indicate methods that could not be run within a reasonable time frame for the given instance. (Color figure online)

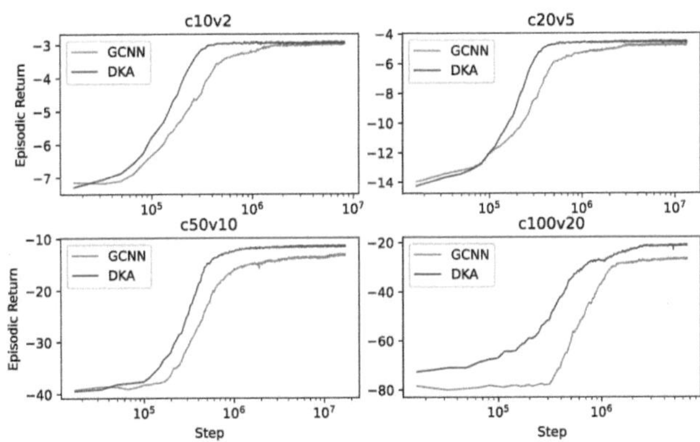

Fig. 6. Training curves comparison between MAXFS agents trained using GCNN and DKA encoders. The x-axis is presented on a logarithmic scale for improved visibility.

performance gap between GCNN and DKA is particularly pronounced in these domains compared to the LF domains. In the latter, the constraint coefficients are scalar, and the SAT domains feature categorical (combinatorial) constraints, which make the problem landscape significantly more complex and challenging to encode. We believe that attention mechanisms act as a superior function approximator in these scenarios, potentially capturing the richer structure and dependencies of the SAT problem space more effectively. However, further investigation is needed to validate this conjecture and better understand the role of attention in these domains.

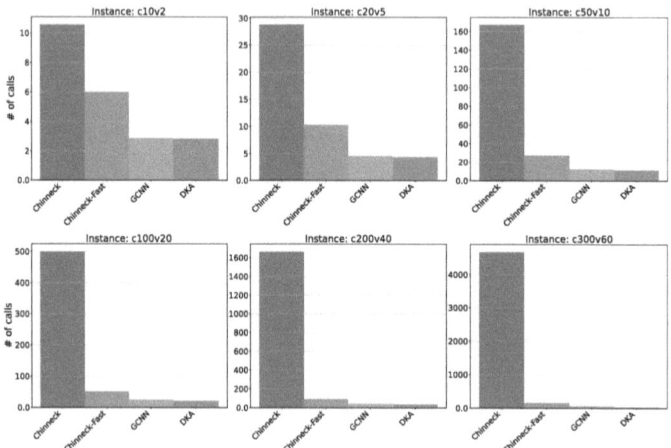

Fig. 7. Number of solver calls for different heuristics across various infeasible RL instances.

Instances Generation. To generate infeasible LF instances, we randomly create a constraint matrix and bound values. Specifically, the coefficients for the matrix and bounds are sampled uniformly from a range of integers between -100 and 100. Randomness in these values often leads to conflicting constraints, resulting in infeasible LFs. For infeasible SAT instances, we generate a random CNF formula by creating clauses using random variables, where each variable is either included as-is or negated. The clauses are designed to collectively create contradictions, ensuring that the SAT instance is unsatisfiable.

Additional Results on the SATLIB. Although the generation of instances for SAT problems was relatively random, we also trained and evaluated our agent using a distribution consistent with SATLIB Uniform Random-3-SAT (see Table 4) . These instances are specifically generated to avoid being trivially unsatisfiable and are sampled from the phase transition region, where the problem instances are the most challenging. More details on the instances and characteristics of the transition phase can be found at the following URL: https:// www.cs.ubc.ca/~hoos/SATLIB/Benchmarks/SAT/RND3SAT/descr.html.

Implementation Details. To account for the removal of a variable, we chose to introduce a binary indicator variable rather than physically removing the

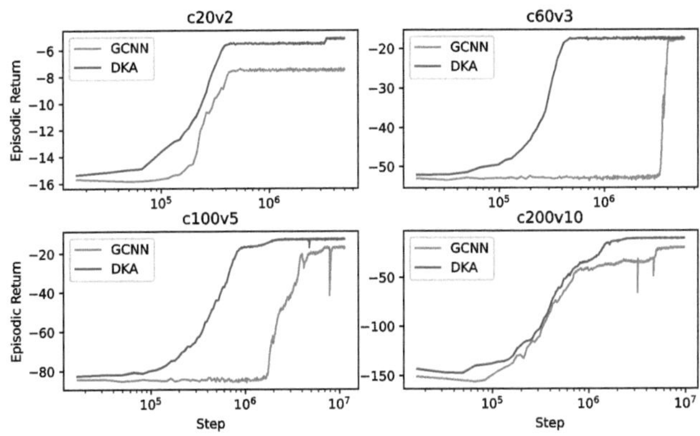

Fig. 8. Training curves comparison between MAXSAT agents trained using GCNN and DKA encoders. The x-axis is presented on a logarithmic scale for improved visibility.

Table 4. Average solution cost for various heuristics across 300 test instances of MAXSAT following the SATLIB Uniform Random-3-SAT distribution; lower values indicate better performance.

	RC2 [17]	Rnd	Ours + GCNN [13]	Ours + DKA [38]
SAT(c19v3)	**1.03 ± 0.18**	4.82 ± 2.66	1.62 ± 0.74	**1.03 ± 0.18**
SAT(c50v10)	**1.06 ± 0.23**	8.15 ± 5.1	2.03 ± 1.2	1.09 ± 0.28
SAT(c75v15)	**1.12 ± 0.32**	10.77 ± 6.1	2.11 ± 1.26	1.32 ± 0.54
SAT(c91v20)	**1.06 ± 0.23**	9.20 ± 6.23	2.55 ± 1.52	1.42 ± 0.63

node from the graph, simplifying the implementation. In Tables 1 and 2, a separate model is trained for each instance distribution and evaluated on 300 unseen instances from the same distribution, resulting in a total of 20 models. In Table 3, a single model is trained on one specific distribution and evaluated on 300 instances from four different distributions, each with a higher number of variables and constraints.

Settings. To check the feasibility of our set of constraints, we used two open-source solvers. HiGHS [15], available in SciPy [35] for LF, and Glucose4 [3] for SAT, available in PySAT [16]. All experiments were run on a desktop machine (Intel i9, 10th generation processor, 64GB RAM) with a single NVIDIA RTX 3090 GPU, using Python. We used PyTorch and PyTorch Geometric [12] to implement our DRL policy and Gymnasium [22] to model the RL environment. The parameters in Table 5 were used to optimize our policy. Table 6 presents the training times in hours.

Table 5. Hyperparameters

Hyperparameter	fValue
Number of layers	4
Number of attention heads per layer	8
Node embeddings dimension	128
Number of parallel environments	64
Episode length	256
Discount factor γ	0.999
Optimizer (θ of PPO)	Adam
Optimizer learning rate	2e-5
PPO epochs	3
PPO clip	0.2
PPO GAE λ	0.95
PPO value loss coefficient	0.5
PPO entropy loss coefficient	0.01
PPO batch size	256

Table 6. Training time (hours) for various instances size of LF and SAT problems

	GCNN	DKA
LF($c10v2$)	0.55	0.16
LF($c20v5$)	1.22	0.36
LF($c50v10$)	10	4.5
LF($c100v20$)	26	23
LF($c150v30$)	71	61
SAT($c20v2$)	10	7
SAT($c40v4$)	1	0.46
SAT($c60v3$)	1.9	0.71
SAT($c100v5$)	6.62	5
SAT($c200v10$)	20	14

5   Conclusion, Limitation and Future Work

In this work, we have investigated how deep RL and GNNs can be leveraged to efficiently repair infeasible problems. Focusing on LF and SAT, which encompass a wide spectrum of problems in OR, we developed DRL policies capable of competing with (and even outperforming) classical heuristics without any expert knowledge. We introduced two agents that employ different GNN encoders: one with reduced memory and computational requirements, and the other based on attention mechanisms that demonstrate superior performance in solution quality, stability, generalization, and sample efficiency. A limitation, common to neural approaches, is that despite promising results, it remains difficult to clearly surpass the most advanced state-of-the-art solvers, such as RC2 in the case of MAXSAT. Moreover, our approach relies on repeatedly checking the feasibility of a CSP, which can become computationally expensive (particularly for complex problems such as MILPs or hard SAT instances, where feasibility checks are themselves costly).

Our framework can be seen as a black-box method that only requires feasibility checks at each step, without the need for access to the internal state of the solver, which makes it very general and widely applicable.

Future work could address several challenges. One promising direction is learning to repair MILP and CP instances, as our GNN framework can also encode these problems. We propose a method for repairing CSPs by removing constraints. Future research could explore alternative methods of problem repair, such as modifying the coefficients of constraints within an LP for continuous repair, or providing recommendations for specific cases, such as determining the appropriate number to place in a cell to restore feasibility in a Sudoku puzzle. Furthermore, approaches that employ a general representation of CSP, as demonstrated by the architecture proposed in [32], may offer a wide applicability beyond specific domains.

Another promising extension of this work is to leverage the internal state of the solver, which checks feasibility, to improve the heuristic. This could result in a hybrid method that combines both learning-based and traditional approaches, thereby transforming it into a white-box method.

Finally, applying our approach to large datasets of real-world infeasible problems could further validate and enhance the practicality of our methods.

Acknowledgment. We acknowledge the support of the IRT-MINDS project.

References

1. Amaldi, E., Kann, V.: The complexity and approximability of finding maximum feasible subsystems of linear relations. Oper. Res. **147**, 181–210 (1994)
2. Amaldi, E.: From finding maximum feasible subsystems of linear systems to feed-forward neural network design, PhD dissertation, Swiss Federal Institute of Technology at Lausanne (EPFL) (1994)
3. Audemard, G., Simon, L.: Glucose: a solver that predicts learnt clauses quality. In: SAT Competition, pp. 7–8 (2009)
4. Bailey, J., Stuckey, P.J.: Discovery of minimal unsatisfiable subsets of constraints using hitting set dualization. In: Practical Aspects of Declarative Languages: 7th International Symposium, PADL 2005, Long Beach, CA, USA, January 10–11, 2005. Proceedings 7, pp. 174–186. Springer (2005)
5. Bello, I., Pham, H., Le, Q.V., Norouzi, M., Bengio, S.: Neural combinatorial optimization with reinforcement learning. arXiv preprint arXiv:1611.09940 (2017)
6. Bennett, K.P., Bredensteiner, E.J.: A parametric optimization method for machine learning. INFORMS J. Comput. **9**(3), 311–318 (1997)
7. Borchers, B., Furman, J.: A two-phase exact algorithm for max-sat and weighted max-sat problems. J. Comb. Optim. **2**, 299–306 (1998)
8. Chen, Z., Liu, J., Wang, X., Lu, J., Yin, W.: On representing linear programs by graph neural networks. arXiv preprint arXiv:2209.12288 (2022)
9. Chinneck, J.W.: An effective polynomial-time heuristic for the minimum-cardinality IIS set-covering problem. Ann. Math. Artif. Intell. **17**(1), 127–144 (1996)
10. Chinneck, J.W.: Fast heuristics for the maximum feasible subsystem problem. INFORMS J. Comput. **13**(3), 210–223 (2001)
11. Chinneck, J.W.: Feasibility and Infeasibility in Optimization: Algorithms and Computational Methods, vol. 118. Springer Science & Business Media (2007)
12. Fey, M., Lenssen, J.E.: Fast graph representation learning with PyTorch Geometric. In: ICLR Workshop on Representation Learning on Graphs and Manifolds (2019)
13. Gasse, M., Chételat, D., Ferroni, N., Charlin, L., Lodi, A.: Exact combinatorial optimization with graph convolutional neural networks. Adv. Neural Inf. Process. Syst. **32** (2019)

14. Gibbons, D., Lim, C.C., Shi, P.: Deep learning for bipartite assignment problems. In: 2019 IEEE International Conference on Systems, Man and Cybernetics (SMC), pp. 2318–2325 (2019). https://doi.org/10.1109/SMC.2019.8914228
15. Huangfu, Q., Hall, J.J.: Parallelizing the dual revised simplex method. Math. Program. Comput. **10**(1), 119–142 (2018)
16. Ignatiev, A., Morgado, A., Marques-Silva, J.: PySAT: a python toolkit for prototyping with SAT oracles. In: International Conference on Theory and Applications of Satisfiability Testing, pp. 428–437. Springer (2018)
17. Ignatiev, A., Morgado, A., Marques-Silva, J.: RC2: an efficient MaxSAT solver. J. Satisfiability Boolean Model. Comput. **11**(1), 53–64 (2019)
18. Joshi, C.K., Cappart, Q., Rousseau, L.M., Laurent, T.: Learning the travelling salesperson problem requires rethinking generalization (2022). https://doi.org/10.4230/LIPIcs.CP.2021.33
19. Kipf, T.N., Welling, M.: Semi-supervised classification with graph convolutional networks. arXiv preprint arXiv:1609.02907 (2016)
20. Kool, W., van Hoof, H., Welling, M.: Attention, learn to solve routing problems! arXiv preprint arXiv:1803.08475 (2019)
21. Kumar, V.: Algorithms for constraint-satisfaction problems: a survey. AI Mag. **13**(1), 32–32 (1992)
22. Kwiatkowski, A., et al.: Gymnasium: a standard interface for reinforcement learning environments. arXiv preprint arXiv:2407.17032 (2024)
23. Liffiton, M.H., Sakallah, K.A.: Algorithms for computing minimal unsatisfiable subsets of constraints. J. Autom. Reason. **40**, 1–33 (2008)
24. Liu, M., et al.: Can graph neural networks learn to solve the MaxSAT problem? (Student abstract). In: Proceedings of the AAAI Conference on Artificial Intelligence, vol. 37, pp. 16264–16265 (2023)
25. Lynce, I., Marques-Silva, J.P.: On computing minimum unsatisfiable cores (2004)
26. Mangasarian, O.L.: Misclassification minimization. J. Global Optim. **5**(4), 309–323 (1994)
27. Parker, M.R.: A set covering approach to infeasibility analysis of linear programming problems and related issues, Ph.D. thesis, University of Colorado at Denver Denver, Colorado (1995)
28. Pfetsch, M.E.: Branch-and-cut for the maximum feasible subsystem problem. SIAM J. Optim. **19**(1), 21–38 (2008)
29. Puterman, M.L.: Markov Decision Processes: Discrete Stochastic Dynamic Programming. Wiley (2014)
30. Schulman, J., Wolski, F., Dhariwal, P., Radford, A., Klimov, O.: Proximal policy optimization algorithms. arXiv preprint arXiv:1707.06347 (2017)
31. Sutton, R.S., Barto, A.G.: Reinforcement Learning: An Introduction. MIT Press (2018)
32. Tönshoff, J., Kisin, B., Lindner, J., Grohe, M.: One model, any CSP: graph neural networks as fast global search heuristics for constraint satisfaction. arXiv preprint arXiv:2208.10227 (2022)
33. Vaswani, A., et al.: Attention is all you need. Adv. Neural Inf. Process. Syst. **30** (2017)
34. Vinyals, O., Fortunato, M., Jaitly, N.: Pointer Networks. arXiv preprint arXiv:1506.03134 (2017)
35. Virtanen, P., et al.: SciPY 1.0: fundamental algorithms for scientific computing in Python. Nat. Methods **17**(3), 261–272 (2020)
36. Xing, Z., Zhang, W.: MaxSolver: an efficient exact algorithm for (weighted) maximum satisfiability. Artif. Intell. **164**(1–2), 47–80 (2005)

37. Zhang, C., Song, W., Cao, Z., Zhang, J., Tan, P.S., Chi, X.: Learning to dispatch for job shop scheduling via deep reinforcement learning. Adv. Neural. Inf. Process. Syst. **33**, 1621–1632 (2020)
38. Zouitine, M., Berjaoui, A., Lagnoux, A., Pellegrini, C., Rachelson, E.: Learning heuristics for combinatorial optimization problems on k-partite hypergraphs. In: International Conference on the Integration of Constraint Programming, Artificial Intelligence, and Operations Research, pp. 304–314. Springer (2024)

Optimal Matched Block Design For Multi-arm Experiments

Nathan Brixius[✉]

Airbnb, Inc., San Francisco, CA, USA
nathan.brixius@airbnb.com

Abstract. We introduce new techniques for matched block design in multi-arm experiments. In matched block design, units with similar covariate values are grouped into blocks, with one unit per treatment in each block. Existing methods for unit-level block design often fail to produce optimal matches for multi-arm experiments. We present a mixed integer programming (MIP) formulation that guarantees optimal solutions for the general multi-arm matched blocking problem using a clique-based equipartitioning approach. For cases where the MIP is computationally infeasible, we introduce heuristics that decompose large problems into tractable subproblems while providing explicit quality-runtime tradeoffs. We demonstrate that our methods significantly outperform existing techniques on a diverse test suite, achieving consistent improvements in block balance quality. Additionally, we show how these methods can be adapted to improve covariate balance across treatment groups. We demonstrate that matched block design presents an interesting application area for metaheuristic and exact optimization methods.

Keywords: matched block design · optimization · mixed integer programming · randomized experiments

1 Introduction

Determining causal effects is fundamental to evidence-based decision making. Randomized experiments, where subjects are assigned to different treatment groups, remain the gold standard for establishing causality.

However, in many real-world settings, subject-level treatment assignment is often impractical or impossible. Instead, subjects must be collected into groups called units, which then become the level of assignment. In a retail experiment, individual stores might serve as units rather than employees [18]; in educational research, schools rather than students [17]; and on digital platforms, geographic regions rather than individual users.

A key goal in experiment design is to achieve balance in unit covariates across treatment groups. In randomized experiments, balance is ensured if there

Submitted February 13, 2025, Revised July 20, 2025

Y. Zhang et al. (Eds.): LION 2025, LNCS 15745, pp. 48–63, 2026.
https://doi.org/10.1007/978-3-032-09192-5_4

are sufficient units. When sample sizes are limited or when greater precision is desired, alternative approaches are needed. Block design addresses this challenge by explicitly considering unit covariates to form matched groups across treatments. In this approach, units with similar characteristics are grouped into blocks, with each block containing one unit for each treatment.

Software tools for block design include `blocktools` [23], `mipmatch` [32], and `brsmatch` [20]. Such implementations account for balance within and across blocks, as well as computational runtime. These considerations in turn depend on factors such as the number of treatments, experimental units, and covariates.

Let n be the number of experimental units, and t be the number of treatments, i.e. block size. Without loss of generality we assume $n \bmod t = 0$ and block count $m = \frac{n}{t}$. Figure 1 shows the relationship between subjects, units, blocks, and treatments.

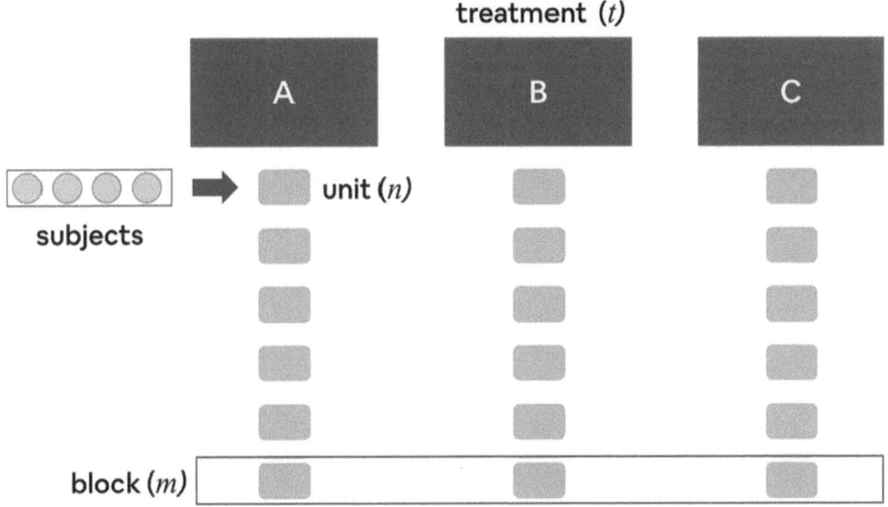

Fig. 1. Subjects, Units, Blocks, and Treatments.

Balance can be assessed by comparing unit covariates within and across blocks. Two types of measures for block design are block balance and treatment balance. By block balance we mean that units in a block are similar, as determined by a covariate distance metric, e.g. Mahalanobis distance. By treatment balance we mean that the covariate distributions among treatment groups are similar.

Our primary focus in this work is block balance, as assessed by the sum of all pairs of distances between matched pairs within blocks. We refer to this problem as Matched Blocking (MB), and the special case of $t = 2$ as Pair Blocking (PB).

The primary contribution of this work is a new mixed integer programming formulation of MB. This formulation can be solved to optimality in many scenarios, as demonstrated across a set of test instances from the literature. While our method produces optimal matches, it is not computationally feasible in all situations. Therefore a second contribution is a heuristic framework capable of balancing runtime and block balance considerations, leveraging the strengths of our MIP formulation. We show that our techniques outperform commonly used methods over a test suite that includes both previously considered and newly created instances.

The remainder of this work is organized as follows. We first consider PB, where efficient methods for optimal matches are well known. We then turn to MB, where this is not the case. While MB approaches often rely on heuristics that do not provide guarantees of match quality, we show how an extension of a mixed integer programming (MIP) formulation for PB can be extended to MB. This formulation can be solved to optimality across most of the instances in our test suite. We then consider a heuristic framework that trades optimality for faster runtime.

While block balance is our primary concern, we address treatment balance in Sect. 5. We show that even a simple postprocessing heuristic approach to treatment balance yields significantly improved covariate balance.

2   Block Design

2.1   Applications

Experimental blocking was introduced in 1935 by Ronald Fisher. The term "blocking" comes from Chap. 22 of [6] concerning an agricultural experiment where an "area is divided into eight blocks... and each of these is divided into five plots". In Fisher's setting, blocks were predefined and not "designed". Subsequently, block design has been applied to a range of domain areas.

Greevy et al. described several medical applications and supporting methods in [8], in particular the efficacy of ACE-inhibitor drugs in pediatric cancer survivors. Rosenbaum considered cadmium and lead in the blood of smokers in the matching study described in [26]. The same author in [27] provides a comprehensive overview of matching algorithms inside and outside the context of block design, using the possible impact of antidepressant medication on bone density as a motivating example.

There are an array of social science applications of matched block design. Bruhn and McKenzie [4] survey several applications. First, Sri Lankan microenterprises from [5] are used to simulate a program boosting firm profits. Second, a sub-sample of the Mexican ENE survey in [14] focuses on heads of household receiving training or nutrition intervention to raise income. Third, the Indonesian Family Life Survey (IFLS) [30] examines children at risk of dropping out of school and households facing increases in expenditure per capita. Finally, the LEAPS data in Pakistan [1] considers youth education and nutrition. The microenterprise dataset is described in [31].

Keele et al. applied match block design to assess the effectiveness of a computer-aided instruction program for improving reading comprehension in [17]. Moore described several political science applications in [24]. Panagopoulos and Green applied MB to a campaign finance analysis in [25].

2.2 Solution Approaches

Various algorithmic approaches for block design have been proposed.

Rosenbaum et al. [28] considered greedy and regression-based techniques for matched block design, with the aim of reducing covariate imbalance. Greevy et al. provided a more formal treatment of such techniques, using Mahalanobis distance and bipartite matching for the $t = 2$ case [8]. In 2009, Bruhn and McKenzie [4] established that matched pair design yielded more balanced covariates than randomized methods, as measured on four panel data sets. This work roughly coincides with the introduction of the `blocktools` package in 2008, see [23].

Imai, King et al. advocate for the use of PB with block randomized experiments in [11], noting: "Since matching prior to random treatment assignment can greatly improve the efficiency of causal effect estimation ([3,8]), and matching in pairs can be substantially more efficient than matching in larger blocks, matched-pair, cluster-randomization (PB) would appear to be an attractive design for field experiments [13]." Matched pair design methodologies were employed and enhanced by in a universal health insurance experiment in Mexico, see [11]. In [12] the authors extended their analysis, considering the impact of the choice of distance metric in matched block design.

Mixed integer programming (MIP) was first applied to matched block design by Zubizarreta in [32], who considered "the total sum of distances and a weighted sum of specific measures of covariate imbalance." The method described in [32] differs from our scenario because it considers a) PB only, b) modeling covariate imbalance directly, c) linearized objectives only. The author's `mipmatch` package is the only previous matched block design implementation to use MIP of which we are aware. Zubizarreta and Keele further considered both block and treatment balance in multilevel matching, again with $t = 2$ [33].

In 2016, Higgins et al. [10] proposed a heuristic approach to the full $t \geq 2$ m atched block design problem, with a guaranteed bound of no worse than 4x from optimal match quality. Further, this approach scales to large problem sizes. This approach could be a useful alternative to those presented here, but we are not aware of a publicly available implementation.

There is also literature on closely related problems. Balanced risk set matching is the problem of matching treated subjects with one or more control subjects. This may occur across multiple groups, but the key difference from block design is that pairs across groups are matched, rather than forming strata (blocks). Li et al. provided the `brsmatch` package as outlined in [20]. Lu and Rosenbaum in [22] gave a globally optimal risk set matching algorithm for three experimental groups, which does not carry over to MB since it involves pairwise matching.

Lu, Greevy et al. consider the related problem of optimally assigning blocked units to treatments in [21]. Their results can be applied once blocks have been determined, i.e. as a postprocessing step to block design, as we discuss in Sect. 5.

3 Pair Blocking

Assume a set of units $u_1, \ldots, u_n$. Let $f \in \mathbb{R}^{n \times v}$ represent unit covariates with $f_{i\alpha}$ the value of covariate α for unit u_i. Let $c \in \mathbb{R}^{n \times n}$ with c_{ij} representing the distance between units i and j, with distances not necessarily satisfying the triangle inequality. Since a unit cannot be matched with itself, $c_{ii} = \infty$. It is common to use the Mahalanobis distance, but we do not depend on this assumption in our analysis.

Assuming $t = 2$, let $x_{ij} \in \{0, 1\}$ represent whether items i and j are matched, that is $x_{ij} = 1$ iff item i and item j form a matched pair. Pair Blocking is then the problem of minimizing total distance between matched pairs, i.e.

$$\min \quad \sum_{i=1}^{n} \sum_{j=i+1}^{n} c_{ij} x_{ij}.$$

3.1 Greedy Pair Blocking

Greedy algorithms are an intuitive approach for solving PB. A greedy approach selects available (i, j) with minimal distance, matches them, and repeats.

Algorithm 1. PB-G(c)

1: let $s = \emptyset, h = \text{HEAP}(\{((i, j), c_{ij}) \mid i = 1, \ldots, n; \ i < j\})$ {heap ordered by c_{ij}}
2: **while not** EMPTY(h) **do**
3:    $(i, j) := \text{POP}(h)$ {extract pair with minimum distance}
4:    $s := s \cup \{(i, j)\}$
5:    REMOVE$(h, \{((k, l), c_{kl}) \mid k \in \{i, j\} \text{ or } l \in \{i, j\}\})$ {no longer consider i, j}
6: **end while**
7: **return** s

PB-G corresponds to `algorithm=optGreedy`[1] in `blocktools`. Observe that PB-G is not guaranteed to yield optimal solutions. Consider:

$$c = \begin{bmatrix} \infty & 1 & 2 & 2 \\ 1 & \infty & 2 & 2 \\ 2 & 2 & \infty & 11 \\ 2 & 2 & 11 & \infty \end{bmatrix}$$

PB-G yields solution $\{(1, 2), (3, 4)\}$ with total cost $= 12$, but $\{(1, 3), (2, 4)\}$ has cost 8.

---

[1] With minor differences in tiebreaking logic.

3.2 Optimal Pair Blocking

Defining x as before, we formulate PB as:

$$\min \quad \sum_{i=1}^{n}\sum_{j=i+1}^{n} c_{ij}x_{ij}$$

$$\text{s.t.} \quad \sum_{j=1}^{n} x_{ij} = 1 \qquad i = 1,\dots,n, \qquad\qquad \text{(PB-O)}$$

$$\sum_{i=1}^{n} x_{ij} = 1 \qquad j = 1,\dots,n,$$

$$x_{ij} \in \{0,1\}.$$

The formulation (PB-O) is equivalent to a linear assignment problem (LAP) with cost matrix c, equivalent to the well-known minimum cost bipartite matching problem, see [21]. Though x is binary, the integrality condition can be relaxed due to Birkhoff [2]. The LAP described by (PB-O) is easily solved for problem sizes considered in this work; the Hungarian method of Kuhn [19] and Jonker-Volgenant approach [16] each have $O(n^3)$ complexity.

In Sect. 6 we compare the match quality of PB-G and PB-O. Despite the guaranteed globally optimal solution quality of (PB-O), there may be legitimate reasons for using a heuristic approach. First, while PB-G and PB-O are both polynomial time algorithms, PB-G is faster in practice. Second, as noted by [11], if it is anticipated that items will be added or removed prior to experimentation, online approaches may be more appropriate. In light of such considerations, `blocktools` offers the choice between PB-O and PB-G as algorithm options `optGreedy` and `optimal`.

In certain cases, minimizing the sum of distances between matched pairs may not be the desired objective, see [32]. The formulation (PB-O) can be adapted for such considerations, but we will not present them here for brevity.

4 Matched Blocking

We now consider extending PB solution methods to multi-arm experiments, i.e. $t \geq 2$. A simple approach for MB, which we refer to as MB-N, is to scan items in order of index, add the unmatched item of least distance to the current block, and repeat. MB-N corresponds to corresponds to `algorithm=naiveGreedy` in `blocktools`.

PB-G is trivially extended to MB by making $t-1$ greedy choices per iteration, which we refer to as MB-G. We omit the details for brevity. Since MB-G also solves PB, MB-G is not guaranteed to produce optimal solutions as previously shown.

4.1   Clique Formulation

Let $G = (V, E)$ be a graph with vertices representing items and edges item distances. MB is equivalent to partitioning G into $m = \frac{n}{t}$ subcliques (blocks) of size t, referred to in [15] as t-way equipartitioning. We seek the equipartitioning that minimizes total distance within blocks.

Formulation (PB-O) can be extended to model t-way equipartitioning. Letting binary decision variable x_{ij} where $x_{ij} = 1$ iff item i and item j are assigned to the same block, we have:

$$
\begin{aligned}
\min \quad & \sum_{i=1}^{n} \sum_{j=i+1}^{n} c_{ij} x_{ij} \\
\text{s.t.} \quad & \sum_{j} x_{ij} = t - 1 & & i = 1, \ldots, n, \\
& \sum_{i} x_{ij} = t - 1 & & j = 1, \ldots, n, \qquad \text{(MB-O)} \\
& x_{ij} + x_{jk} - x_{ik} \leq 1 & & \forall i < j < k, \\
& x_{ij} - x_{jk} + x_{ik} \leq 1 & & \forall i < j < k, \\
& - x_{ij} + x_{jk} + x_{ik} \leq 1 & & \forall i < j < k, \\
& x_{ij} \in \{0, 1\}.
\end{aligned}
$$

This formulation closely resembles that provided by [9] in the context of a generic clustering framework. Note formulation (MB-O) has n^2 variables and $O(n^3)$ constraints. The large number of constraints in this model, necessary to ensure assigned items form cliques, can make MB-O computationally infeasible in certain cases, as we shall demonstrate in Sect. 6. Alternatively MB-O can be reformulated as a bilinear model, but such a reformulation is less amenable to efficient solution by commonly used solvers.

4.2   Divide and Conquer

We now consider approaches that offer explicit tradeoffs between block balance and computational time. Unlike MB-O, these approaches do not guarantee optimal solutions to MB, however they do leverage the strengths of the formulation presented in Sect. 3.2.

We now propose a divide-and-conquer metaheuristic that separates the n items into subpartitions, solves each subpartition (a smaller MB), and returns the union of results. Suppose $s \leq n$ is the desired upper bound on subpartition size. Then clearly $q = s - (s \mod t)$ is the largest subpartition size with equally sized subcliques, and $p = \left\lceil \frac{n}{q} \right\rceil$ is the subpartition count. The size of each subpartition is $s_i = q$ for $i = 1 \ldots p - 1$ and $s_p = q - (s \mod t)$.

Let MB-S be any solution method for MB. Let PARTITION be a function that separates cluster indexes into disjoint g_i, $i \in 1 \ldots p$. Finally, let $c[g_i, g_i]$ denote the submatrix with row and column indexes g_i. The resulting metaheuristic is given in Algorithm 2.

Algorithm 2. MB-DC(c, t, s)

1: $q = s - (s \mod t)$, $p = \left\lceil \frac{n}{q} \right\rceil$
2: $s_i = q$ for $i = 1 \ldots p - 1$, $s_p = q - (s \mod t)$
3: $g = \text{PARTITION}(c, s)$
4: **for** $i = 1 \ldots p$ **do**
5:     $x_i = \text{MB-S}(c[g_i, g_i])$
6: **end for**
7: **return** $\bigcup_i^p x_i$

The resulting block balance yielded by MB-DC depends on the choice of PARTITION, MB-S, and maximum partition size s. We now present two MB-DC variants with different quality-runtime tradeoffs. Observe that MB-DC reduces to MB-G when $s = t$ and to MB-O when $s = n$.

Greedy Partitioning. A simple choice for PARTITION is to use MB-G and divide along block boundaries of the result. If these partitions are then solved using MB-S = MB-O, the solution of each partition is guaranteed to be no worse than the results yielded by MB-G. Therefore the divide-and-conquer solution is guaranteed to be no worse than MB-G. We refer to MB-DC with greedy partitioning as MB-DC-GP.

Assignment Partitioning. MB-DC-GP improves on MB-G only to the extent that items within subpartitions g_i are rearranged. Intuitively, it therefore makes sense to assign blocks to partitions that are close to each other, rather than simply partition in the order provided by MB-G. The idea behind assignment partitioning is to assign blocks to partitions in such a way that minimizes total distance between the centroids of assigned blocks.

As previously noted there are $m = \frac{n}{t}$ blocks in a solution to MB. Consider the m centroids of the items within each block in an MB solution. Let $d \in \mathbb{R}^{m \times m}$ be the centroid-centroid distances. Let $z_{ij} = 1$ iff block i is assigned to partition j. We solve the partitioning assignment problem (PAP):

$$\min \quad \sum_{i=1}^{m}\sum_{j=1}^{p} d_{ij} z_{ij}$$

$$\text{s.t.} \quad \sum_{j=1}^{m} z_{ij} = 1 \qquad i = 1, \ldots, m, \tag{PAP}$$

$$\sum_{i=1}^{m} z_{ij} = s_j \qquad j = 1, \ldots, p,$$

$$z_{ij} \geq 0.$$

The resulting LAP formulation of PAP is easily solved, as was demonstrated in Sect. 3.2. The desired partitions are given by $g_j = \{i \mid z_{ij} = 1\}, j = 1, \ldots, p$. We refer to this variant as MB-DC-PAP.

Extensions. Finally, note that if MB-DC is recursive, that is, MB-S $=$ MB-DC, the resulting algorithm is analogous to nested dissection approaches for sparse matrix factorization, see [7]. We do not explore this connection further here.

5   Treatment Balance

As noted in Sect. 2, certain block design techniques attempt to balance covariate values across treatment groups directly. In [32] the authors consider a weighted sum of covariate imbalance measures as an objective function. These measures include balancing means of treatment groups, multivariate moments, and Kolmogorov-Smirnov statistics.

Block distance is invariant with respect to item swaps within blocks. Therefore we can apply swap-based treatment balance techniques as a postprocessing step to any of the block balance approaches previously described. Let us consider treatment balance using a standard measure, standardized mean difference (SMD). First, let us define $y_{ilk} = 1$ if and only if unit i is assigned to block l and treatment k. Letting

$$F_{k\alpha}(y) = \sum_{i=1}^{n}\sum_{l=1}^{m} f_{i\alpha} y_{ilk},$$

the standardized mean difference between treatments k and l for covariate α is

$$SMD(k, l, \alpha; f, y) = \frac{F_{k\alpha}(y) - F_{l\alpha}(y)}{\sqrt{\sum_{i=1}^{n} f_{i\alpha}^2}}.$$

Taking the mean across all pairs, we define the following block design covariate balance metric:

$$b(f, y) = \sum_{\alpha=1}^{v} \frac{1}{\binom{t}{2}} \sum_{k=1}^{t} \sum_{l=k+1}^{t} SMD(k, l, \alpha; f, y).$$

We then consider the nonlinear assignment problem

$$\min \quad b(f, y)$$

$$\text{s.t.} \quad \sum_{i=1}^{n} y_{ilk} = 1 \quad l = 1, \ldots, m; \ k = 1, \ldots, t,$$

$$\sum_{l=1}^{m} y_{ilk} = 1 \quad i = 1, \ldots, n; \ k = 1, \ldots, t \qquad \text{(COV-OPT)}$$

$$\sum_{k=1}^{t} y_{ilk} = 1 \quad k = 1, \ldots, t; \ l = 1, \ldots, m$$

$$y_{ilk} \in \{0, 1\}.$$

A local improvement approach for COV-OPT is to sort blocks l by SMD and then repeatedly look for swaps within blocks that reduce $b(f, y)$. This approach is employed in the results that follow. While we do not consider optimal solutions for COV-OPT here, we regard this an area for future research.

6 Computational Results

6.1 Test Instances

We have assembled a test suite of matched block design problems, summarized in Table 1. In Table 1, n is the number of items and v is the number of covariates. In certain matched block design scenarios, "grouping" attributes are specified. In such cases the dataset is divided into g independent subproblems; see the `blocktools` documentation [23] for more details.

The sources of the test suite instances are as follows. The Sri Lankan microenterprises, Mexican BNE survey, IFLS, and LEAPS examples from [5] are included. The x100 test problem is a simple synthetic problem included with

Table 1. Block Balance Test Suite Instances

Name	n	v	g	Description
x100	100	2	3	test problem included with `blocktools`
states	51	12	1	synthetic Airbnb experiment data
silver	492	4	1	a publicly available dataset provided by Ian Silver
ifls_expend	30	7	1	a subset of the Indonesian Family Live Survey (IFLS) dataset
ifls_school	30	7	1	a subset of the IFLS dataset
leaps_height	30	7	1	child and household data from the LEAPS project
leaps_test	30	7	1	child and household data from the LEAPS project
mexico_ene	30	7	1	a subset of the Mexican employment survey
sri_lanka	30	7	1	a subset of the Sri Lankan microenterprise dataset

`blocktools`. A publicly available dataset provided by Ian Silver is also included, see [29]. Finally, we also include a synthetic dataset from Airbnb called `states`. In this data set, there are three treatment groups and 51 units (the fifty US states and DC).

6.2 Algorithm Performance

In Table 2 we compare the match quality of `blocktools`, MB-G, MB-DC-G, MB-DC-PAP, and MB-O. The `blocktools` results are produced by taking the

Table 2. Block balance by algorithm relative to best `blocktools` result

Name	t	blocktools Value	MB-G Value	Gap	MB-DC-G Value	Gap	MB-DC-PAP Value	Gap	MB-O Value	Gap
ifls_expend	2	31.5	32.8	4.0%	32.8	4.0%	32.8	4.0%	31.5	0.0%
ifls_expend	3	73.9	73.9	0.0%	73.9	0.0%	73.9	0.0%	72.2	−2.3%
ifls_expend	4	106.3	106.3	0.0%	101.1	−4.9%	106.3	0.0%	99.6	−6.3%
ifls_school	2	29.3	30.8	5.1%	30.8	5.0%	30.8	5.1%	29.3	0.0%
ifls_school	3	74.7	74.7	0.0%	74.7	0.0%	73.0	−2.2%	69.4	−7.0%
ifls_school	4	115.7	115.7	0.0%	111.5	−3.6%	112.5	−2.8%	104.4	−9.8%
leaps_height	2	31.3	33.0	5.5%	33.0	5.5%	33.0	5.5%	31.3	0.0%
leaps_height	3	76.4	76.4	0.0%	73.4	−4.0%	75.8	−0.9%	68.3	−10.7%
leaps_height	4	108.1	108.1	0.0%	102.2	−5.4%	107.2	−0.9%	101.3	−6.3%
leaps_test	2	33.5	36.2	8.2%	36.2	8.2%	35.7	6.6%	33.5	0.0%
leaps_test	3	80.3	80.3	0.0%	79.4	−1.2%	76.7	−4.5%	72.3	−10.0%
leaps_test	4	116.8	116.8	0.0%	112.7	−3.4%	116.2	−0.5%	106.8	−8.6%
mexico_ene	2	27.3	27.8	2.0%	27.8	2.0%	27.6	1.4%	27.3	0.0%
mexico_ene	3	69.0	69.0	0.0%	67.1	−2.8%	69.0	0.0%	63.4	−8.1%
mexico_ene	4	97.4	97.4	0.0%	91.3	−6.3%	96.5	−0.8%	90.5	−7.1%
silver	2	27.7	29.7	7.2%	29.7	7.2%	29.7	7.2%	27.7	0.0%
silver	3	73.1	72.8	−0.4%	72.1	−1.4%	72.1	−1.4%	*	*
silver	4	132.2	139.5	5.5%	139.5	5.5%	137.6	4.1%	*	*
sri_lanka	2	30.6	31.0	1.2%	31.0	1.2%	31.0	1.2%	30.6	0.0%
sri_lanka	3	74.1	74.1	0.0%	73.6	−0.7%	74.0	−0.1%	68.6	−7.4%
sri_lanka	4	114.6	114.6	0.0%	110.7	−3.4%	112.5	−1.9%	99.0	−13.6%
states	2	93.9	99.2	5.6%	99.2	5.6%	99.2	5.6%	93.9	0.0%
states	3	241.4	241.4	0.0%	241.4	0.0%	240.4	−0.4%	217.1	−10.0%
states	4	348.4	348.4	0.0%	348.4	0.0%	345.5	−0.8%	329.4	−5.5%
x100	2	18.9	21.3	12.7%	21.3	12.7%	20.9	10.6%	18.9	0.0%
x100	3	55.9	55.9	0.0%	55.9	0.0%	53.2	−4.8%	49.6	−11.3%
x100	4	116.7	116.7	0.0%	116.2	−0.4%	111.1	−4.8%	89.8	−23.1%

best result of the `naiveGreedy`, `optGreedy`, and `optimal` algorithm options. The newly presented methods MB-G, MB-DC-G, MB-DC-PAP, and MB-O are implemented in Python. Instances of MB-O and PAP arising in MB-DC-G, MB-DC-PAP, MB-O are solved using the Gurobi 14.1 solver.

We report the relative gap in total distance for our new solution methods, using the best results from `blocktools` as a baseline. We do not report MB-O results for the `silver` instances because they are too large to be solved optimally within two hours. For MB-DC-G and MB-DC-PAP, we choose maximum subproblem size $s = 8t$. Note that for $t = 2$, `blocktools` provides optimal results in all cases.

Table 2 shows that MB-O matches or exceeds `blocktools` block balance quality in all test instances, except for `silver` where the method is too computationally expensive. In multi-arm instances we frequently see improvement of 10% or more. The computationally more efficient MB-DC-G and MB-DC-PAP approaches also frequently outperform `blocktools`.

We noted in Sect. 4.2 that the block balance of MB-DC-G and MB-DC-PAP is guaratneed to be no worse than that provided by MB-G. Nevertheless, `blocktools` results are sometimes better than MB-DC-G and MB-DC-PAP. Note that, for $t = 2$, `blocktools` produces optimal results when `algorithm=optimal`. Further, the `blocktools` implementation of MB-G randomly breaks minimum-distance ties, see [23].

In Table 3 we compare algorithm runtime for selected test instances. We see that for problems of modest size (everything other than `silver`), MB-O is solved in well under ten seconds, well within reason for offline use cases.

Table 3. Runtime (in seconds) by problem and algorithm

Name	t	MB-G	MB-DC-G	MB-DC-PAP	MB-O
`silver`	2	1.44	3.00	3.11	1.44
	3	1.61	4.27	5.7	*
	4	1.3	6.62	30.29	*
`states`	2	0.01	0.15	0.15	1.91
	3	0.01	0.56	0.33	2.43
	4	0.01	0.45	0.58	7.21
`x100`	2	0.04	0.32	0.31	1.2
	3	0.03	0.58	0.65	3.22
	4	0.03	2.85	2.39	4.04

In Table 2 we noted that our implementation of MB-O fails to solve the full `silver` instance within two hours. In Table 4 we solve increasingly large subsets of `silver`. An asterisk denotes that the instance could not be solved within two hours. Solution time increases rapidly with problem size and the number of treatments. Alternative approaches to solving MB-O may address this limitation.

Table 4. MB-O runtime of `silver` by problem and number of treatments

n	t=2	t=3	t=4	t=5	t=6
24	0.17	0.18	0.54	0.54	1.64
48	1.51	1.93	2.22	16.91	46.2
72	3.98	161.44	35.45	150.14	2551.25
96	10.55	1096.32	*	*	*

Finally, we report on treatment balance. We apply a local improvement approach to COV-OPT after applying MB-G, MB-DC-G, and MB-DC-PAP. Figure 2 provides an illustrative example of treatment covariate differences across algorithms. The mean SMD μ for each covariate is plotted for each algorithm for each pair of treatments. It is seen that while there are variations across covariates, overall the mean SMD μ for MB-O is approximately 30% smaller than of MB-G, $\mu = 0.09$ and $\mu = 0.129$, respectively.

We summarize covariate balance across algorithms and instances in Table 5, reporting the ratio of the mean SMD of each solution to the best mean SMD each instance. The mean ratio for MB-O is 0.47, the best of all reported methods. While covariate balance for MB-G, MB-DC-G, and MB-DC-PAP varies by problem, they are similar in the aggregate.

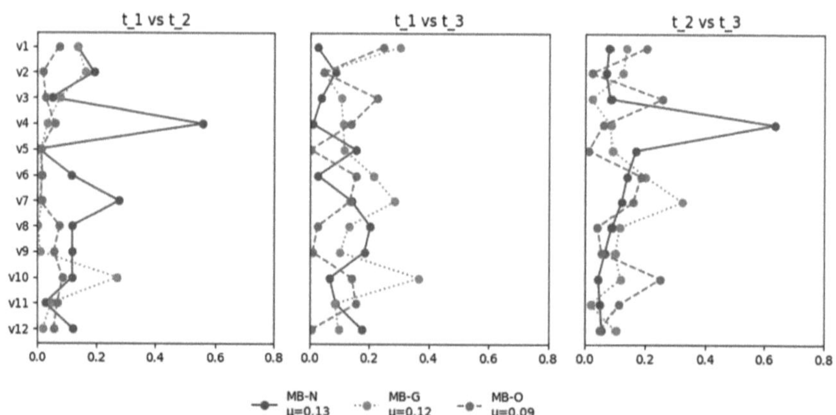

Fig. 2. Mean SMD μ for `states` by treatment and algorithm.

Table 5. Relative gap for covariate mean SMD by instance and algorithm

Name	t	MB-N	MB-G	MB-DC-G	MB-DC-PAP	MB-O
ifls_expend	2	1.00	0.73	0.73	0.73	0.29
	3	1.00	0.77	0.77	0.77	0.55
	4	1.00	0.57	0.36	0.47	0.88
ifls_school	2	1.00	0.44	0.47	0.44	0.26
	3	1.00	0.61	0.61	0.63	0.27
	4	0.89	0.76	0.68	0.63	1.00
leaps_height	2	0.51	1.00	1.00	1.00	0.14
	3	1.00	0.71	0.30	0.74	0.25
	4	0.81	0.99	0.72	0.51	1.00
leaps_test	2	1.00	0.67	0.67	0.62	0.25
	3	1.00	0.78	0.66	0.56	0.32
	4	0.85	0.63	0.65	0.55	1.00
mexico_ene	2	1.00	0.49	0.49	0.57	0.28
	3	0.96	0.96	1.00	0.96	0.23
	4	1.00	0.66	0.65	0.70	0.52
silver	2	1.00	0.50	0.50	0.50	0.34
	3	1.00	0.65	0.48	0.68	0.58
	4	1.00	0.86	0.86	0.60	0.84
states	2	1.00	0.78	0.78	0.78	0.23
	3	1.00	0.67	0.67	0.87	0.37
	4	1.00	0.70	0.70	0.65	0.58
x100	2	1.00	0.18	0.18	0.10	0.01
	3	0.78	0.28	0.09	0.19	1.00
	4	1.00	0.41	0.37	0.26	0.19
μ		0.95	0.66	0.60	0.60	0.47

7 Conclusions

We have presented new approaches for matched block design leveraging mixed integer programming and divide-and-conquer formulations. We have shown that these approaches can provide improved match quality and covariate balance, and that they can be employed to account for large problem sizes. We have also presented a consolidated set of test instances that can be used to evaluate future approaches.

We see several opportunities to build upon the results presented here. First, refinement of the mixed integer formulation and solution approach given in Sect. 4.1 to improve runtime performance. We also believe that metaheuristic approaches are a promising direction for matched block design, especially in conjunction with the divide-and-conquer approach presented in Sect. 4.2. Finally, we believe that there are opportunities to address the covariate balance problem in a more principled manner than that presented in Sect. 5, as well as to consider block and treatment balance in the same modeling framework.

Acknowledgments. We thank Will Betz and Yan Zhang for the work that motivated us to consider block balance more deeply. We thank Ian Silver for providing the dataset used in the `silver` examples, and for his encouragement. Finally, the author thanks the reviewers of this paper for their insightful comments and suggestions, which helped to improve its quality.

References

1. Andrabi, T., Das, J., Khwaja, A.I., Zajonc, T.: Do value-added estimates add value? Accounting for learning dynamics. Am. Econ. J. Appl. Econ. **3**, 29–54 (2011)
2. Birkhoff, G.: Tres observaciones sobre el álgebra lineal [three observations on linear algebra]. Univ. Nac. Tucumán. Rev. A **5**, 147–151 (1946)
3. Bloom, H.: The core analytics of randomized experiments for social research, Technical report, MDRC, New York (2006)
4. Bruhn, M., McKenzie, D.: In pursuit of balance: randomization in practice in development field experiments. Am. Econ. J. Appl. Econ. **1**, 200–232 (2009)
5. De Mel, S., McKenzie, D., Woodruff, C.: Returns to capital in microenterprises: evidence from a field experiment. Q. J. Econ. **123**, 1329–1372 (2008)
6. Fisher, R.: Design of Experiments. Oliver and Boyd, Edinburgh (1935)
7. George, A.: Nested dissection of a regular finite element mesh. SIAM J. Num. Anal. **10** (1973)
8. Greevy, R., Silber, J., Rosenbaum, P.: Optimal multivariate matching before randomization. Biostatistics **5**, 263–275 (2004)
9. Grötschel, M., Wakabayashi, Y.: A cutting plane algorithm for a clustering problem. Math. Program. **45**, 59–96 (1989)
10. Higgins, M.J., Hall, P., Villar, S.: Improving massive experiments with threshold blocking. Proc. Natl. Acad. Sci. USA **113**, 7369–76 (2016)
11. Imai, K., King, G., Nall, C.: The essential role of pair matching in cluster-randomized experiments, with application to the Mexican universal health insurance evaluation. Stat. Sci. **24**, 29–53 (2009)
12. Imai, K., King, G., Nall, C.: Rejoinder: matched pairs and the future of cluster-randomized experiments. Stat. Sci. **24**, 65–72 (2009)
13. Imai, K., King, G., Stuart, E.A.: Misunderstandings among experimentalists and observationalists about causal inference. J. R. Stat. Soc. A. Stat. Soc. **171**, 481–502 (2008)
14. G.E.I. Instituto Nacional de Estadística, Encuesta nacional de empleo 2004 (2004). https://en.www.inegi.org.mx/programas/ene/2004/
15. Ji, X., Mitchell, J.E.: Finding optimal realignments in sports leagues using a branch-and-cut-and-price approach. Int. J. Oper. Res. **1** (2016)
16. Jonker, R., Volgenant, A.: A shortest augmenting path algorithm for dense and sparse linear assignment problems. Computing **38**, 325–340 (1987)
17. Keele, L., Lenard, M., Page, L.: Matching methods for clustered observational studies in education. J. Res. Educ. Effect. **14**, 696–725 (2021)

18. Kesavan, S., Lambert, S.J., Williams, J.C., Pendem, P.K.: Doing well by doing good: improving retail store performance with responsible scheduling practices at the gap, Inc. Manage. Sci. **68**, 7818–7836 (2022)
19. Kuhn, H.W.: The Hungarian method for the assignment problem. Naval Res. Logistics Q. **2**, 83–97 (1955)
20. Li, Y.P., Propert, K.J., Rosenbaum, P.R.: Balanced risk set matching. J. Am. Stat. Assoc. **96**, 870–882 (2001)
21. Lu, B., et al.: Optimal nonbipartite matching and its statistical applications. Am. Stat. **65**, 21–30 (2011)
22. Lu, B., Rosenbaum, P.R.: Optimal pair matching with two control groups. J. Comput. Graph. Stat. **13**, 422–34 (2004)
23. Moore, R.T.: blockTools: blocking, assignment, and diagnosing interference in randomized experiments (2013). https://www.ryantmoore.org/html/software.blockTools.html
24. Moore, R.T.: Multivariate continuous blocking to improve political science experiments. Polit. Anal. **20**, 460–479 (2017)
25. Panagopoulos, C., Green, D.P.: Field experiments testing the impact of radio advertisements on electoral competition. Am. J. Polit. Sci. **52**, 156–168 (2008)
26. Rosenbaum, P.R.: Impact of multiple matched controls on design sensitivity in observational studies. Biometrics **69**, 118–127 (2013)
27. Rosenbaum, P.R.: Modern algorithms for matching in observational studies. Ann. Rev. Stat. Appl. **7**, 143–176 (2020)
28. Rosenbaum, P.R., Rubin, D.B.: The central role of the propensity score in observational studies for causal effects. Biometrika **70**, 41–55 (1983)
29. Silver, I.A.: Blocked randomization with r for randomized controlled trials (2025). https://ianasilver.com/blocked-randomization-with-r-for-randomized-controlled-trials/
30. Strauss, J., Beegle, K., Sikoki, B., Dwiyanto, A., Herawati, Y., Witoelar, F.: The third wave of the Indonesia family life survey (IFLS3): overview and field report. WR-144/1-NIA/NICHD (2004)
31. Woodruff, C., McKenzie, D., de Mel, S.: Returns To Capital In Microenterprises : Evidence from a Field Experiment, The World Bank (2007)
32. Zubizarreta, J.R.: Using mixed integer programming for matching in an observational study of kidney failure after surgery. J. Am. Stat. Assoc. **107**, 1360–1371 (2012)
33. Zubizarreta, J.R., Keele, L.: Optimal multilevel matching in clustered observational studies: a case study of the effectiveness of private schools under a large-scale voucher system. J. Am. Stat. Assoc. **112**, 547–560 (2017). https://doi.org/10.1080/01621459.2016.1240683

CHORUS: Zero-Shot Hierarchical Retrieval and Orchestration for Generating Linear Programming Code

Tasnim Ahmed$^{(\boxtimes)}$ (iD) and Salimur Choudhury (iD)

School of Computing, Queen's University, Kingston, ON, Canada
`{tasnim.ahmed,s.choudhury}@queensu.ca`

Abstract. Linear Programming (LP) problems aim to find the optimal solution to an objective under constraints. These problems typically require domain knowledge, mathematical skills, and programming ability, presenting significant challenges for non-experts. This study explores the efficiency of Large Language Models (LLMs) in generating solver-specific LP code. We propose CHORUS, a retrieval-augmented generation (RAG) framework for synthesizing Gurobi-based LP code from natural language problem statements. CHORUS incorporates a hierarchical tree-like chunking strategy for theoretical contents and generates additional metadata based on code examples from documentation to facilitate self-contained, semantically coherent retrieval. Two-stage retrieval approach of CHORUS followed by cross-encoder reranking further ensures contextual relevance. Finally, expertly crafted prompt and structured parser with reasoning steps improve code generation performance significantly. Experiments on the NL4Opt-Code benchmark show that CHORUS improves the performance of open-source LLMs such as Llama3.1 (8B), Llama3.3 (70B), Phi4 (14B), Deepseek-r1 (32B), and Qwen2.5-coder (32B) by a significant margin compared to baseline and conventional RAG. It also allows these open-source LLMs to outperform or match the performance of much stronger baselines—GPT3.5 and GPT4 while requiring far fewer computational resources. Ablation studies further demonstrate the importance of expert prompting, hierarchical chunking, and structured reasoning.

Keywords: Linear Programming · Large Language Models · Code Generation · Retrieval-Augmented Generation · Gurobi Solver

1 Introduction

Solving Linear Programming (LP) problems, an essential field of applied mathematics, has demonstrated value in areas like supply chain management, energy scheduling, marketing, resource allocation, networking, etc. [1]. Traditionally, this process involves three steps: translating real-world scenarios into mathematical formulations by identifying variables, objectives, constraints, and parameters; encoding these formulations using modeling languages such as Python, R,

© The Author(s), under exclusive license to Springer Nature Switzerland AG 2026
Y. Zhang et al. (Eds.): LION 2025, LNCS 15745, pp. 64–80, 2026.
https://doi.org/10.1007/978-3-032-09192-5_5

or C++; and finally, executing the optimization using solvers [2]. While practitioners proficient in mathematical modeling are generally capable of using solver APIs, the overall process still demands fluency in both the mathematical and programming domains.

Large Language Models (LLMs) have become powerful tools, demonstrating remarkable performance across various tasks, including code generation and reasoning tasks [3]. As LLMs grow in size, they serve effectively as standalone knowledge repositories, with facts encoded in their parameters [4], and can be further enhanced through fine-tuning on specific downstream tasks [5]. However, even large models often lack sufficient domain-specific knowledge for specialized tasks, and their factual accuracy can diminish as the world evolves. Moreover, updating a model's internal knowledge through fine-tuning or pretraining remains challenging, especially with vast and frequently changing text corpora, e.g., code or tool documentation [6]. To address these limitations, retrieval techniques such as Retrieval-Augmented Generation (RAG) have been introduced [7], augmenting model inputs by appending contextually relevant documents retrieved from external knowledge corpora. However, traditional RAG pipelines have limitations. LLMs face challenges processing numerous chunks (e.g., top-100), not only due to efficiency constraints but also because shorter top-k lists (e.g., 5 or 10) improve generation accuracy [8]. With a small k, ensuring high recall of relevant content is crucial. A retrieval model alone may struggle with local alignments across the embedding space, while a cross-encoding ranking model better selects top-k contexts from top-N candidates ($N \gg k$) [9]. Furthermore, fixed-size chunking in code documentation disrupts dependencies, breaking functions, losing syntax, and separating explanations, reducing comprehensibility and leading to ineffective retrieval with incomplete functions and missing variables, increasing hallucination risks.

Efforts to simplify mathematical programming (linear, nonlinear, etc.) using LLMs aim to enhance accessibility for non-experts. Research studies have largely focused on utilizing LLMs to generate mathematical formulations and end-to-end solutions, converting problem descriptions into solver code. Notable contributions in generating mathematical formulation from LP problem descriptions include NL4Opt [2] and LM4OPT [10]. The NL4Opt competition explored converting natural language into structured LP formulations. The dataset samples from this competition include an input paragraph describing an LP problem, with annotations for entity extraction and semantic parsing to generate the representation of the objective and constraints. Using the NL4Opt dataset, the LM4OPT framework was proposed to enhance the performance of smaller LLMs in generating mathematical formulations. The methodology focuses on progressive fine-tuning, starting with broader domain contexts before specializing in formulation task-specific datasets. Tsouros et al. [11] proposed one of the first end-to-end solutions for this task, introducing an LLM-based system that generates mathematical formulations and solver codes from prompts. While demonstrating promise on the NL4Opt dataset, it lacked benchmark comparisons and relied solely on pretrained LLMs. E-OPT [12] provided a benchmark evaluating the mathematical programming code generation capabilities of LLMs across linear, nonlinear, and tabular problems. Teshinizi et al. [13] introduced OptiMUS, a multi-agent LLM framework to both formulate and solve mixed-integer LP problems based on natural language descriptions. It iteratively engages specialized agents, generating multiple responses per problem, leading to very high computational overhead. Another work, OptLLM [14] integrates LLMs with external solvers to automate the modelling and solving of LP problems. The authors curated a dataset with 100 problems from the NL4Opt development set to evaluate the performance of GPT3.5 [15], GPT4 [16], and their fine-tuned Qwen model. In addition, OptiGuide [17] combines GPT4 with plain-text user queries to generate insights into optimization outcomes which enhances decision-making in supply chain management. Researchers have explored LLM-based approaches that solve such problems without external solvers. Yang et al. [18] introduced OPRO, a framework that iteratively refines solutions using a meta-prompt. However, LLMs are not well-suited for such tasks, as their generation relies primarily on next-token prediction rather than structured mathematical reasoning. To summarize, recent advancements have demonstrated that numerous researchers are actively working on generating problem formulations or solver code from LP or similar mathematical programming problem descriptions using LLMs. However, solver codes typically rely heavily on documentation, and popular solvers frequently update their frameworks or tool versions to incorporate new features. These solvers often function as APIs that support multiple programming languages. As a result, fine-tuning a model on a specific dataset for a particular solver ties the model to that specific solver version and programming language. A major version update in the future could render the generated codes prone to errors, necessitating further annotation and re-fine-tuning, a resource-intensive process. Thus, depending on the parametric knowledge of LLMs or fine-tuning for code generation can lead to rigid, narrowly focused solutions that require significant effort to maintain over time. Moreover, while LLMs excel in traditional analytical and algorithmic code generation due to abundant examples and libraries, the scarcity of solver-specific LP code examples limits their domain-specific knowledge, a challenge for both pretrained and fine-tuned LLMs [13].

To address this limitation, our study focuses on providing an adaptive end-to-end solution to generate solver code directly from mathematical problem descriptions, without relying on any specific solver or LLM. We hypothesize that providing an enhanced context is essential to tackle this complex task. Therefore, in this paper, we propose a novel RAG-based framework, CHORUS (Contextual Hierarchical Orchestration for Retrieval-aUgmented code Synthesis), to address this challenge. To the best of our knowledge, this is the first work to introduce a retrieval mechanism for generating optimization solver code directly from natural language problem descriptions. In this study, we generate LP codes for the Gurobi [19] solver API in Python, widely used for its efficiency and robust optimization algorithms. However, our approach remains adaptable to any mathematical problem, given an available solver. Our contributions are as follows:

1. We introduce CHORUS, a novel zero-shot RAG framework that enhances LLMs' retrieval abilities by leveraging hierarchical chunking and contextual metadata from LP solver (Gurobi) documentation for code generation.
2. Proposed two-stage retrieval process with cross-encoder reranking improves code generation accuracy significantly.
3. Expertly crafted prompt and structured parser with reasoning steps improve the reasoning capability of LLMs as well as facilitates automated inference, execution and evaluation.
4. Experiments on the NL4Opt-Code benchmark show significant gains over baseline LLMs and traditional RAG, allowing open-source LLMs to achieve GPT4-level performance with lower computational costs.

2   CHORUS

This section delineates CHORUS, which enhances LLM retrieval by leveraging hierarchical chunking and contextual metadata from LP solver documentation. It describes the retrieval process with cross-encoder reranking and details the use of an expertly crafted prompt with a structured parser and reasoning steps. Figure 1 presents an overview of the proposed pipeline at inference.

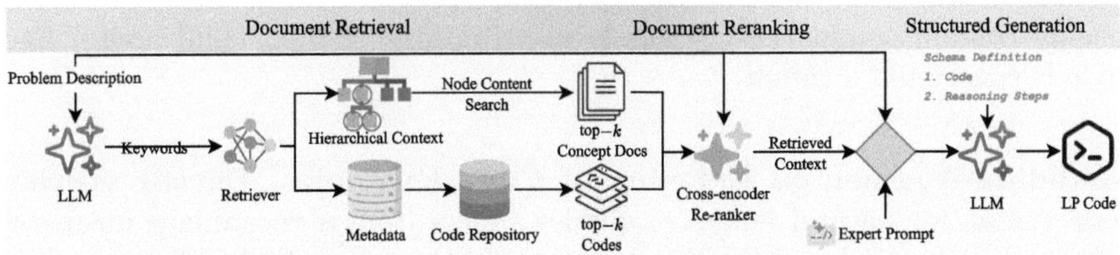

Fig. 1. CHORUS framework for generating Gurobi solver code from LP descriptions. It retrieves and reranks relevant conceptual and implementation-level documents, supplements them with an expert prompt, and employs a structured output parser for automated execution and improved reasoning.

2.1   Document Structuring and Indexing

The Gurobi documentation, serving as the primary source of external context, exhibits a dual nature. The first category comprises theoretical and contextual content that includes descriptions of various models, variables, and constraint types, as well as code snippets. This content is inherently rich in semantic information and provides a deep theoretical foundation necessary for understanding the nuances of different mathematical programming problem formulations. The second category consists solely of complete code examples that demonstrate the practical application of the Gurobi library in solving diverse mathematical programming scenarios. Although these code examples lack explanatory text, they embody critical implementation details that are essential for generating syntactically correct and functionally appropriate code. Given the heterogeneity in the types of content, we adopt a dual approach for storing and retrieving documents.

Hierarchical Tree Indexing of Theoretical Content. Technical documents are inherently structured, and flattening it into equal-length chunks disrupts semantic coherence. By modeling the documentation as a tree, CHORUS mimics human navigation patterns, starting from broad overviews and drilling down to specifics. In a recent study on multi-document QA, Chen et al. [20] proposed a hierarchical multi-route retrieval system that enhances retrieval accuracy and helps reduce hallucinations. Our manual analysis of the retrieved content supports this idea, suggesting that such a retrieval approach helps minimize hallucination by providing the LLM with self-contained and contextually complete technical information. The theoretical documents are processed using the inherent structure of the source PDFs. We extract metadata and content from the PDF files to construct a hierarchical tree representation. At the root of this tree is the entire document, which is then segmented into chapters. Each chapter is further divided into sections, and these sections are subdivided into subsections. The hierarchical tree is designed to preserve both detailed and high-level contextual information. The main bulk of the theoretical content is distributed across the leaf nodes, while the intermediary nodes preserve introductory texts that usually contain summaries or high-level definitions of their child nodes. Each node is considered a chunk.

Metadata-Augmented Indexing of Code Examples. Directly searching code syntax for natural language queries suffers from a vocabulary mismatch. Gao et al. [21] introduced HyDE (Hypothetical Document Embeddings), a dense retrieval method that generates an ideal document, embeds it into a vector, and retrieves similar real documents. This approach enhances retrieval precision by capturing the conceptual essence and mitigating vocabulary mismatches. Inspired by this workflow, our proposed code example retrieval mechanism depends upon enriched metadata to bridge the semantic gap between natural language queries and stored code examples. In contrast to the theoretical contents, the complete code examples are managed as individual chunks, each representing a distinct code snippet. These snippets are stored in a vector database along with generated metadata. For every code example, we generate a set of domain-agnostic keywords (5–7) using an LLM providing an expert prompt. These keywords capture salient features such as the type of variables, the nature of the problem, and the specific constraints used in the model which abstract the functionality of the code. Additionally, a concise 2–3 line natural language synopsis is generated for each code snippet using the same approach. This metadata allows semantic alignment with user queries without requiring direct lexical matches to code syntax, which is critical for interdisciplinary problems where users describe constraints in non-technical terms.

The impact of our document structuring methodology is demonstrated through comparative analyses of chunk characteristics. Figure 2 highlights the natural variation in chunk lengths between theoretical documents and code examples, contrasting with the rigid fixed-length chunking used in traditional RAG systems. Figure 3 validates our metadata augmentation strategy through lexical analysis. The raw code term frequencies reveal dominance of implementation-specific tokens (e.g., "gurobipy", "batchid", "continue") that rarely appear in natural language queries. Conversely, the metadata-derived terminology demonstrates improved alignment with natural query patterns.

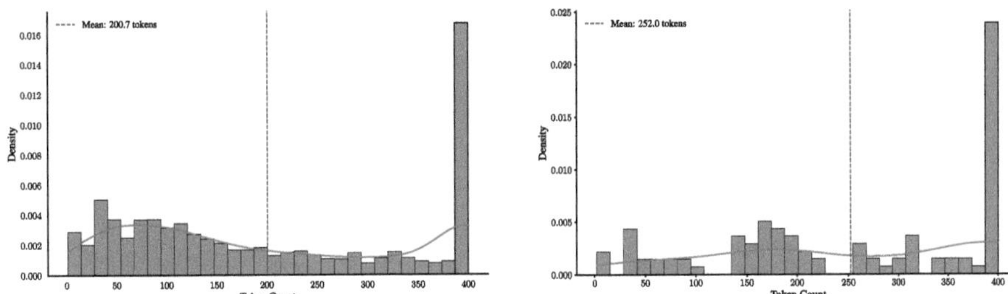

Fig. 2. Token length distributions for (left) theoretical documentation chunks and (right) code examples.

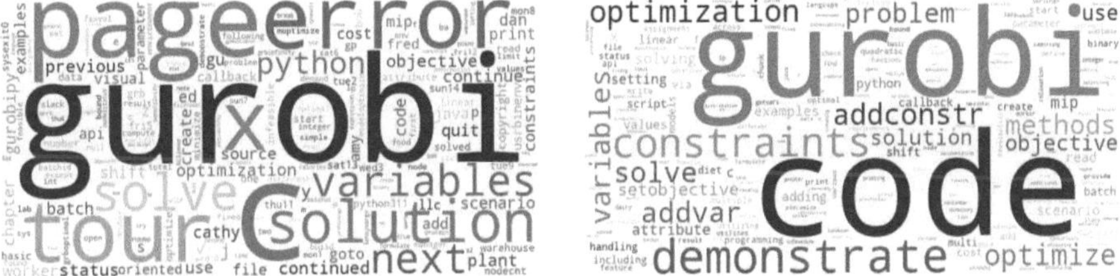

Fig. 3. Word frequency analysis contrasting (left) raw code token prevalence with (right) curated metadata terminology from summaries and keywords.

2.2   Two-Stage Context Retrieval

The retrieval process operates in two phases: conceptual context retrieval from the hierarchical documentation and code example retrieval from the indexed implementations. This separation allows the framework to retrieve both theoretical foundations and practical examples.

Hierarchical Context Retrieval. An LLM starts with extracting a set of keywords from the problem description, focusing on LP-specific concepts (e.g., "binary variables," "supply chain optimization"). Keyword extraction is guided by a prompt engineered to prioritize terms likely to appear in technical documentation. These keywords are used to traverse the hierarchical document tree. For each keyword, top-k most relevant nodes are selected based on the cosine similarity between their embeddings and node content. The retrieved chunk post-processing follows a dynamic, context-aware strategy. To ensure retrieved chunks are contextually self-contained yet concise, adaptive chunk construction dynamically combines sibling or parent content when node sizes fall below the maximum chunk size threshold (400 tokens). For smaller nodes, sibling content is concatenated iteratively until the token limit is reached. Crucially, parent node summaries are prepended to child content to retain hierarchical context. For example, a subsection describing quadratic constraints inherits the introductory context from its parent subsection on constraints, which ensures retrieved chunks provide both specificity and conceptual grounding. This approach mirrors human navigation patterns in technical documents, where practitioners iteratively narrow from broad overviews to granular details.

Code Example Retrieval. The second phase of the retrieval process focuses on the complete code examples. Here, the keywords extracted from the problem description are matched not directly against the code itself but against the metadata that accompanies each complete code example. A similarity search is conducted within the vector database to retrieve top m candidate examples that best match descriptive keywords. The reliance on metadata rather than raw code ensures that the retrieval process remains robust to variations in coding style and syntax. This approach avoids lexical mismatches between natural language queries (e.g., "minimize delivery costs") and code variables (e.g., `model.setObjective(sum(c[i,j]*x[i,j] for i,j in routes), GRB.MINIMIZE)`).

2.3 Cross-Encoder Reranking for Contextual Relevance

After the initial retrieval phase, the system retains two candidate sets: one comprising $k \times n_{keywords}$ conceptual documents from the hierarchical tree and the other comprising m complete code examples. However, to further refine the selection and ensure that only the most relevant documents are utilized during code generation, we introduce a re-ranking mechanism based on a cross-encoder model [22]. While dual-encoders excel at scalable retrieval, their separate encoding of queries and documents limits their ability to capture fine-grained semantic relationships. The cross-encoder model is trained on the MS MARCO passage ranking task[1] and such models have proven to transfer very effectively to new domains, including technical text and code. The re-ranking process involves constructing input pairs for the cross-encoder, where each pair

---

[1] github.com/microsoft/MSMARCO-Passage-Ranking.

consists of the user query, keywords, and the text content (with metadata, in the case of code examples) of a candidate document. The cross-encoder model then computes a relevance score for each pair, effectively quantifying the semantic alignment between the query and the document. The candidate documents are subsequently sorted in descending order of their scores, and a selective truncation is performed wherein the top-3 conceptual documents and the top-2 code examples are retained for further processing.

2.4   LLM Code Generation

The final set of retrieved and re-ranked documents, comprising both conceptual context and code examples, is then integrated into the expert prompt that is fed to the language model. This integration serves a dual purpose—it enriches the input with domain-specific knowledge and ensures that the generated code adheres to both theoretical principles and practical implementation guidelines as prescribed by the Gurobi documentation.

Expert Prompting. To ensure that the generated code rigorously matches each problem specification while maintaining high-quality coding practices, we implement specialized system and user prompts. The system prompt directs the language model to assume the role of an expert in both operations research and Python-based Gurobi development, whereas the user prompt outlines strict requirements such as the function name, error handling strategy, variable naming conventions, and constraint modeling rules. These prompts merge specialized domain guidance with explicit style and accuracy rules, which enforce the LLM to create Python functions that accurately capture the mathematical formulation, limit unneeded output, and include exception handling for potential errors.

Structured Output with Improved Reasoning. A fundamental requirement of CHORUS is its capacity for end-to-end automation, including code inference, evaluation, and execution without reliance on external annotations or human intervention. Preliminary experiments with LLMs revealed a notable inconsistency: despite configuring the sampling temperature to minimize stochasticity, model responses demonstrate significant variability across queries. This observation shows a contradictory strain between the unstructured nature of responses—rooted in their training on vast, heterogeneous corpora—and the structured precision demanded by code generation tasks. To enforce consistency, we implemented a structured output-parsing mechanism, wherein each LLM response is validated against a predefined schema (e.g., a class definition) to extract executable code. However, as demonstrated in [23], strict formatting constraints can impair LLMs' performance on analytical and reasoning tasks, with degradation intensifying as constraints tighten. We assume that enforcing code-only outputs, devoid of any explanatory text, may inadvertently restrict the model's capacity to engage in stepwise reasoning—a critical factor in ensuring code correctness. This hypothesis also aligns with empirical findings from

OpenAI[2], which suggest that LLM output quality improves when models are instructed to provide intermediate reasoning steps within a structured schema. Building on this insight, we augmented our output schema with a dedicated `reasoning_steps` field, explicitly instructing the model to validate the code alignment with the problem specification (e.g., variable definitions, constraint formulations, and objective functions). The schema definition is as follows:

```
class GurobiSolution(BaseModel):
    code: str = Field(description="Complete Python code using
    Gurobi API that solves the LP problem")
    reasoning_steps: str = Field(description="Justification
    mapping code components to problem requirements")
```

3   Experiments

3.1   Task Formulation

Given a natural language problem description X, the CHORUS framework implements a two-stage retrieval process to identify relevant context from structured documentation and code examples. The hierarchical retriever R_H extracts conceptual documents $D_C = \{d_{c1}, ..., d_{ck}\}$ by traversing a tree-structured index of theoretical content, preserving parent-child context relationships. Simultaneously, the metadata-augmented retriever R_M identifies code examples $D_E = \{d_{e1}, ..., d_{em}\}$ through keyword and summary alignment with X. These candidates are reranked by a cross-encoder C to yield optimal context subsets $D_C^* \subseteq D_C$ and $D_E^* \subseteq D_E$. The generator G, conditioned on X, D_C^*, and D_E^*, produces structured output $Y_{\text{structured}}$ containing:

$$Y_{\text{structured}} = \langle \text{code}, \text{reasoning_steps} \rangle$$

where:

- code: Executable Gurobi code adhering to API constraints
- reasoning_steps: Natural language justification mapping code to X

3.2   Dataset and Metrics

To the best of our knowledge, there are currently no publicly available datasets on general problems except the data from the NL4Opt competition. We conducted an empirical study using the NL4Opt-Code dataset, which we curated from the original problem descriptions found in the NL4Opt dataset. The NL4Opt dataset includes LP Word Problems (LPWPs) across six domains: sales, advertising, investment, production, transportation, and sciences. Each LPWP has problem descriptions and annotations for key entities such as constraints, objectives, and variables. We expanded this dataset by generating Gurobi code for each LPWP. The NL4Opt-Code dataset now contains LPWPs, each paired with its corresponding Gurobi solver code.

---

[2] openai.com/index/introducing-structured-outputs-in-the-api.

Dataset Annotation. The dataset annotation was performed by a domain expert (a graduate student who completed an undergraduate course in Optimization and a graduate course in Combinatorial Optimization). The annotation process was carried out in two stages. In the first stage, the annotator derived the LP formulation for each problem, ensuring that the mathematical representation accurately captured the nuances of the problem description. These formulations were then verified by another expert with similar qualifications to ensure their correctness. In the second stage, the domain expert wrote the Gurobi code based on the verified formulations, which was then double-checked by another expert.

Evaluation Metrics. The evaluation metric used in E-OPT [12] uses the output code to produce numerical answers for variables and objectives, which are compared against the ground truth for accuracy. On the other hand, NL4Opt compares the coefficients of variables and decision variables in constraints and the objective function against the ground truth based on an intermediate matrix representation. However, this approach imposes certain limitations, such as requiring all constraints to be in the less-than-or-equal-to form and the objective function to be in the maximization form. As a result, a correct formulation may be considered incorrect if it does not conform to these specific rules. Unlike NL4Opt and ReSocratic, we adopted a more stringent definition for our primary evaluation metric, "accuracy." In addition, we incorporated three supplementary measures that evaluate code quality at varying levels of detail—specifically, "syntactic validity," "semantic similarity," and "edit distance."

Accuracy: A generated response is considered correct only if the optimal objective function value from the generated solver code matches that of the reference solver code. This metric offers an automated and fast method for evaluating the solution, eliminating the need for human evaluators. Additionally, when the objective value matches, it generally indicates that all constraints and objective functions are correctly aligned.

$$\text{Accuracy} = \frac{1}{|X|} \sum_{x \in X} \mathbb{I}\big(f\big(Y_{\text{gen}}(x)\big) = f\big(Y_{\text{ref}}(x)\big)\big) \tag{1}$$

Here,

- X be the set of all problem instances in the test set.
- $Y_{\text{gen}}(x)$ be the generated code for problem instance $x \in X$.
- $Y_{\text{ref}}(x)$ be the reference code for problem instance $x \in X$.
- f be the optimal objective function value produced by executing a solver code.
- $\mathbb{I}$ be the indicator function, where $\mathbb{I}(A) = 1$ if condition A is true, and $\mathbb{I}(A) = 0$ otherwise.

Syntactic Validity: It checks whether the generated code can be parsed and executed without syntax errors. It serves as a sanity check to ensure that the model produces well-formed Python code.

$$\text{Syntactic Validity} = \frac{1}{|X|} \sum_{x \in X} \mathbb{I}\big(\text{parse}\big(Y_{\text{gen}}(x)\big) = \text{True}\big) \tag{2}$$

Semantic Similarity: It quantifies the conceptual similarity between the generated and reference code using embeddings from a pre-trained language model (`all-MiniLM-L6-v2`[3]). It is calculated using cosine similarity between the embedded representations. Jin et al. [24] demonstrated that traditional metrics such as BLEU and BERTScore frequently fall short in capturing nuanced semantic meaning and struggle with domain-specific terminology common in code, whereas LLM-based embeddings show a much stronger alignment with human evaluations.

$$\text{Semantic Similarity} = \frac{1}{|X|} \sum_{x \in X} \cos\big(\text{embed}\big(Y_{\text{gen}}(x)\big), \text{embed}\big(Y_{\text{ref}}(x)\big)\big) \tag{3}$$

Edit Distance: This similarity metric measures the character-level similarity between the generated and reference code using the Levenshtein ratio. Tree-based edit distances closely correlate with execution accuracy and often offer fine-grained insights into code similarity across languages [25].

$$\text{Edit Distance} = \frac{1}{|X|} \sum_{x \in X} \text{SequenceMatcher}(Y_{\text{gen}}(x), Y_{\text{ref}}(x)) \tag{4}$$

3.3   Experimental Setup

We designed the evaluation framework for efficient execution on a local server. To achieve this, we intentionally excluded methodologies that depend on online context and instead relied on open-source models and retrieval processes. Some entities may prefer not to use closed-source online LLMs for security and privacy. In certain scenarios, a solution may need to run on local devices or in remote areas. For instance, a recent country-wide internet outage in a country lasted several days, and if a business or operation depends on LP solving, a standalone local solution can be crucial.

Choice of LLMs. We chose a diverse set of LLMs—Llama3.1 (8B), Llama3.3 (70B) [3], Phi4 (14B) [26], Deepseek-r1 (32B) [27], and Qwen2.5-coder (32B) [28]—and compared them with the well-known baselines, GPT3.5 and GPT4. This selection covers a wide range of parameter sizes and architectures, letting us explore the trade-offs between model complexity, computational efficiency, and code generation quality. For instance, Llama3.1 (8B) gives us insight into how lightweight models perform in resource-limited settings, while Llama3.3 (70B)

---

[3] huggingface.co/sentence-transformers/all-MiniLM-L6-v2.

serves as a benchmark for high-capacity models with stronger reasoning capabilities. Phi4 (14B) and Deepseek-r1 (32B) were selected for their unique training approaches that may enhance reasoning, and Qwen2.5-coder (32B), with its coding focus, is particularly suited for our code generation task. We acknowledge the importance of testing CHORUS with GPT-4 and GPT-3.5. However, our experimental focus on open-source models aligns with the core goal of this work: improving resource-efficient, locally deployable LLMs that support accessibility, privacy, and computational feasibility for a wide range of users. Although GPT-4 and GPT-3.5 perform well, their proprietary nature, high operational costs, and dependence on cloud-based APIs make them less practical for many real-world applications. Additionally, GPT series models' larger parameter size and inference demands make it significantly computationally expensive compared to the open-source models tested [29].

Hardware Configuration. The experiments were conducted on a Linux server equipped with an Intel Xeon Platinum 8358 processor, featuring 38 MB of cache, 32 cores, and a maximum clock speed of 3.2 GHz. An NVIDIA H100 NVL GPU further accelerated the computational tasks with 95.8 GB of GPU memory. The system was supported by 251 GB of RAM.

3.4   Results

Table 1 presents a comparative analysis of the performance of the baseline models and the CHORUS framework across multiple LLMs. While GPT3.5 and GPT4 serve as baseline references, their strong performance (achieving accuracy scores of 0.5260 and 0.6367, respectively) emphasizes the effectiveness of closed-source models in LP code generation. Furthermore, the performance of GPT3.5 and GPT4 suggests that open-source LLMs (e.g., Qwen2.5-coder (32B)) are approaching the baseline performance of commercial, large-scale models. Among the open-source LLMs, the CHORUS framework results in substantial improvements in accuracy relative to the baseline. For example, the accuracy of Llama3.3 (70B) increases from 0.2289 to 0.5675, while Phi4 (14B) improves from 0.1938 to 0.6125. Notably, even smaller or mid-range models (e.g., Llama3.1 (8B) and Deepseek-r1 (32B)) demonstrate significant performance gains, which indicates that the incorporation of retrieval and re-ranking steps effectively enhances the alignment of generated code with the correct objective function and constraints. With the exception of the smallest model in this set (Llama3.1 (8B)), all open-source LLMs surpass GPT3.5 following the integration of the CHORUS framework, achieving performance levels comparable to GPT4. Given that GPT4 is considerably more computationally expensive and energy-intensive, the results demonstrate the effectiveness of CHORUS in optimizing the performance of locally deployed, resource-efficient LLMs. Syntactic validity improves in all CHORUS-based runs, indicating that a more coherent technical context reduces parsing errors. Results show that for the baseline open-source LLMs, the syntactic validity score ranges from 0.69 to 0.96. However, incorporating the CHORUS framework improves the score to 1.00, except for Llama models, where the score is in the range of 0.98. This demonstrates that integrating theoretical documents and code examples enhances syntactic correctness by ensuring the generated code is almost always valid. Such an improvement is particularly significant for code generation tasks, where syntactic accuracy directly impacts downstream usability and execution reliability. Likewise, semantic similarity improves with CHORUS, which implies that the integrated code examples and conceptual documentation help preserve semantic alignment between the generated and reference solutions. However, edit distance, which measures token-level similarity, is not a particularly reliable metric for evaluating code generation tasks. Our experimental results show that edit distance remains nearly similar between the baseline and CHORUS-based runs, despite significant improvements in accuracy. This is because a given problem can be implemented in multiple valid ways, leading to high edit distance scores even for correct solutions. Overall, our evaluation indicates that accuracy is the most critical metric for this task, as it closely mirrors the real-world performance of such a framework in practical applications.

Table 1. Experimental results for both Baseline and CHORUS frameworks across various LLMs. The parameter sizes of the LLMs are detailed in Subsect. 3.3.

Approach	Model	Accuracy ($\uparrow$)	Syntactic Validity ($\uparrow$)	Semantic Similarity ($\uparrow$)	Edit Distance ($\uparrow$)
Baseline	Llama3.1	0.0796	0.8581	0.8365	0.2008
	Llama3.3	0.2289	0.6920	0.9092	0.4185
	Phi4	0.1938	0.2872	0.9153	0.3921
	Deepseek-r1	0.1073	0.9654	0.9186	0.3276
	Qwen2.5-coder	0.4644	0.8618	0.9280	0.4728
	GPT3.5	0.5260	1.0000	0.9159	0.3933
	GPT4	0.6367	0.9931	0.8721	0.3565
CHORUS	Llama3.1	0.1349	0.9827	0.7941	0.2019
	Llama3.3	0.5675	0.9827	0.9291	0.4255
	Phi4	0.6125	1.0000	0.9379	0.4953
	Deepseek-r1	0.5848	1.0000	0.9141	0.3946
	Qwen2.5-coder	0.5986	1.0000	0.9367	0.4747

3.5 Ablation Study

We conduct a comprehensive ablation analysis to quantify the contributions of the core components of our proposed framework—expert prompting, retrieval, and structured reasoning. Table 2 compares five system configurations across five LLMs, which reveals critical insights into optimization code generation dynamics. All configurations use structured output schema for automated evaluation.

Impact of Expert Prompting. The efficiency of expert prompting exhibits significant model-size dependence. While the Deepseek-r1 (32B) model achieves a

386.95% accuracy improvement when using expert prompts (rising from 10.73% to 52.25%), the smaller Llama3.1 (8B) model suffers a 61.3% reduction in accuracy (7.96% to 3.08%). Manual analysis of generated code samples reveals that this divergence derives from differences in instruction-following capacity and parametric knowledge. Larger models demonstrate superior adherence to the implementation and formatting rules of the expert prompt. Expert prompt occupies a significant portion of the context window in smaller models, which overwhelms their limited capacity, leading to syntax errors.

Table 2. Ablation Results (Accuracy): Structured reasoning provides consistent gains across LLMs. Traditional RAG underperforms CHORUS by 46.14–89.33%.

Configuration	Llama3.1	Llama3.3	Phi4	Deepseek-r1	Qwen2.5-coder
Baseline	0.0796	0.2289	0.1938	0.1073	0.4644
Baseline + Expert Prompt	0.0308	0.4833	0.5536	0.5225	0.5640
Traditional RAG	0.0144	0.1578	0.1698	0.1168	0.3224
CHORUS (w/o reasoning)	0.0104	0.5497	0.6020	0.5613	0.5721
CHORUS	0.1349	0.5675	0.6125	0.5848	0.5986

Traditional RAG Limitations. Traditional fixed-length chunking strategy reduces accuracy by 46.14–89.33% compared to CHORUS. This reduction primarily arises from two systemic flaws inherent to traditional retrieval approaches. First, most of retrieved chunks contain fragmented or incomplete contents. Second, semantic dispersion introduces irrelevant content—the majority of the retrieved chunks mismatch problem domains, confusing models with conflicting API examples. These issues disproportionately affect smaller architectures i.e., accuracy of Llamma3.1 plummets to 1.4% with traditional RAG versus 13.5% in CHORUS. Even coding-specialized models like Qwen2.5-coder (32B) exhibit 46.14% reduced accuracy, which demonstrates that domain expertise cannot compensate for poor context selection.

Impact of Structured Reasoning. Omitting the `reasoning_steps` field from CHORUS's structured definition schema reduces accuracy by 1.71–92.29% across model scales. The reasoning field forces models to explicitly map problem elements to code components, a process that improves constraint coverage. This aligns with chain-of-thought [30] but extends its benefits to structured code synthesis through schema enforcement.

4 Conclusion

In this work, we introduced **CHORUS**, a RAG framework for generating Gurobi-based LP code from natural language descriptions. Our methodology ensures semantic cohesion in theoretical documentation through hierarchical chunking and bridges vocabulary mismatches with metadata-augmented retrieval of code examples. Experimental results on the NL4Opt-Code dataset demonstrate that CHORUS consistently outperforms baseline LLMs. Ablation studies further highlight the importance of core components of our proposed framework. One of the key strengths of CHORUS is its flexibility, as it remains both LLM-agnostic and solver-independent. While our experiments focus on generating LP code for Gurobi, the framework can support various mathematical problems, as long as a corresponding solver is available. By decoupling retrieval-augmentation from solver-specific logic, the use of CHORUS allows for seamless adaptation to future solver updates and emerging LLMs, ensuring continuous improvements over standard code generation methods.

4.1   Limitations

Although our framework demonstrates strong performance improvement compared to conventional RAG, certain limitations persist. Code generation remains highly sensitive to prompt engineering, and smaller models struggle to fully incorporate all contextual elements within limited context windows. While our methodology is solver-independent, its performance depends on the quality and relevance of the indexed documentation, a factor that varies across solvers. Additionally, the alignment of LLMs with other optimization topics (e.g., integer linear, mixed, or non-linear problems) is left for future research. Another important consideration is the evolving nature of solver documentation. As solvers frequently introduce new methods and modify existing APIs, this can impact the consistency and completeness of the CHORUS knowledge base. To maintain system accuracy, the documentation parsing and knowledge base construction process would need to be re-executed periodically to reflect these updates. Despite these constraints, CHORUS lays a strong foundation by offering a resource-efficient alternative that allows open-source LLMs to approach GPT4-level performance.

4.2   Future Work

We believe that further exploration into adaptive retrieval strategies, multi-step reasoning, and agentic approach will open avenues for increasingly robust and domain-adapted code synthesis in optimization workflows. In future work, the CHORUS pipeline can be extended by incorporating advanced prompt engineering techniques such as role prompting, reflexion, ReAct, chain-of-thought, and chain-of-experts prompting [31]. Additionally, integrating few-shot learning strategies may further improve model performance in low-resource or domain-specific settings. Furthermore, the generalizability of the framework can be evaluated by applying it to additional mathematical optimization domains and across a broader set of solvers, including OR-Tools and IBM CPLEX.

References

1. Zhang, J., et al.: Solving general natural-language-description optimization problems with large language models (2024). arXiv: 2407.07924 [math.OC]
2. Ramamonjison, R., et al.: NL4Opt competition: formulating optimization problems based on their natural language descriptions. In: Ciccone, M., Solovitzky, G., Albrecht, J. (eds.) Proceedings of the NeurIPS 2022 Competitions Track. Proceedings of Machine Learning Research, pp. 189–203. PMLR (2022)
3. Dubey, A., et al.: The Llama 3 Herd of Models (2024). arXiv: 2407.21783 [cs.AI]
4. Kandpal, N., Deng, H., Roberts, A., Wallace, E., Raffel, C.: Large language models struggle to learn long-tail knowledge. In: Krause, A., Brunskill, E., Cho, K., Engelhardt, B., Sabato, S., Scarlett, J. (eds.) Proceedings of the 40th International Conference on Machine Learning. Proceedings of Machine Learning Research, pp. 15696–15707. PMLR (2023)
5. Roberts, A., Raffel, C., Shazeer, N.: How much knowledge can you pack into the parameters of a language model? In: Webber, B., Cohn, T., He, Y., Liu, Y. (eds.) Proceedings of the 2020 Conference on Empirical Methods in Natural Language Processing (EMNLP), pp. 5418–5426. Association for Computational Linguistics, Online (2020)
6. Mitchell, E., Lin, C., Bosselut, A., Manning, C.D., Finn, C.: Memory-based model editing at scale. In: Chaudhuri, K., Jegelka, S., Song, L., Szepesvari, C., Niu, G., Sabato, S. (eds.) Proceedings of the 39th International Conference on Machine Learning. Proceedings of Machine Learning Research, pp. 15817–15831. PMLR (2022)
7. Lewis, P., et al.: Retrieval-augmented generation for knowledge-intensive NLP tasks. In: Larochelle, H., Ranzato, M., Hadsell, R., Balcan, M., Lin, H. (eds.) Advances in Neural Information Processing Systems, pp. 9459–9474. Curran Associates, Inc. (2020)
8. Xu, Y., et al.: Retrieval meets long context large language models. In: The Twelfth International Conference on Learning Representations (2024)
9. Yu, Y., et al.: RankRAG: unifying context ranking with retrieval-augmented generation in LLMs. In: The Thirty-eighth Annual Conference on Neural Information Processing Systems (2024)
10. Ahmed, T., Choudhury, S.: LM4OPT: unveiling the potential of large language models in formulating mathematical optimization problems. INFOR Inf. Syst. Oper. Res. **62**(4), 559–572 (2024)
11. Tsouros, D.C., Verhaeghe, H., Kadioglu, S., Guns, T.: Holy Grail 2.0: From Natural Language to Constraint Models. arXiv abs/2308.01589 (2023)
12. Yang, Z., et al.: Benchmarking LLMs for optimization modeling and enhancing reasoning via reverse socratic synthesis (2024). arXiv: 2407.09887 [cs.LG]
13. AhmedTeskuna, A., Gao, W., Udell, M.: OptiMuS: optimization modeling using step solvers and large language models. arXiv abs/2310.06316 (2023)
14. Zhang, J., et al.: Solving general natural-language-description optimization problems with large language models. In: Yang, Y., Davani, A., Sil, A., Kumar, A. (eds.) Proceedings of the 2024 Conference of the North American Chapter of the Association for Computational Linguistics: Human Language Technologies (Volume 6: Industry Track), pp. 483–490. Association for Computational Linguistics, Mexico City, Mexico (2024)
15. Ye, J., et al.: A comprehensive capability analysis of GPT-3 and GPT-3.5 series models (2023). arXiv: 2303.10420 [cs.CL]

16. OpenAI, Achiam, J., Adler, S., et al.: GPT-4 Technical Report (2024). arXiv: 2303.08774 [cs.CL]
17. Li, B., Mellou, K., Zhang, B., Pathuri, J., Menache, I.: Large language models for supply chain optimization (2023). arXiv: 2307.03875 [cs.AI]
18. Yang, C., et al.: Large language models as optimizers. arXiv abs/2309.03409 (2023)
19. Gurobi Optimization, LLC, Gurobi Optimizer Reference Manual (2024). https://www.gurobi.com
20. Chen, X., Gao, P., Song, J., Tan, X.: HiQA: a hierarchical contextual augmentation RAG for multi-documents QA (2024). arXiv: 2402.01767 [cs.CL]
21. Gao, L., Ma, X., Lin, J., Callan, J.: Precise zero-shot dense retrieval without relevance labels. In: Rogers, A., Boyd-Graber, J., Okazaki, N. (eds.) Proceedings of the 61st Annual Meeting of the Association for Computational Linguistics (Volume 1: Long Papers), pp. 1762–1777. Association for Computational Linguistics, Toronto, Canada (2023)
22. D jean, H., Clinchant, S., Formal, T.: A thorough comparison of cross-encoders and LLMs for reranking SPLADE (2024). arXiv: 2403.10407 [cs.IR]
23. Tam, Z.R., Wu, C.K., Tsai, Y.L., Lin, C.Y., Lee, H.Y., Chen, Y.N.: Let me speak freely? A study on the impact of format restrictions on large language model performance. In: Dernoncourt, F., Preoţiuc-Pietro, D., Shimorina, A. (eds.) Proceedings of the 2024 Conference on Empirical Methods in Natural Language Processing: Industry Track, pp. 1218–1236. Association for Computational Linguistics, Miami, Florida, US (2024)
24. Jin, X., Lin, Z.: SimLLM: calculating semantic similarity in code summaries using a large language model-based approach. Proc. ACM Softw. Eng. 1(FSE) (2024). https://doi.org/10.1145/3660769
25. Song, Y., Lothritz, C., Tang, X., Bissyandé, T., Klein, J.: Revisiting code similarity evaluation with abstract syntax tree edit distance. In: Ku, L.-W., Martins, A., Srikumar, V. (eds.) Proceedings of the 62nd Annual Meeting of the Association for Computational Linguistics (Volume 2: Short Papers), pp. 38–46. Association for Computational Linguistics, Bangkok, Thailand (2024). https://doi.org/10.18653/v1/2024.acl-short.3
26. Abidin, M., et al.: Phi-4 technical report (2024). arXiv: 2412.08905 [cs.CL]
27. DeepSeek-AI, Guo, D., Yang, D., et al.: DeepSeek-R1: incentivizing reasoning capabilities in LLMs via reinforcement learning (2024). arXiv: 2501.12948 [cs.CL]
28. Hu, B., et al.: Qwen2.5-Coder Technical Report (2024). arXiv: 2409.12186 [cs.CL]
29. Kalyan, K.S.: A survey of GPT-3 family large language models including ChatGPT and GPT-4. Nat. Lang. Process. J. **6**, 100048 (2024). https://doi.org/10.1016/j.nlp.2023.100048
30. Wei, J., et al.: Chain-of-thought prompting elicits reasoning in large language models. In: Proceedings of the 36th International Conference on Neural Information Processing Systems. NIPS 2022. Curran Associates Inc., LA, USA (2022)
31. Schulhoff, S., Ilie, M., Balepur, N., et al.: The prompt report: a systematic survey of prompt engineering techniques (2025). arXiv: 2406.06608 [cs.CL]. https://arxiv.org/abs/2406.06608

Taxi Re-positioning Considering Driver Compliance

Cebrina Lindstroem[1][(✉)], Stefan Ropke[1], and Reza Pourmoayed[2]

[1] DTU Management, Technical University of Denmark, Akademivej Building 358,
2800 Kgs., Lyngby, Denmark
ceblin@dtu.dk
[2] BNR A/S, Rolighedsvej 32, 8240 Risskov, Denmark

Abstract. Managing taxi fleets in large cities is challenging, especially when drivers operate independently. This study improves taxi repositioning by predicting demand and providing smart recommendations. Using real-world data from a Scandinavian taxi service, we employ neural networks with LSTM layers to forecast demand and test different strategies, like a simple greedy algorithm and a more structured matching-based approach, to guide taxis to high-demand areas. Since drivers ultimately decide whether to follow these suggestions, we also model their behavior using a probabilistic acceptance strategy. Through a simulation of a real day, we analyze how different approaches impact passenger wait times and overall efficiency. The results show that proactive re-positioning significantly reduces wait times but can increase total driving distance. The greedy algorithm tends to perform better in quickly getting taxis to passengers, while the matching model is more efficient in minimizing unnecessary travel. However, increased mobility comes at a cost, as rerouting leads to longer driving distances. Additionally, driver behavior plays a crucial role, with lower acceptance rates reducing the effectiveness of predictive recommendations. Overestimating demand in such cases helps mitigate inefficiencies.

Keywords: Taxi re-positioning · LP based heuristics · Multivariate time series forecasting

1 Introduction

Efficiently allocating taxis in large Scandinavian cities is a difficult challenge. The aim is to minimize passenger wait times while optimizing the transport system, driver earnings, the number of trips serviced by the same cars, and inefficient wait times. Much of the existing literature assumes this problem to have centralized authority that can dispatch the taxis and direct taxis to specific zones or passengers. Yet, in multiple real-life settings, taxi dispatching is decentralized with each taxi working as an individual agent. This shifts the problem from optimizing a centralized system to guiding and incentivizing a collection of autonomous agents.

© The Author(s), under exclusive license to Springer Nature Switzerland AG 2026
Y. Zhang et al. (Eds.): LION 2025, LNCS 15745, pp. 81–96, 2026.
https://doi.org/10.1007/978-3-032-09192-5_6

In both the centralized and de-centralized versions of this vehicle routing problem, the dispatching occurs based on some assumptions or predictions for future demand. To predict future demand, recurrent neural networks with LSTM cells are a strong tool, as used in this study. However, with these predictions, there is still a need to understand how they can be used to guide taxis working as individual agents efficiently.

We tackle this challenge by integrating demand prediction with a simulation-based framework to explore different aspects of decentralised taxi routing. Our simulation models a scenario where individual drivers receive zone recommendations based on predicted demand. The drivers have full control over their routing decisions. To generate recommendations, we test different approaches, such as greedy algorithms and solving an assignment problem.

In this paper, we examine this challenge using a real-life case study from a Scandinavian taxi service. Currently, the system does not provide any guidance to the taxis, so the drivers make completely independent decisions regarding their movement. However, there is interest in implementing a guidance system to help drivers make more efficient decisions, which motivates this investigation of integrating demand prediction with decentralized routing decisions.

Furthermore, we model driver behavior through a probabilistic acceptance strategy to see how these behaviors affect different outcomes for both drivers and passengers.

Our key contributions are 1) providing a simulation-based framework for evaluating decentralized taxi rerouting in systems where drivers have full control over their rerouting decisions and 2) examining the dual aspects of recommendation generation and driver response, providing insights into how system performance is influenced by drivers who choose not to follow the guidance.

2   Related Work

The problem of repositioning taxis and similar services is well-studied. Research in this area follows two main approaches: classical operations research (OR) techniques and deep reinforcement learning.

In the OR stream, Hao et al. [8] address taxi relocation in Singapore after the morning rush hour, when taxis tend to cluster in low-demand downtown areas. Their model, solved once daily, uses demand prediction based on weather data, applying a regression tree with precipitation-based splits. A distributionally robust model plans repositioning, outperforming simpler approaches.

Tavor and Raviv [11] assume fully automated taxis and solve repositioning continuously. They propose three LP-based models: one maintaining a fixed number of cars per zone and two using demand forecasts. These models are solved every 15 min and tested on New York City taxi data, showing that demand-aware approaches balance empty mileage and customer wait times effectively.

Guo et al. [7] use the smart-predict-then-optimize approach proposed by Elmachtoub and Grigas [4] for taxi re-positioning. They consider a taxi-like service provided by Didi Chuxing in Chengdu, China, and provide several mathematical models for relocating vehicles. They conclude that their smart-predict-then-optimize approach outperforms simpler benchmarks as well as other mathematical models defined in the paper.

The problem of re-positioning taxis has been addressed through reinforcement learning in, for example, [6,9], and [10]. The approach by Gammelli et al. [6] is of special interest as it uses graph reinforcement learning combined with linear programming to reposition taxis. The graph neural network suggests how taxis should be distributed among zones, and the linear program (LP) finds a cost-efficient way to obtain this distribution from the current one. This simplifies the reinforcement learning algorithm as it does not have to consider the feasibility of its proposed distribution and does not have to specify the precise movements of taxis, the LP takes care of these parts.

The issue of drivers not adhering to repositioning suggestions has not received much attention in the literature. Wang and Wang [12] studies relocating vehicles using data from Didi Chuxing in China. It is acknowledged that vehicle relocation tasks will only be a suggestion to the drivers, and drivers are not required to follow the suggestion, however, this fact is, to the best of our understanding, not embedded in the solution method or in the simulation results. The solution method is based on mathematical models that aim at matching up as many vehicles with transport requests. Very recently, Brar et al. [2] and Chen et al. [3] studied the question further. Brar et al. [2] model driver behavior based on repositioning preferences and confidence level in the recommendations provided and incorporate the driver model in a repositioning algorithm. Chen et al. [3] approach the issue of driver behavior from two angles, one is to analyze past data to learn how a driver *cruise* while being idle, and the other is to use a survey to understand what makes it likely for a driver to accept a repositioning request. The model for driver behavior is then integrated into a repositioning algorithm based on reinforcement learning. Our work differs from both Brar et al. and [3] by focusing on a fully decentralised fleet, thus evaluating the effects of driver uncertainty on system-wide outcomes rather than mitigating the effects of incomplete driver compliance.

3 Problem Description

The inspiration for this problem stems from a collaboration with BNR A/S and the challenges faced by Scandinavian taxi centrals. This section provides context on the issue.

Taxis are vital to urban transport, and in many Scandinavian cities, they operate independently while the central dispatches rides. Although trips are assigned via a queue system, drivers retain full autonomy over repositioning between trips and accepting assigned rides. This study assumes full trip acceptance, focusing on the effects of repositioning while maintaining driver control. Taxi demand fluctuates based on temporal, spatial, and contextual factors like weather and events. Currently, repositioning is left to the driver's intuition, making the system reactive. The goal is to develop a proactive guidance system. Trips are allocated based on queue order within zones. If a zone queue is empty, the trip is assigned from the nearest available queue. Each trip has a fixed pickup and drop-off, without ridesharing. While some trips originate from street hails, the simulation assumes all trips are centrally booked.

In the simulation, taxis act as independent agents, making movement decisions freely. They enter new queues when switching zones and receive rerouting recommendations but decide whether to follow them. Priority is given to taxis already waiting in a zone over those en route. A probabilistic acceptance strategy is applied, with drivers accepting rerouting suggestions at rates of 100%, 70%, 50%, or 30%. This approach was chosen due to no available data regarding historical acceptance behaviour as no such recommendation system is currently implemented. Additionally, the study aims to focus on understanding the impact of decentralized drivers within a system, making this approach well-suited to the research objective. The goal is to minimize passenger wait times while balancing the additional kilometers driven due to rerouting.

4   Datasets and Demand Prediction

A central part of the simulation is the ability to predict demand arising in different zones and, based on this, reroute the taxis to zones where there is a supply deficit. The prediction is trained and used on a dataset collected over multiple years.

4.1   Data

The data for this study comes from a real Scandinavian taxi central and includes all trips from August 2016 to December 2024. This dataset serves as the foundation for training the demand prediction model, containing details on trip type, price, pickup/delivery times, and locations.

Notably, the dataset only includes completed trips, omitting declined or unrecorded ones, which limits insight into total demand.

For prediction, we used a zone-based approach, clustering historical trips with k-means clustering based on the locations of the trips. As city-supported zones evolved over time, this method ensured a stable, flexible framework. We selected 35 zones based on the elbow method [5]. We found 35 zones provided the best granularity, balancing spatial detail with model simplicity.

Figure 1 provides insight into the dataset used in this paper and the clustering. It shows a bar chart of the zones (ordered numerically) based on the total demand within each zone. There is a clear variance in the number of trips across the zones, reflecting the differences in demand between high- and low-density areas.

4.2   Demand Prediction Using Neural Networks with LSTM-Cells

To optimize taxi routing toward high-demand areas, our solution relies on a demand prediction model. We use a neural network with Long Short-Term Mem-

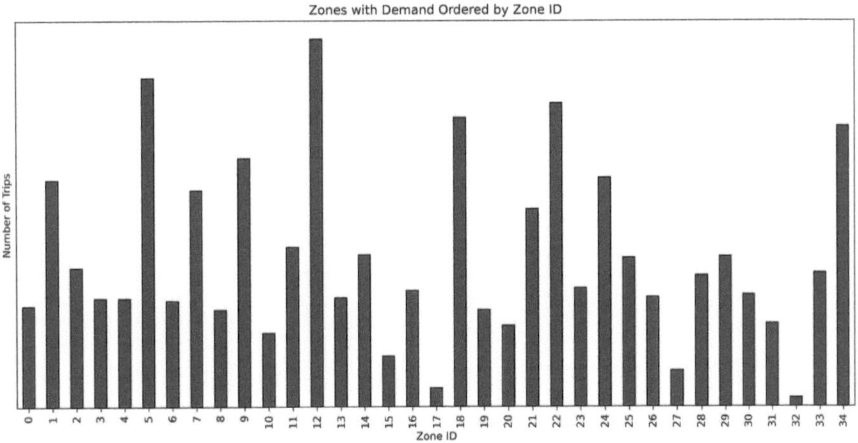

Fig. 1. A bar chart respresenting the number of trips per zone, ordered by zone ID.

ory (LSTM) cells, which effectively handle time-series data by capturing long-term patterns and mitigating vanishing gradients. This capability is crucial for demand forecasting, where historical trends strongly shape future outcomes.

The model predicts demand across all zones simultaneously as a multivariate time-series problem, rather than 35 separate univariate time-series as the zone demands are highly correlated. This also ensures more consistent and coherent predictions across the city. Input data consists of summed historical demand, with predictions based on lag variables representing past demand. We used 10-minute time steps and found the best balance between accuracy and efficiency with 144 lag variables (covering two days). We experimented with incorporating external variables such as weather conditions, COVID-19 case counts, pre-bookings, and holidays. However, these features offered minimal gains in prediction accuracy while substantially increasing computational complexity and runtime. The only added inputs were seasonality/cyclic inputs, specifically the month, weekday as well as the period within the hour. These inputs were cyclically encoded to preserve the cyclic nature of the data, while also reducing the number of input features compared to alternatives like one-hot encoding [1].

The model provides both point predictions and confidence intervals by estimating quantiles, where the 90th percentile serves as the upper bound and the 10th percentile as the lower bound. The network is trained through quantile regression, which minimizes a quantile loss function tailored to each specified quantile. We predicted three time periods into the future, thus as any point we predict the demand for the next 30 min divided into 10-minute intervals.

Our architecture includes two shared LSTM layers (256 and 128 units) and an individual LSTM layers (128 units each) for median, upper, and lower quantile predictions. A dropout rate of 0.3 is applied between layers. Figure 2 shows an overview of the architecture.

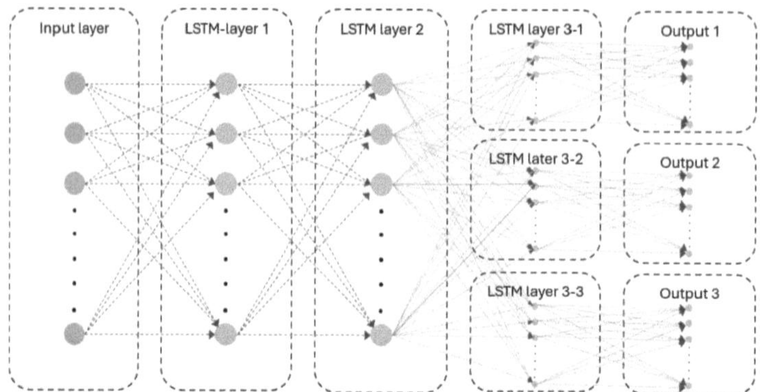

Fig. 2. The architecture of the neural network used for predicting demand.

The demand prediction is programmed in Python using the Python package Keras. It is trained using a training-validation-testing split of 70%-10%-20%, and the day of the simulation is part of the testing data (and occurs at the end of the time series). Thus the model is not trained on the data for the simulation, but it has been evaluated based on this data before use. The model achieved an R-squared value of approximately 0.74 on validation and testing data.

We compared the prediction performance of our multivariate model to traditional models such as Auto-ETS, which were run separately on each zone. Auto-ETS achieved an average R-squared of 0.48 across all zones on the test set, with values ranging from 0.17 to 0.66. The multivariate LSTM model achieved a significantly higher average R-squared of 0.74 across all zones on the test set, ranging from 0.30 to 0.94. It outperformed the Auto-ETS models in all but three zones.

In Fig. 3, the predicted and actual values for each zone are displayed, along with the confidence intervals for the predictions, for a single day between the times 8 and 18. The model demonstrates strong performance in many zones, accurately predicting demand patterns. However, in certain cases, such as Zone 10, the model struggles due to significant fluctuations in the actual values. This uncertainty is reflected in larger confidence intervals, whereas zones like Zone 7 has smaller confidence intervals and still manage to capture key demand peaks effectively. A similar plot for a low-demand day is shown in the appendix to illustrate the models ability to handle diverse demand situations.

5   Simulation and Suggestion Methods

Based on the prediction model obtained as described in Sect. 4.2, we modeled a simulation framework to "replay" an 8-hour interval of a real day's demands.

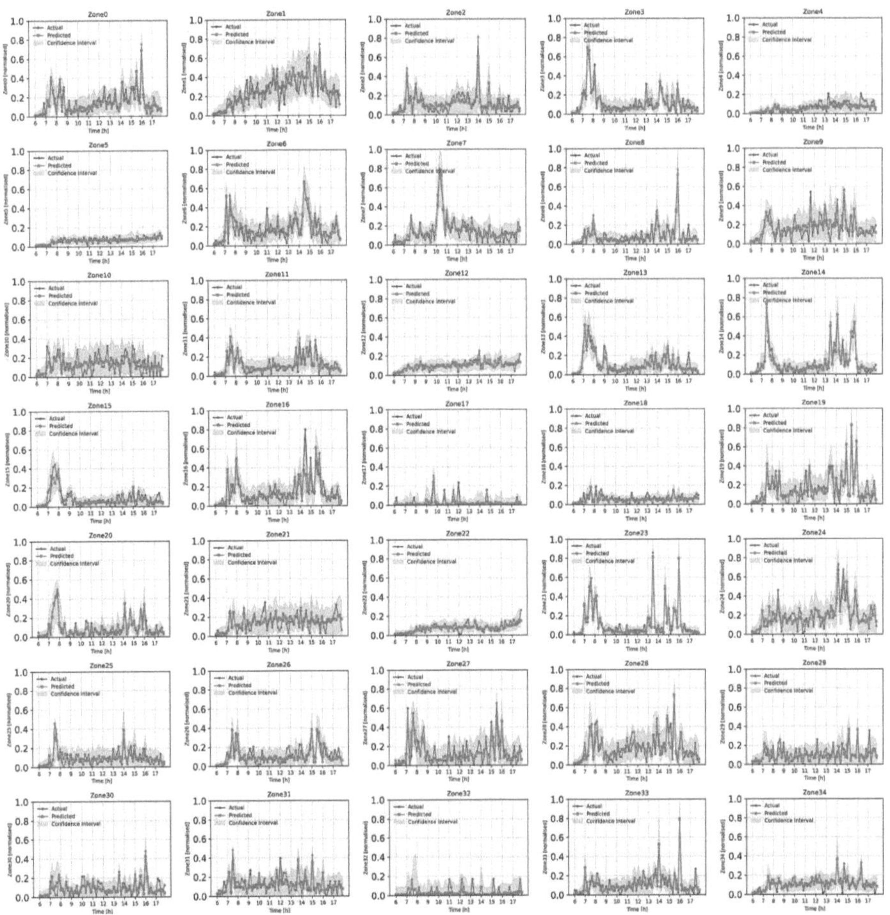

Fig. 3. Each plot shows the actual and predicted values for a single zone, with actual values as blue dots and predictions as orange x. Confidence intervals are shown with light blue shading. The x-axis shows time from 6 to 18 h, and the y-axis represents normalized demand.(Color figure online)

5.1 Simulation Framework

To evaluate the effectiveness of different rerouting techniques, we implemented a discrete event-based simulation. This approach lets up replay an actual day. The discrete event-based framework was chosen for its ability to model systems where changes occur at irregular intervals as is the case of taxi demand and routing. It is implemented as a priority queue, so new events can be added (e.g. a end-of-trip-event added whenever a new begin-trip-event has been handled. Key events for the simulation are:

1. New Trip: Trip requests generated based on a single day. These are part of the event queue from the beginning. When a new trip arises, it is assigned

to a taxi, which in turn generates a new trip end event that is added to the event queue.

2. Trip ends: When a trip ends, the status of the assigned taxi is updated. The taxi becomes available and is added to the dispatching queue of the end zone of the trip, where it currently is. A routing suggestion event is also added to the event queue, enabling the system to provide a relocation recommendation.

3. Prediction: Demand forecasts are periodically updated and inserted into the event queue at fixed intervals. In this setup, predictions are generated every 10 min, matching the model's forecast interval. These events are used during routing suggestion events but do not themselves generate additional events.

4. Routing Suggestions: Relocation recommendations are issued to drivers either when a trip ends or when a taxi has remained idle for an extended period. If accepted, the taxi begins moving to the suggested zone, an end-reroute event is added to the event queue, and the taxi is assigned a queue number in the new zone (though dispatching prioritizes taxis already present in the zone, if such taxis exist). If the suggestion is rejected, a new routing suggestion event is scheduled after 10 min in case relocation becomes preferable.

5. End Reroute: The rerouting process concludes and the taxi's status is updated. A new routing suggestion event is added to the event queue after 10 min, in case the taxi remains unassigned and relocation becomes necessary.

A pseudocode of the simulation framework can be found in Algorithm 1.

5.2 Suggestion Methods for Rerouting

A central part of the simulation comes from the routing suggestions for the drivers. In this part of the simulation the drivers are presented to a new zone they should relocate to, and then they decide whether or not to follow the systems recommendation. We worked with three different suggestion methods: (1) A static strategy, where we did not move taxis at all, thus providing a baseline for our data, (2) A simple greedy algorithm calculating the deficit in each zone, i.e. the difference between predicted demand and available taxis, and reroutes drivers to the zone with the largest deficit, and (3) A matching problem solution, where we solve a matching problem every few minutes and reroute based on this.

We also tried different suggestion intervals, as we found it beneficial to let a taxi stay in one place right after it finished a trip. Thus, if the suggestion interval was 0 min, we would suggest a new zone for a taxi right after they finished a trip, while if the suggestion interval was 10 min, we would wait 10 min before suggesting a new trip and thus let the taxi get into the queue of their finishing zone. Waiting briefly before suggesting a relocation proved helpful in some cases, as the current location might turn out to be a hotspot, allowing the taxi to get a trip without needing to relocate.

Each suggestion algorithm had some common implementations. First, if we did not have a zone recommendation for a taxi, we let the taxi stay and gave them a new recommendation after the next prediction. Secondly, if a taxi chooses to decline our suggestion, we give them a new suggestion after 10 min, unless they

---

Algorithm 1. Simulation framework

1: input: a set of taxis V, a set of events E, a prediction model m and a suggestion algorithm alg_{sug}

2: **while** E is not empty **do**

3:     get next event e from E

4:     **if** e is of type new-trip **then**

5:         Assign the demand of e to the first available taxi v from V in queue in the zone. If no queue in the zone assign to the nearest available taxi v from V. If no taxi available, wait.

6:         If a taxi is assigned a trip add an end-of-trip event to E

7:     **else if** e is of type end-of-trip **then**

8:         Complete trip and let taxi v be available for new trips

9:         Add routing suggestion event to E

10:     **else if** e is of type prediction **then**

11:         Predict from current state the demand into the next number of time periods based on prediction model m

12:     **else if** e is of type routing-suggestion **then**

13:         Suggest to vehicle from event e the next zone to travel to based on alg_{sug}. If the vehicle accept begin reroute, add the taxi to the dispatching queue of the suggested zone, and add end reroute event to E. Otherwise add routing suggestion event to E for the vehicle in 10 minutes.

14:     **else if** e is of type end reroute **then**

15:         The state of taxi is updated

16:         Add routing suggestion event to E

17:     **end if**

18: **end while**

---

get assigned a trip before this time. Finally, in both the greedy algorithm and the matching problem, we subtracted the actual demand since the last prediction from the predicted demand. Thus, if $pred_j$ defines the predicted demand in zone j, and act_j is the actual realized demand that has occurred in zone j since the last prediction, the predicted demand used in the suggestion would be $pred_j - act_j$.

Greedy Algorithm. The greedy algorithm calculates the deficit in all zones based on the latest prediction and reroutes taxis according to the largest deficit. We explored both global and local approaches: the global version calculated the deficit across all zones and made decisions based on this, while the local version focused only on nearby zones for its calculations. We found that the global version often resulted in taxis being dispatched to distant zones, leading to inefficiencies. As a result, we opted for the local version of the greedy algorithm.

In this local version, the algorithm considers only the nearby zones (defined as those that can be reached within 8 min) and calculates the deficit for each. It then suggests the zone with the largest deficit. To prioritize more immediate demand, the algorithm places greater weight on deficits occurring within the next 10 min. If the predicted demand for the next 10 min is satisfied, the algorithm extends the horizon to 20 min, and later to 30 min, in line with our three-period prediction model.

If a deficit was predicted in a zone, we refrained from suggesting a new zone for the taxis, as it would be counterproductive to relocate vehicles to different areas when other taxis needed to return to the original location to meet demand.

Matching Problem. The matching problem algorithm solves a matching problem at specific time steps, where each predicted demand for each zone must be met by the same number of taxis. Specifically, we let V be the set of taxis and J be the set of zones. We then define $x_{ij} = 1$ if vehicle $i \in V$ is matched to zone $j \in J$, and 0 otherwise. Define d_{ij} as the distance from vehicle $i \in V$ to zone $j \in J$. Furthermore, we let z_j be a continuous positive slack variable, defined as a demand not met in a zone. To heavily prioritize satisfying all demand, we impose a penalty on z_j with a value of $\alpha = 1,000,000$. We let $pred_j$ describe the predicted demand in zone $j \in J$. Unlike the greedy algorithm, we do not restrict how far taxis can travel, as our goal is to find the best global solution. Travel distance is already penalized in the objective function. Thus the problem we solve at a specific time step is:

$$\min \sum_{i \in V} \sum_{j \in J} d_{ij} x_{ij} + \sum_{j \in J} \alpha z_j$$

s. t.

$$\sum_{j \in J} x_{ij} \leq 1 \qquad \forall i \in V \qquad (1)$$

$$\sum_{i \in V} x_{ij} + z_j \geq pred_j \qquad \forall j \in J \qquad (2)$$

$$x_{ij} \in \{0,1\} \qquad \forall i \in V, j \in J \qquad (3)$$

$$z_j \geq 0 \qquad \forall j \in J \qquad (4)$$

We found that the optimal time frame for solving the matching problem was every 5 min, as this frequency strikes a balance that prevents recalculations that lead to conflicting solutions or contradictory recommendations. We used Cplex to solve the linear problem.

6   Computational Results

For the numerical study, we ran the simulation described in Sect. 5.1 to replay an 8-hour interval of a real day on this Scandinavian taxi central. Specifically, we took all actual trips run between 8:00 and 16:00 on this day and let them appear dynamically throughout the simulation. This 8-hour interval had 6112 actual trips, and in the simulation, we serviced these trips with a fleet of 300 vehicles, where we assume all of them are in operation all of the time, as opposed to real life where they take natural breaks throughout the day. We used OpenStreetMap to calculate distances between any two points. We treated all taxis and trips within a zone as clustered together, considering only distances between zones.

6.1 Varying the Suggestion Methods and Changing the Acceptance Rates

We ran the simulation with the different suggestion methods as well as different acceptance rates, i.e. the chance a taxi accepts our recommendation. Specifically, we ran the model with acceptance rates of 100%, 70%, 50%, and 30%. The results can be found in Table 1. The columns show the used suggestion algorithm, taxi acceptance percentage, average duration to the pickup location in seconds, percentage of trips serviced in their starting zone, percentage of taxis receiving trips in their relocation zone, and finally, the total driven kilometers.

As can be seen rerouting the taxis to zones before their suggestion greatly decreases the average duration from a taxi to their assigned pickup trip. Please note that the average duration is quite low, as we only consider distances and durations between zones, and do not consider wait times for acceptance. Compared to the static strategy, where we do not move any of the taxis preemptively, the greedy algorithm (if followed completely by all taxis) can decrease the average duration by 71%, whereas the matching solution can decrease the average duration by 55% when followed completely. It thereby proves a significant improvement in passenger wait times and proximity to dynamically occurring trips, by using a prediction to reroute the taxis. However, in this case, it seems the greedy algorithm performs better than solving the LP of a matching problem, which can be explained by the imbalance of solving an exact problem based on predicted (i.e. not exact) numbers. The cost of using a proactive rerouting strategy can, however, be seen in the total driven kms. When using the greedy algorithm completely we see an increase in the total driven km compared to the static strategy of 16%, whereas using the matching problem only increases

Table 1. Results from varying degrees of acceptance rates across different suggestion methods. Successful reroute is defined as the percentage of taxis that receive a trip within the zone they were rerouted to.

Algorithm	Acceptance percentage (%)	Average duration to pickup (s)	Trips serviced within a zone (%)	Successful reroutes (%)	Total kms driven (km)
Static strategy	-	177	69	-	56,783
Greedy	100	51	87	76	65,661
Greedy	70	55	85	75	63,249
Greedy	50	60	82	74	61,931
Greedy	30	71	79	74	59,857
Matching	100	80	80	68	59,036
Matching	70	82	79	67	58,461
Matching	50	83	79	69	57,983
Matching	30	89	76	69	57,389

the driven kms by 4%. Thus, there is a definitive trade-off between an increased proximity to pickup locations and the total kms driven.

When comparing different acceptance percentages, we see an increase in the average duration to pickup when the acceptance percentage falls for both suggestion methods. However, there is still a great decrease of the average duration to pickup, as can be seen, when we only accept 30% of the suggestion methods, the average duration for the greedy algorithm is still decreased 60% compared to the static strategy, while the matching problem decreases average duration by 50%. These numbers will slowly increase until they reach the static strategy when no rerouting suggestions are accepted. The reason the number only slowly approaches the average duration in the static strategy is explained by the fact, that when the acceptance percentage falls, the simulation ends up suggesting new zones more often to the taxis, as the ones that decline, will get a new suggestion 10 min later, when a new prediction has been made. For example, we make 2485 rerouting suggestions in the greedy algorithm when all suggestions are accepted, whereas we make 3666 suggestions when only 30% of suggestions are accepted.

6.2 Results of the Algorithm When We Have Complete Knowledge of the Future

We ran the simulation with the same algorithm strategies, except we assumed complete knowledge of the future, i.e. we knew exactly how many trips would occur in a zone in the next three time periods when rerouting. The results can be found in Table 2.

As can be seen, having complete knowledge of the future further decreases the average duration to the pickup locations. We see the greedy algorithm, when there is a perfect acceptance rate of 100%, decreases the average duration to pickup by 12% compared to using the predicted values, whereas in the matching

Table 2. Results from the simulation with complete knowledge of the future.

Algorithm	Acceptance percentage (%)	Average duration to pickup (s)	Trips serviced within a zone (%)	Successful reroutes (%)	Total kms driven (km)
Greedy	100	45	91	75	70,370
Greedy	70	52	90	77	67,212
Greedy	50	59	88	78	65,327
Greedy	30	70	85	78	62,787
Matching	100	43	86	84	59,324
Matching	70	54	82	83	59,069
Matching	50	62	80	79	58,504
Matching	30	83	75	80	57,443

Table 3. Results from the simulation with different prediction scenarios.

Algorithm	Acceptance percentage (%)	Scenario	Average duration to pickup (s)	Trips serviced within a zone (%)	Successful reroutes (%)	Total kms driven (km)
Greedy	100	low	85	77	45	58,771
Greedy	70	low	89	76	43	58,243
Greedy	50	low	93	74	42	57,657
Greedy	30	low	102	74	44	57,665
Greedy	100	high	57	86	77	77,648
Greedy	70	high	57	86	79	73,704
Greedy	50	high	58	85	80	69,218
Greedy	30	high	67	83	79	65,780

problem, we got a 46% decrease, supporting the explanation that this method is more sensitive to inaccurate predictions. We actually now see that the matching problem outperforms the greedy algorithm when we have complete acceptance. However, this is quickly negated when we change the acceptance rates, as already when we have an acceptance rate of 70%, the greedy algorithm again outperforms the matching problem. This indicates that the matching problem is also more sensitive to the acceptance rate since the solution relies on all taxis following the suggestions. In contrast, the greedy algorithm only considers the current state and recommends the best option for an individual taxi. However, even with e.g. 50% acceptance rate, we find an improvement of the matching problem over the static strategy of 53% when knowing the future, but only with a cost increase of kms of 3%. Therefore, the algorithm may be advantageous if more accurate predictions can be obtained, as opposed to the greedy approach, which results in a higher cost increase.

6.3   Performance in Different Scenarios

The prediction model, as described in Sect. 4.2, outputs three different values, one for the actual prediction and an upper and lower confidence interval. In Table 3 the results from running the Greedy algorithm on both the high and low prediction scenario is seen, where the high and low prediction compares to higher and lower predicted values in each zone corresponding to upper and lower quantiles, respectively.

As seen in the results, both the high and low scenarios perform worse than the median prediction when the acceptance rate is 100%. There is also a significant difference in total kilometers driven: the low scenario results in minimal additional driving, as it takes a conservative approach and only relocates taxis when there is a high certainty of a demand deficit. In contrast, the high scenario leads to a substantial increase in kilometers driven, as it tends to overestimate the needed demand, causing more frequent vehicle movement. It is worth noting that all of the results in these scenarios are better than when not moving the taxis at all.

Interestingly, when the acceptance rate drops to 50% or 30%, the high scenario outperforms the median prediction. This is likely because, with fewer taxis accepting re-routing suggestions, more taxis need to be directed to high-demand zones. By overestimating demand, the high scenario compensates for the lower acceptance rate, ensuring better overall coverage.

7   Conclusion

This study investigates taxi re-repositioning by integrating demand prediction with uncertain driver compliance. Our simulations demonstrate that proactive rerouting significantly reduces passenger wait times, although it may increase total kilometers driven. The greedy algorithm proves particularly effective at minimizing wait times, while the matching approach balances efficiency with lower additional travel costs.

One key finding is that driver acceptance rates play a crucial role in overall system performance. When acceptance rates drop, overestimating demand (high scenario) can help maintain efficiency by ensuring enough taxis relocate to high-demand areas. This highlights the challenge of balancing system-level interventions with driver autonomy in decentralized environments.

Future research could explore alternative suggestion and prediction methods, such as Markov decision processes. Additionally, more advanced driver behavior models may provide further insights into drivers' choices. Another interesting direction is to investigate how drivers might be incentivized to accept the suggestions, as this study shows that system-wide benefits arise when all drivers follow the recommendations. This could also involve examining the outcomes for drivers who follow the suggestions compared to those who do not. Finally, examining the impact of exact distances within zones would be valuable, as it improves real-world applicability but also requires considering where taxis should stay within the zone.

Acknowledgments. This study was funded by BNR A/S. We thank Filipe Rodrigues for advice and discussions on demand prediction using recurrent neural networks.

Disclosure of Interests. C. Lindstroem is an industrial PhD student at BNR A/S, and R. Pourmoayed is employed by BNR A/S. S. Ropke has no affiliations with the company. The authors declare that they have no competing interests that are relevant to the content of this article.

A   Appendix

Figure 4 shows the predicted zone plots for a low-demand day (weekend). Actual values are blue dots, predicted values are orange x. Confidence intervals marked with light blue shading. The x-axis shows time from 6 to 18 h, and the y-axis represents normalized demand.

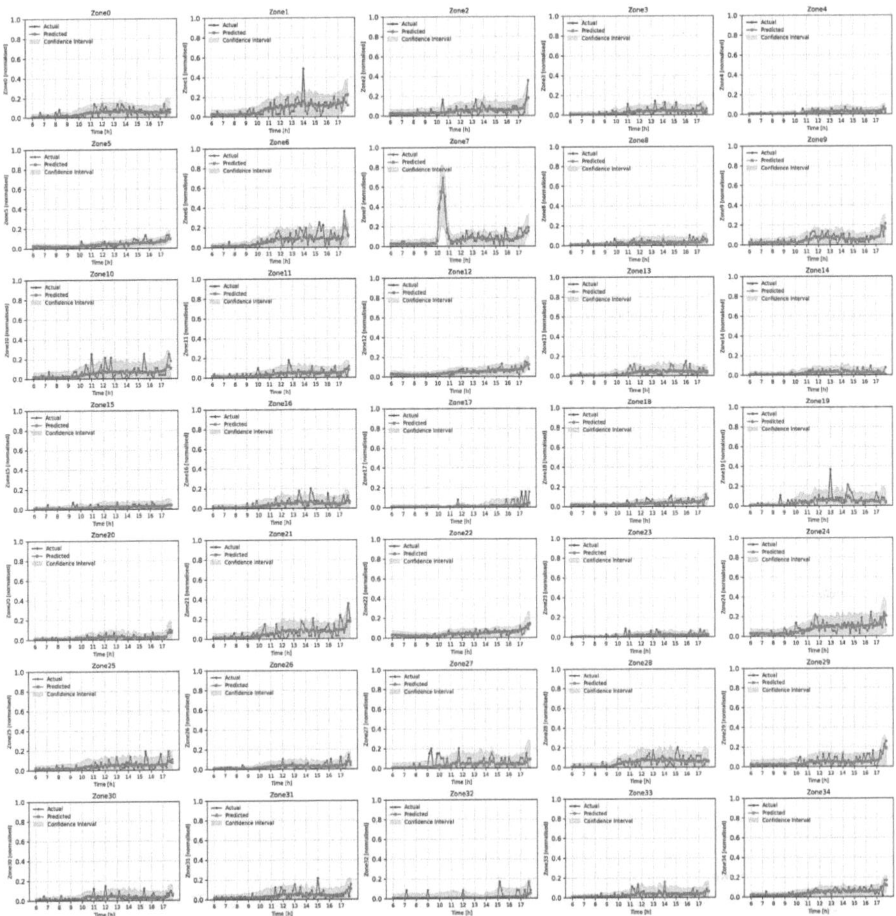

Fig. 4. Predicted and actual values with confidence intervals for all 35 zones on a low-demand day.

References

1. Adams, A., Vamplew, P.: Encoding and decoding cyclic data. S. Pac. J. Nat. Sci. (1998)
2. Singh Brar, A., Su, R., Zardini, G.: Vehicle rebalancing under adherence uncertainty. arXiv preprint arXiv:2412.16632 (2024)
3. Chen, H., et al.: I-rebalance: Personalized vehicle repositioning for supply demand balance. In: Proceedings of the AAAI Conference on Artificial Intelligence, vol. 38, pp. 46–54 (2024)
4. Elmachtoub, A.N., Grigas, P.: Smart predict then optimize. Manag. Sci. **68**(1), 9–26 (2022)
5. Marisa, F., Wardhani, A.R., Purnomowati, W., Vitianingsih, A.V., Maukar, A.L., Puspitarini, E.W.: Potential customer analysis using k-means with elbow method. JIKO (Jurnal Informatika Dan Komputer) **7**, 307 (2023)

6. Gammelli, D., Harrison, J., Yang, K., Pavone, M., Rodrigues, F., Pereira, F.C.: Graph reinforcement learning for network control via bi-level optimization. arXiv preprint arXiv:2305.09129 (2023)
7. Guo, Z., Bin, Yu., Shan, W., Yao, B.: Data-driven robust optimization for contextual vehicle rebalancing in on-demand ride services under demand uncertainty. Transp. Res. Part C Emerg. Technol. **154**, 104244 (2023)
8. Hao, Z., He, L., Zhenyu, H., Jiang, J.: Robust vehicle pre-allocation with uncertain covariates. Prod. Oper. Manag. **29**(4), 955–972 (2020)
9. Holler, J., et al.: Deep reinforcement learning for multi-driver vehicle dispatching and repositioning problem. In: 2019 IEEE International Conference on Data Mining (ICDM), pp. 1090–1095. IEEE (2019)
10. Mao, C., Liu, Y., Shen, Z.J.M.: Dispatch of autonomous vehicles for taxi services: a deep reinforcement learning approach. Transp. Res. Part C **115**, 102626 (2020)
11. Tavor, S., Raviv, T.: Anticipatory rebalancing of RoboTaxi systems. Transp. Res. Part C Emerg. Technol. **153**, 104196 (2023)
12. Wang, H., Wang, Z.: Short-term repositioning for empty vehicles on ride-sourcing platforms (2020)

A Shared Memory Optimal Parallel Redistribution Algorithm for SMC Samplers with Variable Size Samples

Alessandro Varsi[1]([✉])[iD], Efthyvoulos Drousiotis[1][iD], Paul G. Spirakis[2][iD], and Simon Maskell[1][iD]

[1] Department of Electrical Engineering and Electronics, University of Liverpool, Liverpool L69 3GJ, UK
`{A.Varsi,E.Drousiotis,SMaskell}@liverpool.ac.uk`
[2] Department of Computer Science, University of Liverpool, Liverpool L69 3GJ, UK
`P.Spirakis@liverpool.ac.uk`

Abstract. Sequential Monte Carlo (SMC) samplers are Bayesian inference methods to draw N random samples from challenging posterior distributions. Their simplicity and competitive accuracy make them popular in various applications of Machine Learning (ML), Bayesian Optimization (BO), and Statistics. In many applications, run-time is critical under strict accuracy requirements, making parallel computing essential. However, an efficient parallelization of SMC depends on how effectively its bottleneck, the redistribution step, is parallelized. This is hard due to workload imbalance across the cores, especially when the samples are of variable-size. A parallel redistribution for variable-size samples was recently proposed for Shared Memory (SM) architectures. This method resizes all samples to the size of the biggest sample, $\overline{M}$, and constrains the samples to be indivisible, i.e., forces the cores to redistribute whole samples. This leads to inefficient run-time, and a sub-optimal time complexity, $O(\overline{M} \log_2 N)$. This study addresses the challenge of Optimal Parallel Redistribution (OPR) for variable-size samples. We first prove that OPR for indivisible variable-size samples is NP-complete. Then, we present an OPR algorithm for SM that does not resize the samples and allows cores to redistribute either whole samples or fractions of them. We prove theoretically that this approach achieves optimal $O(\hat{M} \log_2 N)$ time complexity, where $\hat{M}$ is the average size of the redistributed samples. We also show experimentally that the proposed approach is up to $10\times$ faster than the reference method on a 32-core SM machine.

Keywords: Parallel Algorithms · Sequential Monte Carlo · Parallel Redistribution · Shared Memory · Parallel Methods for ML and BO

1 Introduction

1.1 Motivation

Sequential Monte Carlo (SMC) samplers are commonly used Bayesian inference methods to make estimations of the state of a statistical model, $\mathbf{X}$, given some

© The Author(s), under exclusive license to Springer Nature Switzerland AG 2026
Y. Zhang et al. (Eds.): LION 2025, LNCS 15745, pp. 97–112, 2026.
https://doi.org/10.1007/978-3-032-09192-5_7

data, $\mathbf{Y}$. The key idea is to generate randomly N weighted, and statistically independent samples from the posterior distribution of the model, $\pi(\mathbf{X}, \mathbf{Y})$, and use the samples to make estimates. The simplicity and the state-of-art performance of SMC samplers make them a popular choice in many application domains, e.g. Machine Learning (ML) [1], Bayesian Optimization (BO) [2,3], and Medicine [4].

Several ideas have been explored to make the estimations of SMC samplers more accurate. Examples include using better proposal distributions [5–7], tempering [8], or simply increasing N [9,10]. However, while these ideas are all valid and generally applicable, they share a common side effect: the SMC sampler becomes more computationally intensive. This side effect is more problematic in time-crucial applications, such as Epidemiology [11] (e.g., when computing the R-number of viruses, such as for COVID-19) or Crime Prediction [12]. Hence, parallel computing becomes crucial to trade-off accuracy and run-time.

1.2   Problem Definition and Related Work

Since the samples are statistically independent, they can be generated randomly in embarrassingly parallel fashion, which makes SMC an appealing alternative to other Bayesian inference methods, such as Markov Chain Monte Carlo [13]. However, at some point, the samples may experience a numerical error, called degeneracy, which makes the samples (and the state estimates) diverge from the true state of interest. This error is corrected by using a resampling algorithm, which replaces the samples that have diverged by redistributing copies of the samples that have not. Resampling is widely known to be the parallelization bottleneck of SMC samplers. This is due to the issues encountered when parallelizing the redistribution sub-task of resampling, in such a way that the workload is optimally distributed across P processing elements (or cores, the terms are used interchangeably here). For clarity, we provide the following definition:

Definition 1 (Optimal Parallel Redistribution (OPR)). *The problem of redistributing N samples with P parallel cores is solved optimally if and only if the maximum workload on any core is minimized, where the workload of a core is defined as the sum of the copying costs for the samples assigned to that core.*

In the special case where all samples have the same size, M, parallel redistribution has been extensively studied and solved optimally, achieving lower bound time complexity, $O(\log_2 N)$. Examples can be found for both Distributed Memory (DM) [14,15] and Shared Memory (SM) architectures [9,16,17].

In the most general case, the samples may have variable sizes. This occurs in several areas of ML and BO, where the model is designed to sample and estimate abstract data types, such as Decision Trees (DTs) [18–20], Additive Structures [21], and Time Series Structures [22]. In this scenario, OPR is significantly more complicated to achieve than in the fixed-size scenario, as the workload is even more unbalanced. Indeed, variable-size samples introduce an additional layer of complexity: balancing workloads now depends not only on the number of samples assigned to each core but also on the cost of copying each individual sample, which can vary significantly. The work in [23] describes a parallel SMC sampler for DTs on SM, in which the redistribution step is centralized to a single core, achieving little to no speed-up. In [24], a parallel redistribution for variable-size samples is proposed for SM and then ported to DM [25]. This approach first resizes all samples until they match the size of the biggest sample, $\overline{M}$, and then performs parallel redistribution using conventional parallel techniques for redistributing N fixed-size samples, where the cores redistribute an equal (integer) number of samples. The approach in [24] achieves $O(\overline{M} \log_2 N)$ time complexity, and is, therefore, sub-optimal since the cost of computation per sample is bound to the worst case, $O(\overline{M})$. This consequently poses the following question: is OPR achievable when redistributing N variable-size samples?

1.3 Paper Contribution and Outline

In this paper, we first prove that if all cores are constrained to redistribute an integer number of variable-size samples, OPR is NP-complete. In other words, under this constraint, OPR for variable-size samples is achievable in polynomial time only if P=NP. This is the case for DM systems and SM approaches like [24]. Then we propose a novel SM parallel redistribution algorithm for variable-size samples in which the samples are divisible (i.e., the cores may either redistribute whole samples or fractions of them) and do not need to be resized. Our theoretical analysis proves that our approach achieves optimal $O(\hat{M} \log_2 N)$ parallel time complexity, where $\hat{M}$ is the average size of the samples to be redistributed.

We also present experimental results comparing two versions of the same SMC sampler, differing only in their parallel redistribution approach. The results demonstrate that our approach achieves a significantly faster run-time than the approach described in [24], by roughly a $10\times$ factor.

The rest of the paper is organized as follows: in Sect. 2, we first describe SMC samplers in general, and then we briefly explain how SMC is parallelized in [24]. Section 3 proves that OPR is NP-complete if the cores are constrained to redistribute an integer number of variable-size samples. In Sect. 4, we present and analyze our approach in detail. Section 5 shows the numerical results. Section 6 concludes the paper and proposes possible future research directions.

2 Background and Previous Work

In this section, we first introduce the SMC sampler. For details, the reader is referred to [26]. Then we provide brief details on how SMC is parallelized in [24].

2.1 Sequential Monte Carlo Samplers

The SMC sampler generates N random samples of $\pi(\mathbf{X}, \mathbf{Y}) = p(\mathbf{Y}|\mathbf{X})p(\mathbf{X})$, i.e., the posterior distribution of the model, and $p(\mathbf{X})$ and $p(\mathbf{Y}|\mathbf{X})$ are the prior and the likelihood, respectively. SMC performs K iterations, for $k = 0, 1, \ldots, K-1$, to generate and update N random samples of the posterior, $\mathbf{x}_k^i$, for $i = 0, 1, \ldots, N-1$. However, since $\pi(\cdot, \cdot)$ is often impossible to be sampled from directly, the SMC sampler uses a convenient and arbitrary proposal distribution (or simply proposal) to generate the samples, and then gives each sample a weight, $\mathbf{w}_k^i$, which quantifies how well $\mathbf{x}_k^i$ approximates the true state, $\mathbf{X}$.

At the first iteration, $k = 0$, each sample, $\mathbf{x}_0^i$, is sampled from the prior, and its weight is initialized to $\mathbf{w}_0^i = \pi(\mathbf{x}_0^i, \mathbf{Y})/p(\mathbf{x}_0^i)$. After that, for all $k > 0$, the SMC sampler draws new samples and updates their weights as follows:

$$\mathbf{x}_k^i \sim q(\cdot|\mathbf{x}_{k-1}^i), \tag{1a}$$

$$\mathbf{w}_k^i = \mathbf{w}_{k-1}^i \frac{\pi\left(\mathbf{x}_k^i, \mathbf{Y}\right)}{\pi\left(\mathbf{x}_{k-1}^i, \mathbf{Y}\right)} \frac{q\left(\mathbf{x}_{k-1}^i|\mathbf{x}_k^i\right)}{q\left(\mathbf{x}_k^i|\mathbf{x}_{k-1}^i\right)}, \tag{1b}$$

where $q(\cdot|\cdot)$ is the proposal. We note that Equation (1) is often referred to as Importance Sampling (IS) in the literature.

After (1), the weights need to be normalized as follows:

$$\tilde{\mathbf{w}}_k^i = \frac{\mathbf{w}_k^i}{\sum_{j=0}^{N-1} \mathbf{w}_k^j}, \tag{2}$$

so that $\sum_i \tilde{\mathbf{w}}_k^i = 1$ and $\tilde{\mathbf{w}}_k^i$ can approximate expectations of the posterior.

In theory, this approach correctly represents the posterior as $N \to \infty$. In practice, IS suffers from degeneracy, a numerical error caused by having the samples generated from the proposal, instead of directly from the posterior. This leads to all weights but one to drop to 0 within a few iterations, making the samples diverge from the posterior. To correct this error, the typical approach is to perform the resampling algorithm if the effective sample size,

$$N_{\text{eff}} = \frac{1}{\sum_{i=0}^{N-1} \left(\tilde{\mathbf{w}}_k^i\right)^2}, \tag{3}$$

drops below an arbitrary threshold, normally set to $0.5N$. Resampling replaces the samples with low weights with copies of the samples with high weights.

Many resampling schemes exist in the literature. Here, we use systematic resampling [27], arguably the most common scheme. This resampling variant regenerates the sample population by performing the following three steps.

Step 1 - Choice. This step determines how many times each sample, $\mathbf{x}_k^i$, will be copied. In systematic resampling, this choice involves computing $\mathbf{cdf} \in \mathbb{R}^N$, the cumulative sum (or prefix sum, the terms are used interchangeably) of the normalized weights as follows:

$$\mathbf{cdf}^i = N \sum_{j=0}^{i-1} \tilde{\mathbf{w}}_k^j, \quad \forall i = 0, 1, 2, \ldots, N-1. \tag{4}$$

Each sample, $\mathbf{x}_k^i$, for $i = 0, 1, \ldots, N-1$, will then be copied as many times as

$$\mathbf{ncopies}^i = \lceil \mathbf{cdf}^i + N\tilde{\mathbf{w}}_k^i - u \rceil - \lceil \mathbf{cdf}^i - u \rceil, \tag{5}$$

where $u \sim \mathtt{Uniform}[0, 1)$, and $\lceil \cdot \rceil$ is the ceiling operator. From (4) and (5), it can be inferred that $\mathbf{ncopies} \in \mathbb{N}_0^N$, where $\mathbb{N}_0 = \mathbb{N} \cup \{0\}$, and complies with

$$\sum_{i=0}^{N-1} \mathbf{ncopies}^i = N. \tag{6a}$$

$$0 \leq \mathbf{ncopies}^i \leq N, \tag{6b}$$

Step 2 - Redistribution. This step is in charge of actually replicating each sample as many times as in (5), such that the samples for which $\mathbf{ncopies}^i = 0$ will be removed. Algorithm 1 describes a textbook implementation of a Sequential Redistribution (S-R). We note that every resampling variant requires redistri-

---

Algorithm 1. Sequential Redistribution (S-R)

Input: x, ncopies, N
Output: $\mathbf{x}_{new}$
  $i \leftarrow 0$
  for $j \leftarrow 0;\ j < N;\ j \leftarrow j+1$ **do**
    for $copy \leftarrow 0;\ copy < \mathbf{ncopies}^j;\ copy \leftarrow copy + 1$ **do**
      $\mathbf{x}_{new}^i \leftarrow \mathbf{x}^j$
      $i \leftarrow i + 1$
    end for
  end for

---

bution. Therefore, since this paper presents an OPR for variable-size samples, the novelty we propose is generally applicable to any resampling variant.

Step 3 - Reset. After redistribution, the weights are simply reset to $1/N$, and a new iteration starts over from (1).

After K iterations, the samples are assumed to have converged to the true posterior, and the true state can be estimated as a weighted mean of the samples.

2.2 Previous Work on Parallel SMC for Variable Size Samples

Equations (1), (5), and the *Reset* step in resampling are embarrassingly parallel, as these operations are all element-wise tasks. On SM, these tasks can be parallelized by equally dividing the iteration space across P SM threads (i.e., the processing elements), such that each thread performs $O(\frac{N}{P})$ iterations.

Equations (2), (3) both require the computation of a vector sum. This operation is notoriously parallelizable by using reduction, which is well-known to achieve $O(\frac{N}{P} + \log_2 P)$ time complexity.

The prefix sum in Equation (4) can also be performed in $O(\frac{N}{P} + \log_2 P)$ by using prefix reduction [28–30].

Algorithm 1 takes $O(N)$ sequentially, as Equation (6a) holds. However, this algorithm is impossible to be parallelized if using embarrassingly parallel techniques. The main reason is that the workload associated with replicating each sample, $\mathbf{x}^i$, a total of $\mathbf{ncopies}^i$ times is inherently unbalanced as Equation (6b) holds. Furthermore, the workload may be even more unbalanced in the general case in which the samples have variable sizes, which is the scenario we focus on in this paper. A SM parallel redistribution for variable-size samples has been proposed in [24] and is illustrated in Algorithm 2. Brief details follow.

Algorithm 2 makes all cores redistribute the same (integer) number of samples, $\frac{N}{P}$. Therefore, this means that the samples must be assumed to be indivisible, whose definition is formally given as follows.

Assumption 1 (Indivisibility). *Each variable-size sample, $\mathbf{x}^i$, must be assigned entirely to a single core, i.e., cannot be split or divided across the memory spaces of multiple cores. In other words, if sample $\mathbf{x}^i$ of size $\mathbf{M}^i$ is assigned to core id, then the entirety of the $\mathbf{M}^i$ sample values contributes to the workload of core id.*

Algorithm 2. Reference Parallel Redistribution for Variable Size Samples

---

Input: x, ncopies, N, P, $n - \frac{N}{P}$
Output: x

$\overline{M} \leftarrow \texttt{Max}(\underline{\texttt{dims}}(\mathbf{x}), P)$ //dims returns the dimension of $\mathbf{x}^i$ $\forall i$
$\mathbf{x} \leftarrow \texttt{Pad}(\mathbf{x}, \overline{M}, P)$ //Appends up to $\overline{M} - 1$ Not a Numbers to each sample
$\mathbf{csum} \leftarrow \texttt{Prefix Sum}(\mathbf{ncopies}, P)$, Begin redistribution
Spawn P threads with $id = 0, 1, \ldots, P - 1\{$

   $p \leftarrow \texttt{Binary Search}(\mathbf{csum}, \mathbf{ncopies}, n, P)$
   $cp \leftarrow \min\left(\mathbf{csum}^p - id \times n, n\right)$
   $\mathbf{x}^{n:n \times cp-1} \leftarrow \mathbf{x}^p$
   $\mathbf{x} \leftarrow \texttt{S-R}(\mathbf{x}^{p+1:N-1}, \mathbf{ncopies}^{p+1:N-1}, n - cp)$

$\}$ //End redistribution
$\mathbf{x} \leftarrow \texttt{Restore}(\mathbf{x}, \overline{M}, P)$

Under these conditions, the workload is likely unbalanced due to the uneven sample sizes. Algorithm 2 then bypasses this problem by resizing all the samples to size, $\overline{M}$, i.e., the size of the biggest sample. This step is parallelized in $O(\overline{M} \log_2 N)$. After that, the samples can be redistributed in $O(\overline{M} \log_2 N)$ by using known parallel redistribution algorithms for fixed-size samples [17]. More precisely, the P threads first compute $\mathbf{csum} \in \mathbf{N}_0^N$, the prefix sum of $\mathbf{ncopies}$. Then, each thread can use $\mathbf{csum}$ to identify a subset of $\frac{N}{P}$ samples to redistribute in $O(\frac{N}{P})$ with a constant time per sample of $O(\overline{M})$ by using Algorithm 1. Ultimately, the original size of the redistributed samples is restored in $O(\overline{M}\frac{N}{P})$ parallel time. Overall, [24] proves Algorithm 2 achieves a sub-optimal $O(\overline{M} \log_2 N)$

time complexity because the computational cost per sample is bounded to the worst case, $O(\overline{M})$. This conclusion leads to the following question: is it possible to achieve OPR on N variable-size, indivisible samples in polynomial time? The next section provides an answer to this question.

3 OPR for Variable Size Indivisible Samples Is NP-Complete

Theorem 1. *Under Assumption 1, the problem of Optimal Parallel Redistribution (OPR), defined as distributing N variable-size samples across $P \leq N$ cores such that the maximum workload among all cores is minimized, is NP-complete.*

Proof. To prove Theorem 1, we show OPR under Assumption 1 is both in NP and NP-hard.

OPR is in NP: By Assumption 1, each sample is assigned entirely to a single core. Given a proposed solution (redistribution of samples across P cores, for $id = 0, 1, \ldots P - 1$), we can compute the workload for each core:

$$\mathbf{W}^{id} = \sum_{i\text{-th samples assigned to } id} \mathbf{ncopies}^i \mathbf{M}^i,$$

where $\mathbf{M}^i$ is the size of the i-th sample. The maximum workload is:

$$\overline{W} = \max_{0 \leq id \leq P-1} \mathbf{W}^{id}.$$

Verifying whether $\overline{W}$ is less than or equal to a given threshold can be done in polynomial time. Thus, OPR is in NP.

OPR is NP-hard: We reduce the *Partition Problem*, a known NP-complete problem [31], to OPR. The Partition Problem is defined as follows: given a set of positive integers $\mathbf{S} = \{a_0, a_1, \ldots, a_{N-1}\}$, determine if there exists a partition of $\mathbf{S}$ into two subsets, $\mathbf{S}_1$ and $\mathbf{S}_2$, such that:

$$\sum_{x \in \mathbf{S}_1} x = \sum_{y \in \mathbf{S}_2} y = \frac{1}{2} \sum_{z \in \mathbf{S}}.$$

Given an instance of the Partition Problem, we construct an OPR instance with $P = 2$ cores, sample sizes $\mathbf{M}^i = a_i$, for $i = 0, 1, \ldots, N - 1$, and $\mathbf{ncopies}^i = 1$ (i.e., each sample is replicated exactly once). The goal in OPR is to minimize:

$$\overline{W} = \max\left(\sum_{x_i \in \mathbf{S}_1} \mathbf{M}^i, \sum_{x_j \in \mathbf{S}_2} \mathbf{M}^j \right).$$

If the Partition Problem has a solution, there exists a distribution of samples across two cores such that $\overline{W} = \frac{1}{2} \sum_{z \in \mathbf{S}}$. Conversely, if OPR achieves $\overline{W} = \frac{1}{2} \sum_{z \in \mathbf{S}}$, the Partition Problem has a solution. Thus, OPR is NP-hard for $P = 2$.

For $P > 2$, OPR generalizes the Partition Problem to the *Multiprocessor Scheduling Problem*, where the objective is to minimize $\overline{W}$ across P cores, subject to Assumption 1. The Multiprocessor Scheduling Problem is known to be NP-hard, and the indivisibility constraint preserves the hardness of OPR.

Since OPR is in NP and NP-hard, we conclude that OPR is NP-complete. $\square$

4   A SM OPR Algorithm for Variable Size Samples

Theorem 1 proves that OPR for N variable-size, indivisible samples cannot be solved in polynomial time, unless P=NP. However, while Assumption 1 is indeed mandatory on DM (as the samples are by definition allocated in different memories), it is optional on SM (as the samples are all stored in the same physical memory). In this section, we prove that on SM it is possible to optimally redistribute N variable-size samples in parallel, if Assumption 1 is relaxed.

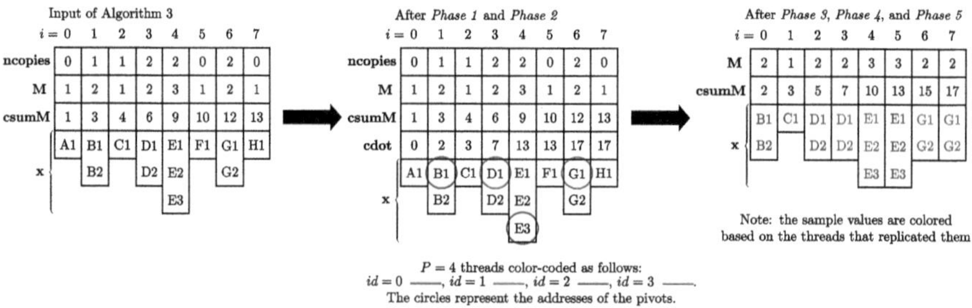

Fig. 1. Algorithm 3 - example for $N = 8$ and $P = 4$. Each sample value is encoded with a letter and a number for brevity.

To achieve this goal, the key idea is to allow the SM threads to redistribute whole samples and/or (potentially) fractions of certain samples. More precisely, if a M-dimensional sample, $\mathbf{x}^i$, needs to be copied by a certain thread with identification number, id, depending on how the workload is distributed across the P threads, thread id might have to either copy the whole sample, $\mathbf{x}^i$, or copy only $M_s < M$ sample values of $\mathbf{x}^i$, such that thread $id + 1$ will have to copy the remaining $M - M_s$ sample values of $\mathbf{x}^i$.

The first thing we need to do is to restructure $\mathbf{x}$ as a one-dimensional (1-D) array of total size $\sum_{i=0}^{N-1} \mathbf{M}^i$, instead of being an array of N samples, each sample being a list of $\mathbf{M}^i$ values, as in Algorithm 2. This means that, once redistribution is complete, $\mathbf{x}$ will have a new total size equal to

$$W_{\text{tot}} = \sum_{i=0}^{N-1} \mathbf{ncopies}^i \mathbf{M}^i, \tag{7}$$

which can also be viewed as the total workload (or total number of memory copies) to redistribute $\mathbf{x}$. Since we work with a 1-D array, we need to be able to address each (variable-size) sample in $O(1)$. We do this by storing in memory $\mathbf{csumM} \in \mathbb{N}_0^N$, i.e., the cumulative sum of $\mathbf{M}$, such that

$$\mathbf{csumM}^i = \sum_{j=0}^{i} \mathbf{M}^j, \quad \forall i = 0, 1, 2, \ldots, N-1, \tag{8}$$

can be considered the base address of i-th sample in a 1-D array of variable-size samples. Broadly speaking, if $\mathbf{x}$ is an array of variable-size lists, $\mathbf{x}^{i,m}$ is the m-th value of the i-th sample, while the same value will be found in index $\mathbf{csumM}^i + m$ if $\mathbf{x}$ is restructured as a 1-D array. With these given inputs, it is possible to prove that redistribution can be parallelized in $O(\frac{N}{P} + \log_2 N)$ with a much lower cost of computation per sample compared to Algorithm 2. Further details follow.

Phase 1 - Prefix Dot Product. Equation (7) expresses the total workload to redistribute $\mathbf{x}$ as the dot product of $\mathbf{ncopies}$ and $\mathbf{M}$. This suggests that we could compute $\mathbf{cdot} \in \mathbb{N}_0^N$, the cumulative dot product of $\mathbf{ncopies}$ and $\mathbf{M}$, such that each $\mathbf{cdot}^i$, for all $i = 0, 1, \ldots, N-1$, will contain the required workload, as a function of the size of each sample, to redistribute the samples up to the i-th index. More precisely, we compute:

$$\mathbf{cdot}^i = \sum_{j=0}^{i-1} \mathbf{ncopies}^j \mathbf{M}^j, \quad \forall i = 0, 1, 2, \ldots, N-1. \tag{9}$$

From (7) and (9), it is relatively straightforward to infer that $W_{\text{tot}} = \mathbf{cdot}^{N-1}$.

Phase 2 - Workload Split. We now want to divide the workload as evenly as possible across the P parallel threads, $id = 0, 1, \ldots, P-1$. Since the threads need to redistribute W_{tot} sample values in total, in theory, every thread could search for a pivot index, p, which is the first index such that $\mathbf{cdot}^p \geq id \frac{W_{\text{tot}}}{P}$, and then redistribute $\frac{W_{\text{tot}}}{P}$ sample values starting from its pivot. However, in practice, W_{tot} might not be an integer factor of P, since the samples have variable size. Therefore, we decide to split the workload in such a way each thread will either redistribute $\lfloor \frac{W_{\text{tot}}}{P} \rfloor$ or $\lfloor \frac{W_{\text{tot}}}{P} \rfloor + 1$ sample values, where $\lfloor \cdot \rfloor$ is the floor operator, such that $\lfloor \frac{W_{\text{tot}}}{P} \rfloor$ is the quotient of $W_{\text{tot}} \div P$. More precisely, the workload, $\mathbf{W}^{id} \in \mathbb{N}_0$, to be done by the thread id is computed as follows:

$$\mathbf{W}^{id} = \begin{cases} \lfloor \frac{W_{\text{tot}}}{P} \rfloor + 1, & \text{if } id + 1 \leq W_{\text{tot}} \bmod P, \\ \lfloor \frac{W_{\text{tot}}}{P} \rfloor, & \text{if } id + 1 > W_{\text{tot}} \bmod P, \end{cases} \tag{10}$$

where $\bmod$ is the remainder of $W_{\text{tot}} \div P$. In order to find the pivots, each thread, id, must compute, $\mathbf{cw}^{id}$, the cumulative sum of all sample values being redistributed by all threads having a lower identification number than id. One could use the exclusive prefix sum of $\mathbf{W}$ to compute $\mathbf{cw}^{id}$. However, the same could be computed in $O(1)$ by using the information in Equation (10) as follows:

$$\mathbf{cw}^{id} = \begin{cases} id \times \mathbf{W}^{id}, & \text{if } id + 1 \leq W_{\text{tot}} \bmod P, \\ \lfloor \frac{W_{\text{tot}}}{P} \rfloor (\mathbf{W}^{id} + 1) + (id - \lfloor \frac{W_{\text{tot}}}{P} \rfloor) \mathbf{W}^{id}, & \text{if } id + 1 > W_{\text{tot}} \bmod P. \end{cases} \tag{11}$$

Each thread, *id*, can then independently search for the first index, p, such that

$$\mathbf{cdot}^p \geq \mathbf{cw}^{id}. \tag{12}$$

Binary Search (BS) can be used to search for the pivots since **cdot** is inherently sorted as the output of every prefix reduction is always monotonically increasing.

Phase 3 - Copying Samples. Each thread *id* can now independently copy the $\mathbf{W}^{id}$ sample values starting from the pivot index found in the previous step. At any given iteration, a sample value will actually be copied if and only if it happens to be a quantity of any i-th sample such that $\mathbf{ncopies}^i > 0$.

Phase 4 - Redistribution of Sample Sizes. After the previous step, the samples have been redistributed. However, we also need to guarantee that the IS step, at the next iteration, is able to address each sample in $O(1)$ as in (8). This requires redistributing the values in $\mathbf{M}$ according to **ncopies**, which can be done by repeating three analogous phases to *Phase 1*, *Phase 2*, and *Phase 3*. More precisely, we first compute the prefix sum of **ncopies** to obtain $\mathbf{csum} \in \mathbb{N}_0^N$. Then each thread, *id*, searches for a pivot index, p, i.e., the first index such that $\mathbf{csum}^p \geq id\frac{N}{P}$. After that, each thread can independently redistribute $\frac{N}{P}$ elements in $\mathbf{M}$ by using Algorithm 1 starting from index p.

Phase 5 - Update **csumM**. The previous step is computed with a view to updating the values of **csumM**. After *Phase 4*, this can be done by computing (8). For completeness, we note that, having restructured $\mathbf{x}$ to be a flattened 1-D array, also means that the IS step is required to update **csumM** as in (8), since the proposal $q(\cdot|\cdot)$ may increase or decrease the size of each and every sample.

Algorithm 3 summarizes the previous steps, and Fig. 1 illustrates an example for $N = 8$ and $P = 4$. The following lemma analyzes the time complexity.

Lemma 1. *Let* $\mathbf{x}$ *be a 1-D array that contains N samples, each with size $\mathbf{M}^i$, for all $i = 0, 1, 2, \ldots, N-1$, which could be different from any $\mathbf{M}^j$ with $j \neq i$. Let* $\mathbf{ncopies} \in \mathbf{N}_0^N$ *be an array of integers that complies with (6). Algorithm 3 redistributes the samples in $\mathbf{x}$ according to $\mathbf{ncopies}$ in $O(\hat{M} \log_2 N)$ time complexity, where $\hat{M} = \frac{1}{N}\sum_i \mathbf{ncopies}^i \mathbf{M}^i$, i.e., the average size of new samples.*

Proof. To prove Lemma 1, we first analyze the time complexity of each step of Algorithm 3 individually. *Phase 1*, and *Phase 5* can be parallelized by using prefix reduction, which is well known to scale as $O(\frac{N}{P} + \log_2 P)$, with a fast constant time that does not depend on the size of the samples.

In *Phase 2*, each thread performs a BS individually, which takes $O(\log_2 N)$ computations, which also are independent of the sizes of the samples.

Phase 4 redistributes the elements of $\mathbf{M}$ according to **ncopies**. This can be parallelized in $O(\frac{N}{P} + \log_2 N)$ because it consists of one prefix reduction, one BS, and Algorithm 1 being used to redistribute $\frac{N}{P}$ elements of $\mathbf{M}$. The constant time of this phase will be negligible, since $\mathbf{M}$ is a 1-D array of integers, and also independent of the size of the samples.

Algorithm 3. Novel Parallel Redistribution for Variable Size Samples

---

Input: x, ncopies, csumM, N, P, $n = \frac{N}{P}$
Output: x, csumM

  cdot $\leftarrow$ `Prefix Dot Product`(ncopies, M, P) //*Phase 1*
  Spawn P threads with $id = 0, 1, \ldots, P - 1$ { //*Beginning of Phase 2*

      $W_{\text{tot}} \leftarrow \mathbf{cdot}^{N-1}$
      if $id + 1 \leq W_{\text{tot}} \bmod P$ **then**
        $\mathbf{W}^{id} \leftarrow \lfloor \frac{W_{\text{tot}}}{P} \rfloor + 1$, $\mathbf{cw}^{id} \leftarrow id \times \mathbf{W}^{id}$
      else
        $\mathbf{W}^{id} \leftarrow \lfloor \frac{W_{\text{tot}}}{P} \rfloor$, $\mathbf{cw}^{id} \leftarrow \lfloor \frac{W_{\text{tot}}}{P} \rfloor \times (\mathbf{W}^{id} + 1) + (id - \lfloor \frac{W_{\text{tot}}}{P} \rfloor) \times \mathbf{W}^{id}$
      end if
      $p \leftarrow$ `Binary Search` (cdot, ncopies, M, $\mathbf{cw}^{id}$, P) //End of *Phase 2*
      $\mathbf{x}^{n:n \times cp-1} \leftarrow \mathbf{x}^{p}$ //*Beginning of Phase 3*
      $\mathbf{x} \leftarrow$ `S-R`($\mathbf{x}^{p+1:N-1}$, $\mathbf{ncopies}^{p+1:N-1}$, $n - cp$)

  } //End of *Phase 3*
  csum $\leftarrow$ `Prefix Sum`(ncopies, P) //*Beginning of Phase 4*
  Spawn P threads with $id = 0, 1, \ldots, P - 1$ {

      $p \leftarrow$ `Binary Search`(csum, ncopies, n, P)
      $cp \leftarrow \min{(\mathbf{csum}^{p} - id \times n, n)}$
      $\mathbf{M}^{n:n \times cp-1} \leftarrow \mathbf{M}^{p}$
      $\mathbf{M} \leftarrow$ `S-R`($\mathbf{M}^{p+1:N-1}$, $\mathbf{ncopies}^{p+1:N-1}$, $n - cp$)

  } //End of *Phase 4*
  csumM $\leftarrow$ `Prefix Sum`(M, P) //*Phase 5*

In *Phase 3*, each thread performs either $\lfloor \frac{W_{\text{tot}}}{P} \rfloor + 1$ or $\lfloor \frac{W_{\text{tot}}}{P} \rfloor$ memory copies, by Equation (10). Due to Equation (7), $\frac{W_{\text{tot}}}{P}$ in (10) can be rearranged as follows:

$$\frac{W_{\text{tot}}}{P} = \frac{\sum_i \mathbf{ncopies}^i \mathbf{M}^i}{P} = N \times \frac{\sum_i \mathbf{ncopies}^i \mathbf{M}^i}{N \times P} = \frac{N}{P} \times \hat{M}$$

Therefore, as $P \to N$, this splitting strategy is analogous to having each thread work on copying $O(1)$ sample, with size $\hat{M} = \frac{1}{N} \sum_i \mathbf{ncopies}^i \mathbf{M}^i$, i.e., equal to the average size of the new samples.

Overall, all steps combined amount to a total time complexity equal to $O(\hat{M} \log_2 N)$ as $P \to N$. In other words, Algorithm 3 scales logarithmically with N and with a computational cost per sample equal to $O(\hat{M})$. $\square$

Theorem 2. *Algorithm 3 divides the workload optimally across the SM threads.*

Proof. To prove this theorem, one needs to prove that the time complexity of Algorithm 3, $O(\hat{M} \log_2 N)$ as proven in Lemma 1, is optimal both with respect to number of samples, N, and the size of the samples.

With respect to N, Algorithm 3 scales logarithmically, i.e., the same time complexity as the prefix sum in (4), which is proven to be optimal [29].

With respect to the sample sizes, each thread in Algorithm 3 performs as many memory copies per sample as in (10). Equation (10) guarantees that the cost of computation per sample performed by each thread is at most within one memory copy of the ideal average, $\frac{W_{tot}}{P}$. It is then straightforward to infer that Algorithm 3 always achieves the optimal workload balance, as any other strategy which assigns more than $\lfloor \frac{W_{tot}}{P} \rfloor + 1$ memory copies to any thread, would necessarily underwork/overwork a subset of the threads by at least two memory copies with respect to the ideal average. $\square$

5   Numerical Results

In this section, we want to compare two versions of the same parallel SMC sampler sampling from a posterior distribution of variable-size samples: one version utilizing Algorithm 2 to parallelize the redistribution step, and the other version utilizing our proposed approach, i.e., Algorithm 3. For transparency, we note that all codes are implemented in C++ and parallelized with OpenMP 4.5. Implementation details are omitted for brevity due to the page limit, but the reader can access the code from the following link: https://github.com/AVarsi88/Parallel_SMC_on_Shared_Memory.

From the results in Sect. 4, we expect both SMC samplers to scale progressively as we increase the number of threads, P. However, given the improved cost of computation per sample, we expect SMC with Algorithm 3 to be much faster than SMC with Algorithm 2 when $\hat{M} \ll \overline{M}$, and we expect the two approaches to achieve similar performance when $\hat{M} \approx \overline{M}$. To test this, we generate samples of variable size by sampling from the following statistical model:

$$\pi(\cdot, \cdot) = \begin{cases} 0.5e^M(1 - \alpha)(\mathbf{m}_s^1 - \mathbf{m}_s^0 + 1), & \text{if } \mathbf{m}_s^0 \leq M \leq \mathbf{m}_s^1 \\ 0.5e^M \alpha(\mathbf{m}_b^1 - \mathbf{m}_b^0 + 1), & \text{if } \mathbf{m}_b^0 \leq M \leq \mathbf{m}_b^1. \end{cases} \tag{13}$$

This model is well-suited for our testing purposes because it gives us precise control over the sample sizes and their variability. Indeed, the samples generated by an SMC sampler sampling from (13) will contain a varying number, M, of coin flip results (i.e., either 0 or 1). Most importantly, each sample will either have a large size in the range $\mathbf{m}_b$ with a probability $p(M \in \mathbf{m}_b) = \alpha$, or a small size in the range $\mathbf{m}_s$ with a probability $p(M \in \mathbf{m}_s) = 1 - \alpha$. This is illustrated in Fig. 2. In our test, we arbitrarily choose $\mathbf{m}_s = \{1, 3\}$ dimensions for the small samples, and a (significantly larger) range of $\mathbf{m}_b = \{698, 700\}$ dimensions for the big samples, such that $\overline{M} = 700$. Then we control the average size of the samples, by varying α. More precisely, $\hat{M} \ll \overline{M}$ when α is small, and $\hat{M} \approx \overline{M}$ when $\alpha \rightarrow 1.0$. Indeed, we use $\alpha = \{0.01, 0.5, 1.0\}$.

The experimental results are provided in Figs. 3, and 4. These results are taken on a workstation that mounts a 2 Xeon Gold 6138 CPU which provides up to 40 physical cores, and 384GB of memory. Therefore, we use power-of-two numbers up to $P = 32$ SM threads for the experiments as both Algorithms 2, and 3 use the divide-and-conquer paradigm. We also use power-of-two numbers for N, to make it easier to balance the workload across the SM threads. More precisely, we use $N = \{2^{10}, 2^{15}, 2^{20}\}$. We also note that all reported run-times

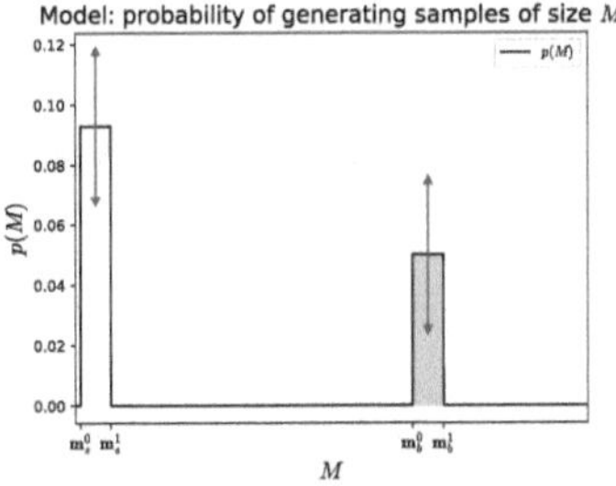

Fig. 2. Model: $p(M)$. The shaded region is α, the total probability of drawing big samples. The red arrows symbolize the possibility of having varying α. (Color figure online)

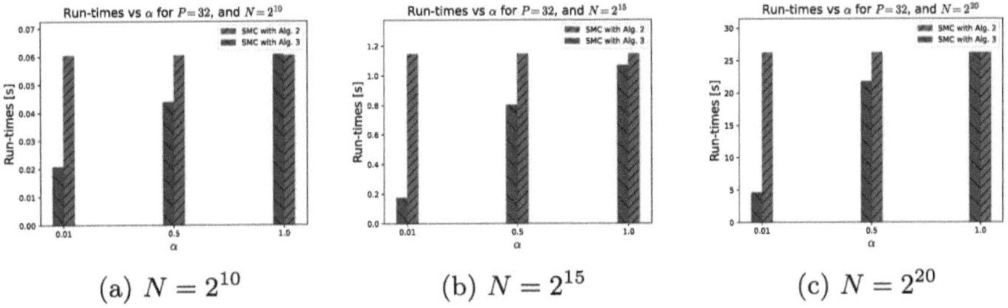

(a) $N = 2^{10}$ (b) $N = 2^{15}$ (c) $N = 2^{20}$

Fig. 3. Run-times as function of α, N, and for $P = 32$.

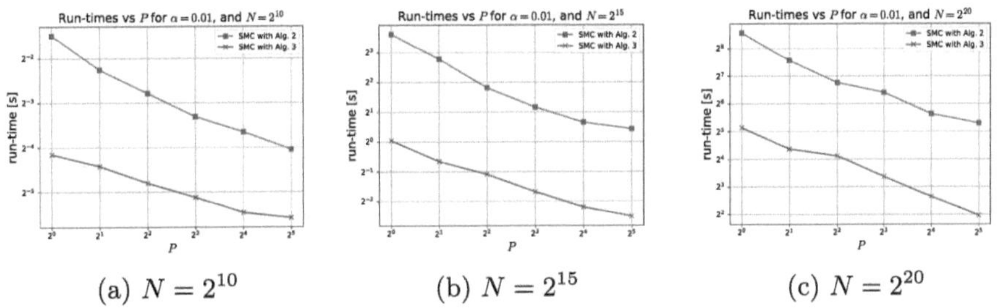

(a) $N = 2^{10}$ (b) $N = 2^{15}$ (c) $N = 2^{20}$

Fig. 4. Run-times as function of N, P and for $\alpha = 0.01$.

are average results for 10 Monte Carlo (MC) runs, and, for each MC run, both versions of the SMC samplers are set to perform $K = 10$ iterations.

Figure 3 shows the run-times of both SMC samplers when α increases. As theory claims, when the samples' sizes are roughly constant (i.e., $\alpha = 1.0$) the run-times of the two SMC samplers are comparable, independently of N. However, when α decreases, an SMC sampler with Algorithm 3 becomes up to $10\times$ faster than the SMC sampler with Algorithm 2. Due to page limitations, we only report the most informative run-times, i.e., for $P = 32$, as the same run-times for any other P would be qualitatively similar. Figure 4 shows how the run-times of both SMC samplers scale for increasing P. As we anticipated, both SMC samplers scale progressively with P, but the SMC with Algorithm 3 has faster run-times due to the improved cost of computation per sample.

6   Conclusion and Future Work

In this paper, we tackle on SM the Optimal Parallel Redistribution (OPR) problem in the context of SMC samplers drawing N variable-size samples. We first prove that if the samples are assumed to be indivisible (i.e., every core must hold an integer number of samples), OPR is NP-complete. We then prove that, if the indivisibility assumption is relaxed, it is possible to construct a parallel redistribution algorithm that divides the workload optimally between the processing elements, as it achieves optimal $O(\hat{M} \log_2 N)$ time complexity, where $\hat{M}$ is the average size of the redistributed samples. The baseline for comparison is presented in [24], which describes a parallel redistribution algorithm for variable-size, indivisible samples which achieves a sub-optimal $O(\overline{M} \log_2 N)$ time complexity, where $\overline{M}$ is the size of the biggest sample (i.e., the worst-case scenario). The numerical results on a 32-core SM machine show that an SMC sampler equipped with our proposed parallel redistribution is up to $10\times$ faster than an equivalent SMC sampler equipped with the parallel redistribution in [24].

Despite the encouraging results, several future avenues remain to be explored. First, the run-time of the proposed method could be accelerated by using GPUs instead of CPUs. Another usual practice in supercomputing is to combine DM and SM, e.g., with MPI+X programming models. This would require an effective DM parallel redistribution for variable-size samples. While OPR for variable-size samples on DM is NP-complete (as discussed in this paper), it may still be possible to design an effective (albeit sub-optimal) algorithm. Developing such a redistribution component is, to the best of our knowledge, an open research question.

References

1. Ma, X., Karkus, P., Hsu, D., Lee, W.S.: Particle filter recurrent neural networks. Proc. AAAI Conf. Artif. Intell. **34**, 5101–5108 (2020)
2. Drousiotis, E., Varsi, A., Spirakis, P.G., Maskell, S.: An SMC sampler for decision trees with enhanced initial proposal for stochastic metaheuristic optimization. In: Learning and Intelligent Optimization Festa, P., Ferone, D., Pastore, T., Pisacane, O. (eds.) Cham, pp. 123–137, Springer Nature Switzerland (2025)
3. Habibi, S., Drousiotis, E., Varsi, A., Maskell, S., Moore, R., Spirakis, P.G.: Conditional importance resampling for an enhanced sequential monte Carlo sampler. In: Learning and Intelligent Optimization Festa, P., Ferone, D., Pastore, T., Pisacane, O. (eds.) Cham, pp. 169–184, Springer Nature Switzerland (2025)
4. Barazandegan, M., Ekram, F., Kwok, E., Gopaluni, B., Tulsyan, A.: Assessment of type ii diabetes mellitus using irregularly sampled measurements with missing data. Bioprocess Biosyst. Eng. **38**, 615–629 (2015)

5. Naesseth, C.A., Lindsten, F., Schn, T.B.: High-dimensional filtering using nested sequential monte Carlo. IEEE Trans. Sig. Process. **67**, 4177–4188 (2019)
6. Drousiotis, E., Phillips, A.M., Spirakis, P.G., Maskell, S.: Bayesian decision trees inspired from evolutionary algorithms. In: Learning and Intelligent Optimization Sellmann, M., Tierney, K. (eds.) Cham, pp. 318–331, Springer International Publishing (2023)
7. Varsi, A., Devlin, L., Horridge, P., Maskell, S.: A general-purpose fixed-lag no-u-turn sampler for nonlinear non-gaussian state space models. IEEE Trans. Aerosp. Electron. Syst. 1–16 (2024)
8. Svensson, A., Schön, T.B., Lindsten, F.: Learning of state-space models with highly informative observations: a tempered sequential monte Carlo solution. Mech. Syst. Signal Process. **104**, 915–928 (2018)
9. Lopez, F., Zhang, L., Beaman, J., Mok, A.: Implementation of a particle filter on a GPU for nonlinear estimation in a manufacturing remelting process. In: 2014 IEEE/ASME International Conference on Advanced Intelligent Mechatronics, pp. 340–345 (2014)
10. Lopez, F., Zhang, L., Mok, A., Beaman, J.: Particle filtering on GPU architectures for manufacturing applications. Comput. Ind. **71**, 116–127 (2015)
11. Sheinson, D., Meng, Y., Elsea, D.: Pin67 real-time analysis of COVID-19 data using sequential monte Carlo methods. Value Health **24**, S118 (2021)
12. Ramaraj, G.D., Venkatakrishnan, S., Balasubramanian, G., Sridhar, S.: Aerial surveillance of public areas with autonomous track and follow using image processing. In: 2017 International Conference on Computer and Drone Applications (IConDA), pp. 92–95. IEEE (2017)
13. Brooks, S.: Markov chain monte Carlo method and its application. J. R. Stat. Soc. Ser. D (the Statistician) **47**(1), 69–100 (1998)
14. Varsi, A., Maskell, S., Spirakis, P.G.: An O (log2N) fully-balanced resampling algorithm for particle filters on distributed memory architectures. Algorithms **14**(12), 342 (2021)
15. Rosato, C., Varsi, A., Murphy, J., Maskell, S.: An O (log2 N) SMC2 algorithm on distributed memory with an approx. optimal l-kernel. In: 2023 IEEE Symposium Sensor Data Fusion and International Conference on Multisensor Fusion and Integration (SDF-MFI), pp. 1–8 (2023)
16. Murray, L.M., Lee, A., Jacob, P.E.: Parallel resampling in the particle filter. J. Comput. Graph. Stat. **25**(3), 789–805 (2016)
17. Varsi, A., Taylor, J., Kekempanos, L., Pyzer Knapp, E., Maskell, S.: A fast parallel particle filter for shared memory systems. IEEE Sig. Process. Lett. **27**, 1570–1574 (2020)
18. Lakshminarayanan, B., Roy, D., Whye Teh, Y.: Top-down particle filtering for Bayesian decision trees. In: Proceedings of the 30th International Conference on Machine Learning Dasgupta, S., McAllester, D. (eds.) vol. 28 of Proceedings of Machine Learning Research, (Atlanta, Georgia, USA), pp. 280–288, PMLR, 17–19 (2013)
19. Lakshminarayanan, B.: Decision trees and forests: a probabilistic perspective. PhD thesis, UCL (University College London) (2016)
20. Quadrianto, N., Ghahramani, Z.: A very simple safe-Bayesian random forest. IEEE Trans. Pattern Anal. Mach. Intell. **37**(6), 1297–1303 (2015)
21. Gardner, J., Guo, C., Weinberger, K., Garnett, R., Grosse, R.: Discovering and exploiting additive structure for Bayesian optimization. In: Artificial Intelligence and Statistics, pp. 1311–1319. PMLR (2017)

22. Saad, F., Patton, B., Hoffman, M.D., Saurous, R.A., Mansinghka, V.: Sequential Monte Carlo learning for time series structure discovery. In: Proceedings of the 40th International Conference on Machine Learning Krause, A., Brunskill, E., Cho, K., Engelhardt, B., Sabato, S., Scarlett, J. (eds.) vol. 202 of Proceedings of Machine Learning Research, pp. 29473–29489. PMLR (2023)
23. Drousiotis, E., Spirakis, P.G., Maskell, S.: Parallel approaches to accelerate Bayesian decision trees. arXiv preprint arxiv.org/abs/2301.09090 (2023)
24. Drousiotis, E., Varsi, A., Spirakis, P.G., Maskell, S.: A shared memory SMC sampler for decision trees. In: 2023 IEEE 35th International Symposium on Computer Architecture and High Performance Computing (SBAC-PAD), pp. 209–218 (2023)
25. Drousiotis, E., Varsi, A., Phillips, A.M., Maskell, S., Spirakis, P.G.: A massively parallel SMC sampler for decision trees. Algorithms **18**(1) (2025)
26. Moral, P.D., Doucet, A., Jasra, A.: Sequential monte Carlo samplers. J. R. Stat. Soc. Ser. B (Statistical Methodology) **68**(3), 411–436 (2006)
27. Hol, J.D., Schon, T.B., Gustafsson, F.: On resampling algorithms for particle filters. In: 2006 IEEE Nonlinear Statistical Signal Processing Workshop, pp. 79–82 (2006)
28. Ladner, R.E., Fischer, M.J.: Parallel prefix computation. J. ACM **27**, 831–838 (1980)
29. Santos, E.E.: Optimal and efficient algorithms for summing and prefix summing on parallel machines. J. Parallel Distrib. Comput. **62**(4), 517–543 (2002)
30. Kozakai, S., Fujimoto, N., Wada, K.: Efficient GPU-implementation for integer sorting based on histogram and prefix-sums. In: Proceedings of the 50th International Conference on Parallel Processing, ICPP '21, (New York, NY, USA), Association for Computing Machinery (2021)
31. Garey, M.R., Johnson, D.S.: Computers and intractability: A Guide to the Theory of NP-Completeness, vol. 174. freeman San Francisco (1979)

A Hybrid Quantum-Inspired and Deep Learning Approach for the Capacitated Vehicle Routing Problem with Time Windows

Jorin Dornemann[1]($\boxtimes$) (ID), Salwa Shaglel[2] (ID), Martin Kliesch[2] (ID), and Anusch Taraz[1] (ID)

[1] Institute of Mathematics, Hamburg University of Technology, 21073 Hamburg, Germany
`jorin.dornemann@tuhh.de`

[2] Institute for Quantum Inspired and Quantum Optimization, Hamburg University of Technology, 21073 Hamburg, Germany

Abstract. This paper introduces a hybrid approach to address the Capacitated Vehicle Routing Problem with Time Windows by integrating quadratic unconstrained binary optimization (QUBO) hardware with deep learning-assisted heuristics. The proposed three-phase heuristic leverages the strengths of QUBO-solving hardware while mitigating its limitations, aiming at offering better scalability to larger problem instances. In the first phase, a deep learning-enhanced QUBO formulation is employed to partition the vertices into clusters. The second phase uses deep learning-assisted tree searches to generate candidate routes within each cluster. These candidate routes are combined in the third phase into a feasible global solution by solving a quadratic unconstrained binary set partition problem. This framework ensures compliance with capacity and time window constraints while maintaining computational efficiency. Computational results indicate that the hybrid approach is promising to potentially scale well for larger problem cases while respecting hardware limitations, offering a viable approach for leveraging quantum-inspired hardware in combination with advanced heuristics for solving complex combinatorial optimization problems.

Keywords: Vehicle Routing · Deep Learning · Quantum-inspired Computing

1 Introduction

The capacitated vehicle routing problem with time windows (CVRPTW) is an extension of the travelling salesman problem (TSP), where the goal is to find a set of routes, such that all vertices except for a depot node are visited exactly once and the capacity and time window constraints are met. The objective is to find routes to minimize the total distance traveled. First introduced by Solomon

[25], the problem has been extensively studied since and has found application in various practical scenarios. Given a fixed number of vehicles, even finding a feasible solution is NP-hard [4]. Therefore, various approaches have been proposed to address the CVRPTW over the years. Recent developments in both software and hardware have inspired novel approaches to address this problem. Advancements in machine learning have provided valuable tools for extracting essential graph properties in routing problems [13,27]. Moreover, specialized hardware is being developed to address combinatorial optimization problems, including quantum computers. Oftentimes, quantum and quantum-inspired algorithms and devices are designed to solve quadratic unconstrained binary optimization (QUBO) problems. Many combinatorial optimization problems can be formulated as QUBOs [17], hence, it is increasingly intriguing to explore ways to utilize specialized hardware for solving combinatorial optimization problems formulated as QUBOs. However, since these hardware systems can handle only a limited number of variables so far, it is essential to develop new scalable methods for integrating this hardware into a solving framework.

In this work, we propose a hybrid approach to approximately solve the CVRPTW. The proposed 3-phase heuristic combines quadratic unconstrained binary optimization hardware with a deep learning-assisted heuristic. The goal is to leverage the strengths of QUBO-solving hardware while ensuring that the hardware limits are not exceeded. For this purpose, we first use a deep learning-complemented QUBO formulation to cluster the set of vertices into subsets, generate sets of candidate routes for each cluster using deep learning-assisted tree searches that are then combined into a complete solution by a quadratic unconstrained binary set partition problem.

We solve the QUBO problems using Fujitsu's Digital Annealer (DA) [18] in its third generation (v3). The DA uses classical dedicated hardware with the purpose to heuristically solve QUBO problems as fast as practically possible. Ultimately, the goal is to use quantum computers for this task and the DA is often considered to be an intermediate quantum-inspired solution that allows to assess some of the potential of quantum computers already today. Technically, the DAv3 uses an application-specific integrated circuit (ASIC) implementing a Markov Chain Monte Carlo (MCMC) method. This method is then called by a global tabu search that can handle QUBO instances up to 100,000 bits. A major advantage of the DA is its massively parallel implementation and an innovative MCMC sampling technique. In comparison to standard simulated annealing, its update steps typically have a much smaller rejection rate speeding up the optimization, and one update step is made within one clock cycle of the ASIC. The DA also supports parallel tempering, which improves the dynamic properties of the Monte Carlo method. Moreover, it uses dynamic off-setting of the objective function to escape local minima in order to explore the optimization landscape more comprehensively. This explains why the DA can outperform other known approaches to solving QUBO problems [15].

2   Related Work

For the CVRPTW, there exist only a limited number of approaches that cluster the set of customers while taking into account the time window constraints. Most methods focus on the spatial characteristics of the customer's features. The sweep-based heuristic [9,25] clusters the customers based on their polar coordinates based on a center of gravity, which in the context of vehicle routing is the depot. An imaginary ray originating from the depot is swept counterclockwise over all vertices. The demand of each swept vertex is accumulated. Once the accumulated demands reach the capacity for one vehicle, or if including the next vertex exceeds the capacity, the current nodes are included in a cluster. Other spatial approaches include [1,5,19]. Qi et al. [22] use time geography theory to represent time and space in the same coordination system defining a spatiotemporal distance, which is used to build clusters minimizing this distance. To our knowledge, the only literature that utilizes machine learning assistance within a clustering algorithm for the CVRPTW is the one by Poullet [21]. They develop an optimal classification tree method [2] to predict the number of vehicles.

The utilization of specialized QUBO hardware for solving vehicle routing and similar problems is still in its early stages of exploration. Most of these approaches have a hybrid structure, as the physical limitations of specialized hardware require dividing the problem into smaller subproblems that can be handled. Tran et al. [26] propose a hybrid quantum-classical tree search, where a classical processor maintains a global search tree and enforces constraints on relaxed sub-problems, and a quantum annealer is applied to obtain strong candidate solutions by sampling from the configuration space of the relaxed problem. Rieffel et al. [23] investigate the effectiveness of a quantum annealer in solving small instances within families of hard operational planning problems. Feld et al. [8] follow the *Cluster First, Route Second* [16] approach to solve the capacitated vehicle routing problem (CVRP) using D-Waves quantum annealer. In their 2-Phase heuristic they divide the set of customers into clusters by formulating this as a QUBO knapsack problem with additional distance minimization. Each cluster is then mapped to a TSP QUBO formulation and subsequently solved using the quantum annealer. Irie et al. [12] develop a QUBO formulation for the CVRP that incorporates constraints with time, state and capacity and conduct experiments on D-Waves quantum annealer, but their formulation quickly exceeds the hardware limits even for smaller instances.

3   Three-Phase Heuristic

Solving the CVRPTW formulated as a QUBO directly using quantum(-inspired) computing is currently not a feasible option because the number of binary variables required to represent an instance quickly surpasses the capabilities of dedicated hardware like the DA. Therefore, we propose an extended version of the *Cluster First, Route Second* method. The three phases are:

1. **Split** the set of vertices into a number of k clusters by solving a cluster optimization problem, complemented with edge weights given by a deep neural network, formulated as a quadratic unconstrained binary optimization problem on the DA. Repeat this for different values of k to diversify the set of clusters.
2. **Generate** a set of different feasible candidate routes for the CVRPTW on each of the found clusters by applying a deep learning-assisted tree search heuristic.
3. **Solve** a set partition problem (SPP) formulated as a QUBO on the DA to compose a complete solution by choosing a subset of the set of candidate routes generated in Step 2 minimizing the total costs.

This approach offers the advantage of significantly reducing the size of the QUBO matrix solved by the DA compared to solving a CVRPTW instance directly on it. The heuristic applied in step 2 is highly flexible, as it does not attempt to solve each cluster as a single-route instance but instead finds the best set of routes within each cluster and can build different sets of routes for each cluster in parallel. This flexibility allows us to vary the number of clusters in step 1 by a parameter λ that we call *cluster range*, forcing different cluster assignments and thereby creating a more diverse set of candidate routes for inclusion in step 3. In this way, we leverage the strengths of the DA, particularly its ability to handle binary quadratic unconstrained optimization problems with a manageable number of binary variables and delegate the handling of the hard capacity and time window constraints to a heuristic for more efficient processing.

3.1 Problem Definition

A CVRPTW instance is given as a directed complete graph $G = (V \cup \{0\}, E)$ with $n+1$ nodes, where node 0 is the special depot node. Additional constraints are imposed by a demand and time window attached to each node i, given as d_i and $[a_i, b_i]$, respectively, with $[a_0, b_0]$ representing the planning horizon regarding the earliest possible departure from and latest possible return to the depot. Each vehicle has a capacity of Z, which the sum of the demands along the vehicle's route must not exceed. The provided time window $[a_i, b_i]$ at each node $i \in V$ represents the time interval in which the service at node i is allowed to start. An arrival at node i before a_i is allowed, the vehicle then has to wait until a_i to start the service. Furthermore, each node $i \in V$ requires a specific service duration h_i. We assume a homogeneous fleet of vehicles so that the capacity and travel time are equal for all vehicles. Furthermore, the edge weights c_{ij} for each edge $e = (i, j)$ represent the transit costs from node i to node j and without loss of generality include the service duration h_i of node i. The objective is to minimize the total distance traveled over all vehicles.

3.2 Graph Convolutional Network

For our approach, we use a Residual Gated Graph Convolutional Neural Network (GCN) [3,13], which is adapted for the CVRPTW in [6] by modifying the layers to account for additional constraints and briefly described in the following.

The neural network assigns a value to each edge in the graph to predict which edges are most promising to include in a solution. These associated edge values can be interpreted as probabilities. In the following, we describe each of the neural network's layers.

Input Layer. The input for the node features is five-dimensional. For node i we have the two-dimensional coordinates $x_i \in [0,1]^2$, the time window given as $[a_i, b_i]$ and the normalized demand d_i/Z, where we set $d_0 = 0$ for the depot. These features are concatenated to the five-dimensional input feature vector y_i and are then embedded to an $\frac{h}{2}$-dimensional representation, where h denotes the hidden dimension of our network that represents the number of parameters each hidden state within the network stores. The depot node gets a separate learned initial embedding parameter in order to mark the depot node as special for the network. For that, define $\hat{y}_0 \in \{0,1\}^{n+1}$ to be the unit vector with entry one at the first position and zeros otherwise. This is put together as the node input feature as follows:

$$\alpha_i = A_1 y_i \oplus A_2 \hat{y}_0, \tag{3.1}$$

where $A_1 \in \mathbb{R}^{\frac{h}{2} \times 5}$, $A_2 \in \mathbb{R}^{\frac{h}{2} \times (n+1)}$ are the weight parameters that are learned during the training procedure and $\cdot \oplus \cdot$ is the concatenation operator.

For the input edge feature, the edge values c_{ij} are embedded as an $\frac{h}{2}$-dimensional feature vector. We use an indicator function δ_{ij} of an edge which has the value one for edges connecting nodes i and j, with $i \neq j$ and i, j not the depot, and value two for edges connecting nodes with itself. To tag the depot as a special node, the indicator function δ_{ij} furthermore has a value of 3 for edges to and from the depot and a value of 4 for the depot self-loop. Together, the edge input feature is given as:

$$\beta_{ij} = A_3 c_{ij} \oplus A_4 \delta_{ij}, \tag{3.2}$$

with $A_3 \in \mathbb{R}^{\frac{h}{2} \times 1}$ and $A_4 \in \mathbb{R}^{\frac{h}{2} \times 1}$ as above are weight parameters to be learned. These $\frac{h}{2}$-dimensional edge and node feature representations are then concatenated to form the h-dimensional input for the first graph convolution layer and are subsequently passed through all graph convolution layers before producing our desired output within our classifier layers.

Graph Convolution Layer. In each of the graph convolution layers the model updates the edge and node embeddings. We leverage the design of the residual gated graph convolutional network developed in Bresson et al. [3] by adding an edge feature representation. Let ℓ be the current layer and for node i and edge

(i, j), let x_i^ℓ be the node features vector and e_{ij}^ℓ the edge features vector. We define the features for layer $\ell + 1$ in the following way:

$$x_i^{\ell+1} = x_i^\ell + \mathrm{ReLU}\left(\mathrm{BN}\left(W_1^\ell x_i^\ell + \sum_{j \in N(i)} \eta_{ij}^\ell \odot W_2^\ell x_j^\ell\right)\right), \tag{3.3}$$

$$e_{ij}^{\ell+1} = e_{ij}^\ell + \mathrm{ReLU}\left(\mathrm{BN}\left(W_3^\ell e_{ij}^\ell + W_4^\ell x_i^\ell + W_5^\ell x_j^\ell\right)\right), \tag{3.4}$$

where $W_k^\ell \in \mathbb{R}^{h \times h}$ for $k \in [5]$ again are the weight parameters to learn, with the notation $[n] = \{1, \ldots, n\}$ for any positive integer n, ReLU being the rectified linear unit, BN stands for batch normalization, $N(i)$ denotes the neighborhood of node i, moreover $\odot$ denotes the Hadamard product operator and η_{ij}^ℓ being defined as $\eta_{ij}^\ell = \frac{\sigma(e_{ij}^\ell)}{\sum_{j' \in N(i)} \sigma(e_{ij'}^\ell) + \varepsilon}$ with σ being the sigmoid function and ε a small value. For the input layer, we set $x_i^0 = \alpha_i$ and $e_{ij}^0 = \beta_{ij}$. We implement W_5^ℓ as a separate parameter to allow the model to distinguish different directions of edges since in the context of CVRPTW we have directed edges in our solutions.

MLP Classifier. A multi-layer perceptron (MLP), which is a fully connected feedforward neural network with a number ℓ_M of hidden layers, is used for generating the desired output, a finite measure that represents probabilities over the edges of our fully connected graph. For each edge embedding e_{ij}^L of the last graph convolution layer L, the MLP outputs the *probability* p_{ij} that this edge is included in the tours of the CVRPTW solution: $p_{ij} = \mathrm{MLP}(e_{ij}^L)$. The edge representations are linked to the ground-truth tour through a softmax output layer, which allows us to train the model parameters end-to-end by minimizing the cross-entropy loss via gradient descent.

3.3 Vertex Clustering

In each iteration of step 1, we split the set of vertices into a varying number of k subsets to vary the assignment of vertices to the clusters and subsequently obtain a more diverse set of candidate routes in step 2. To achieve this, we bound k from above by the minimal number of vehicles needed to solve the instance, as a number of clusters exceeding this bound could potentially exclude the overall optimal solution to this instance by forcing more routes than necessary. The bound $\bar{k}$ is given as the sum over all demands divided by the capacity of a vehicle. We define our decision variables x_{vm} to be

$$x_{vm} = \begin{cases} 1 & \text{if vertex } v \text{ is assigned to cluster } m, \\ 0 & \text{otherwise.} \end{cases} \tag{3.5}$$

With this notation, we can formulate the clustering problem as a quadratic integer program as follows:

$$\min \quad \sum_{m=1}^{k}\sum_{i=1}^{n}\sum_{j=1}^{n} c_{ij}x_{im}x_{jm} \quad \text{s.t.} \tag{3.6}$$

$$\sum_{m=1}^{k} x_{im} = 1 \quad \forall i \in V,$$

$$x_{im} \in \{0,1\} \quad \forall i \in V, m \in [k],$$

where we minimize the sum of costs between all vertex pairs in each cluster under the constraint that each vertex belongs to exactly one cluster. The equivalent QUBO formulation is given as:

$$\min \quad \sum_{m=1}^{k}\sum_{i=1}^{n}\sum_{j=1}^{n} c_{ij}x_{im}x_{jm} + P\left(\sum_{i=1}^{n}\left(\sum_{m=1}^{k} x_{im} - 1\right)^2\right), \tag{3.7}$$

where the constraint is added as a penalty term with a penalty factor $P \geq 0$. This QUBO problem is subsequently solved using the DA [18]. In contrast to other formulations, we do not include inequality constraints ensuring that the sum of the demands of each cluster does not exceed the capacity of the vehicles as well as taking into account the time windows within the clusters, as we do not solve each cluster with one route. This aligns with the strengths of the DA, as inequalities and integer variables cause difficulties.

By augmenting the objective function with additional terms, we can incorporate additional information gained by the deep neural network (cf. Section 3.2) into the cluster decision process instead of relying solely on the costs c_{ij}. For this purpose, let p_{ij} be the probability to be included in the correct solution for each edge (i,j), assigned by the neural network. Note that this associated probability value is directional, i.e., $p_{ij} \neq p_{ji}$. For the clustering procedure, however, the directions are not relevant and would even distort the result, as a large probability for one direction does not mean that the probability for the other direction is also large. We therefore take the maximum value of the two directions: $\overline{p_{ij}} = \max\{p_{ij}, p_{ji}\}$. A shortcoming of using these probabilities straightforwardly for clustering is that it may overlook important relationships between nodes. If we have edges (v_1, v_2) and (v_2, v_3) with high probabilities, but the neural network assigns a low probability to the edge (v_1, v_3) because it recognizes the benefit of visiting v_2 between v_1 and v_3, we want to incorporate the insight that v_1 and v_3 should be assigned to the same cluster into the optimization problem. We do this by replacing the probabilities p_{ij} in (3.9) by *path probabilities*, which we define as:

$$p_{ij}^{\text{path}} := \max\left\{\prod_{(v,w)\in P} p_{vw} : P \text{ is a path from } i \text{ to } j\right\}. \tag{3.8}$$

Then the clustering problem can be formulated as follows:

$$\min \quad \alpha_{\text{dist}} \sum_{m=1}^{k} \sum_{i=1}^{n} \sum_{j=1}^{n} c_{ij} x_{im} x_{jm} \tag{3.9}$$

$$+ \alpha_{\text{prob}} \sum_{m=1}^{k} \sum_{i=1}^{n} \sum_{j=i+1}^{n} (1 - \overline{p_{ij}^{\text{path}}}) x_{im} x_{jm} \quad \text{s.t.}$$

$$\sum_{m=1}^{k} x_{im} = 1 \quad \forall i \in V,$$

$$x_{im} \in \{0,1\} \quad \forall i \in V, m \in [k],$$

where α_{dist} and α_{prob} denote the tuning parameters assigned to each part of the objective function to weigh the different arguments.

Example 1. Figure 1 shows an example solution of the clustering problem (3.9) for an instance with 50 nodes. Figure 1(a) shows the path probability heatmap, where only edges with an associated value greater than 0.25 are displayed, where a darker red indicates a higher value. Figure 1(b)-(d) show the clusters built by solving (3.9), where each cluster is depicted as a different node color, for the different values of $k = 3, 4, 5$.

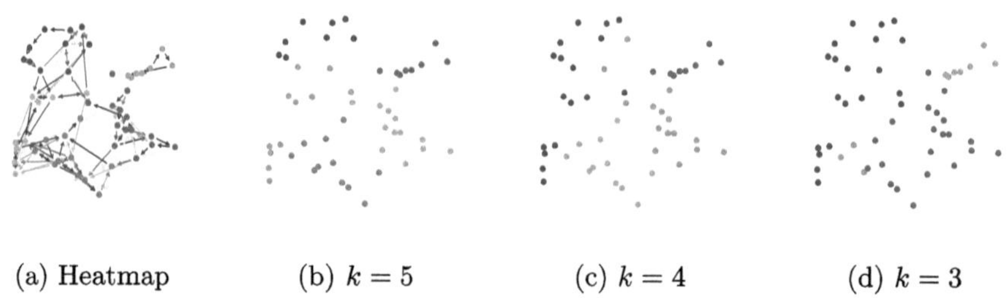

(a) Heatmap   (b) $k = 5$ (c) $k = 4$ (d) $k = 3$

Fig. 1. Path probability heatmap and solutions for clustering problem (3.7) for different numbers k of clusters.

3.4   Candidate Route Generation

Each cluster $C_i, i \in [k]$, is solved as a CVRPTW instance. Note that a CVRPTW solution can consist of several routes. Since the purpose of this step is to generate candidate routes to be included in the SPP in step 3, which then selects routes from the set of generated candidate routes in step 2 to form a feasible solution to the complete instance, it is beneficial to not only create one set of candidate routes per cluster.

For this, we adapt a deep neural network-assisted beam search [6] that utilizes the GCN presented in Sect. 3.2. The beam search is designed to build a number of solutions in parallel. In [6] it is shown that this heuristic works quite well for smaller instances. In the context of our 3-phase heuristic we use the beam search to generate a set candidate routes by considering all individual routes of all built solutions by the beam search, instead of choosing one solution built by the beam search and disregarding all others, as done in [6]. Furthermore, by applying the beam search only to the smaller clusters generated in step (1), we mitigate the inaccuracies of the beam search for larger instances, as shown in [6].

A beam search with beam width b is a limited-width breadth-first search, such that b sets of routes are built simultaneously. It builds a search tree iteratively, where each tree node represents a partially constructed solution. This heuristic uses the values p_{ij} associated with each edge $e = (i, j)$ of the graph, as described in Sect. 3.2, to build solutions that are most promising according to the values p_{ij}.

Starting from the root node that holds the partial solution $[0]$, i.e., being at the depot, in each layer of the beam search tree, only the b most promising partial solutions with respect to a scoring policy S are further explored. The descendants of each search tree node are those that add a vertex to the partial solution of its parent that is compliant with the side constraints of the CVRPTW. The scoring policy S of a (partially) built solution $g = [v_1 \ldots, v_m]$ is given as

$$S(g) = \prod_{i=1}^{m-1} p_{v_i v_{i+1}},$$

where a partial solution g can be interpreted as an array starting at the depot node 0 and then following a trail in the graph where the only node to be included more than once in g is the depot node.

Let $\tilde{n}$ be the number of vertices in one of the clusters including the depot for which we want to build a set of solutions by applying the beam search. Let $S^\ell \in \mathbb{R}^{b \times \tilde{n}}$ be the scoring matrix at the ℓ−th iteration of the beam search, which is subsequently calculated in each iteration of the beam search, and $g_{b'} = [v_1, \ldots, v_\ell]$ be one of the b partially built solutions after ℓ iterationssteps. The w−th entry in the b'−th row of the scoring matrix $S^{\ell+1}$ then reflects the score if the partial solution $g_{b'}$ is continued by adding the vertex w. That is, $S^{\ell+1}$ is given as the policy score $S(g_{b'})$, multiplied by the probability given by the neural network of getting from v_ℓ to w inwith the probability given by the neural network to transit the next step for all nodes $w \in [\tilde{n}]$:

$$S_{b'w}^{\ell+1} = S(g_{b'}) \cdot p_{v_\ell w}.$$

Then, the invalid expansions, including the already visited nodes, are excluded within the scoring matrix $S^{\ell+1}$ by masking the values that correspond to infeasible continuations of the partial solutions to be zero. From the $b \cdot \tilde{n}$ values in the masked scoring matrix the b highest scores are chosen, and the b partial solutions represented by these b search tree nodes are continued while the other are discarded. The beam search stops when b complete solutions are built.

3.5 Set Partition Problem

As described in [4], the CVRPTW can be formulated as a set partition problem (set partition problem (SPP)), which we will define in the following. The SPP is the basis of the currently best exact method for the CVRPTW, the branch-cut-and-price algorithm [4]. It uses the SPP as the master problem, as the solution of the linear relaxations of the SPP provides better lower bounds than, for example, the linear relaxation of the mixed integer linear program (MILP) formulation of the CVRPTW.

For this purpose, let R be the set of all feasible routes on subsets of vertices. For $r \in R$ denote with c_r the cost of route r, let δ_{vr} be a binary coefficient that is equal to one if and only if vertex $v \in V$ is in route r. Define the decision variable y_r to be

$$y_r = \begin{cases} 1 & \text{if route } r \text{ is used in the solution,} \\ 0 & \text{otherwise.} \end{cases} \tag{3.10}$$

Then the set partition problem can be formulated as an integer program

$$\min \ \sum_{r \in R} c_r y_r \quad \text{s.t.} \tag{3.11}$$

$$\sum_{r \in R} \delta_{vr} y_r = 1, \quad v \in V$$

$$y_r \in \{0, 1\} \quad \forall r \in R,$$

where we minimize the costs of the chosen subsets of routes under the constraint that each vertex appears in exactly one route. For a given solution y of this problem we can link together all routes indicated by y to obtain a feasible solution to the instance with respect to the side constraints such that every node is visited and every route starts and ends at the depot. This integer program translates to the following QUBO formulation:

$$\min \ \sum_{r \in R} c_r y_r + P \left(\sum_{v \in V} \left(\sum_{r \in R} \delta_{vr} y_r - 1 \right)^2 \right), \tag{3.12}$$

with P being a penalty factor. This QUBO problem is then solved by the DA. With R being the set of all feasible routes, this corresponds to the optimal solution of the CVRPTW. But in general, finding the optimal solution for this problem is only practically possible for small instances if all feasible routes are included in R, as the number of routes grows exponentially fast. For our approach, we do not choose R as the set of all possible routes but only include the subset of routes generated as described in Sect. 3.4 in the consideration, such that the number of routes included in the SPP is easily controllable via the beam width b.

4   Computational Experiments

This section evaluates the overall performance of our approach by comparing its results to those of other methods, including state-of-the-art solutions for the CVRPTW.

All models are implemented in Python 3.10 and run under Windows 10. The neural network architecture is implemented using PyTorch version 2.0.0 to use GPU computation with Cuda version 11.8. The network consists of $\ell_{\mathrm{GCN}} = 30$ hidden layers and $\ell_M = 3$ layers in the MLP. We use a hidden dimension $h = 300$ in each of the layers. For each problem size, an individual neural network is trained on 1 million randomly sampled instances based on the distribution given in the R201 instance of [25], which consists of randomly generated geographical data, a long scheduling horizon and short to medium-sized time windows allowing only a few customers per route. The instances were solved to optimality by Gurobi [10] for 20 node instances, whereas the instances with 50 and 100 customers, respectively, were solved using one run of LKH [11] and are therefore not necessarily optimal. We apply a supervised learning procedure, where, given as input a graph with the additional node features time windows and demands, the model is trained to output a probability matrix by minimizing the cross-entropy loss via gradient descent with respect to the adjacency matrix corresponding to the target solution. We utilize the Adam optimizer [14] along with a gradual decrease in the learning rate for smoother convergence, starting at a rate of 10^{-3}. We train the nets using a batch size of 24 for 1500 epochs with 500 randomly chosen batches and select the point of training with the lowest validation loss. The training procedure is executed on machines with two CPUs of type Intel Xeon E5-2680v3 @ 2,50GHz with 12 Cores and four NVidia Tesla K80 GPUs with 12GB RAM.

The path probabilities p_{ij}^{path} are calculated by applying the Floyd-Warshall algorithm to the graph (G, ω) with the weight function $\omega : E \mapsto \mathbb{R}$, $\omega((i, j)) = -\log(p_{ij})$ to find shortest paths. With the weight function ω defined like this, a shortest path in (G, ω) corresponds to maximizing (3.8).

To find the best configuration of weights α_{dist} and α_{prob} in the objective function (3.9), each combination of weights in $\{0.0, 0.1, \ldots, 0.9, 1.0\}$ summing up to one is tested for combinations of the distance c, the edge probabilities p given by the neural network and the path probabilities p^{path}. The best weighting of the objective function given by the manual experiments is given as $\alpha_{\mathrm{dist}} = 0.2, \alpha_{\mathrm{prob}} = 0.8$ using the distance and path probabilities. The accuracy of determining clusters based solely on distance differences is lower because the time window constraints in CVRPTW significantly impact route feasibility, making spatial arguments less relevant.

To set the beam width as well as the time limit for calculations using the DA, we manually conduct experiments on small sets of instances for all problem sizes to test different impacts. First, given the clusters that correspond to the optimal routes in the best solution, we test the impact of beam width and time limit for the SPP optimization problem. Second, given the clusters found by solving (3.7) with the DA and the routes built by the beam search for each cluster, we include the routes of the correct solution to find the average calculation time it takes the DA to find this solution when solving the SPP (3.12). Based on these experiments, for the beam width, we choose $b = 50, 150$ and 300 for $n = 20, 50, 100$, respectively. For $n = 20$, we choose a cluster range $\lambda = 2$, for the other problem sizes $\lambda = 3$. Even one to two seconds is enough for the DA to solve the SPP (3.12) to optimality for 20 node instances. To account for the increase of variables when solving (3.7) for multiple values of k, we choose a time limit of 5, 15 and 50 s for $n = 20, 50, 100$, respectively, for the DA to solve the SPP.

In Table 1, we compare our approach to the commercial exact MILP solver Gurobi version 10.0.0 [10] as well as to the best known exact solution method for vehicle routing problems, branch-cut-and-price (BCP) algorithm [7, 24]. We further compare results to the heuristic LKH3 [11] and Google's OR-Tools (GORT) version 9.5.2237 library [20], both of which frequently serve as heuristic baselines in related literature. The comparison is done by calculating the percentage *gap* of the cost of the found solution to the cost of the best known solution for each instance. The runtime for our approach is measured as the time for the execution of all three steps without the communication and idle time with the DA. The values are averages over 100 instances for each problem size n.

For GORT, one can choose different configurations regarding the underlying local search heuristic. We chose guided local search (GLS) to be best suited for our needs. As a time limit is needed for the GLS, we have chosen time limits of 10, 100, and 300 s for the different problem sizes. These time limits are selected to align with the runtime of our method, which allows for a fair comparison between the approaches. Gurobi is applied to the 2-index Integer Linear Program formulation of the CVRPTW and is given a time limit of 10, 100 and 300 s per instance for the different problem sizes, respectively, to allow for a comparison with similar running times. We evaluate the baseline models on machines with two Intel Xeon E5-2680v3 @ 2.50GHz CPUs with 12 cores and 128GB RAM available at the High Performance Computing Cluster of Hamburg University of Technology, while the execution of our 3-phase heuristic is done on a local machine with an Intel Core i7-1165G7 CPU @ 2.80GHz and 16GB RAM running on Windows 11. The reason for this is that communication with Fujitsu's DA is limited to selected local machines, which requires the entire heuristic, including the beam search, to be run on this machine. This in turn affects the runtime of our approach and complicates the comparison with the other approaches in terms of execution time. Therefore, we mark our time in italics in Table 1 and also report results for LKH3, GORT-GLS and BCP on our local machine used for the 3-phase heuristic to provide some more comparability regarding the computational time. The evaluation of our approach as well as the baseline models is done on datasets containing 100 randomly generated instances following the same data distribution as the training data.

Our 3-phase heuristic demonstrates optimal performance for small instance sizes. For instances with 50 nodes, the heuristic yields results comparable to GORT-GLS and LKH executed on a high-performance computer, while outper-

Table 1. Mean cost, gap and time (sec.) per instance for different problem sizes

Model	$n = 20$			$n = 50$			$n = 100$		
	cost	gap (%)	time	cost	gap (%)	time	cost	gap (%)	time
3-Phase Heuristic	5.95	0.00	*10.18*	10.22	0.30	*80.08*	15.10	1.45	*153.53*
Gurobi	5.95	0.00	0.60	10.27	1.63	51.61	17.92	18.47	300.00
BCP	5.95	0.00	1.07	10.10	0.00	10.68	14.89	0.00	119.87
GORT-GLS	5.95	0.00	10.00	10.13	0.33	100.00	15.12	1.56	300.00
LKH3	5.95	0.00	6.20	10.13	0.30	11.74	15.08	1.28	19.69
GORT-GLS local	5.95	0.00	20.00	10.16	0.58	100.00	15.68	3.99	300.00
LKH3 local	5.95	0.00	10.41	10.13	0.30	23.75	15.08	1.28	40.96
BCP local	5.95	0.00	3.31	10.10	0.00	19.30	14.89	0.00	227.17

forming GORT-GLS on the same local machine in terms of solution quality given the same computation time. For medium-sized instances, Gurobi already cannot solve most of the instances to optimality within the given time limit. The BCP method outperforms all heuristics for medium instances. However, it shows a larger relative increase in running time for large instances, whereas LKH and our heuristic exhibit much slower scaling in computation time. For $n = 100$ nodes, our heuristic surpasses GORT-GLS in both running time and quality on both local and high-performance setups, albeit producing slightly inferior results compared to the LKH heuristic. Gurobi is generally unable to solve large instances within the time limit and has a significantly worse optimality gap compared to the heuristics. While BCP continues to produce the best results, the computation time of our heuristic scales noticeably slower with respect to the increase of the input size, which shows the potential of our approach to offer good scalability for larger instances.

5   Discussion

With our proposed 3-phase heuristic, we provide a proof of concept of how to use specialized quantum-inspired computing hardware such as Fujitsu's Digital Annealer in combination with deep learning to solve combinatorial optimization problems with constraints that are difficult to solve directly with the current state of quantum(-inspired) computing hardware. Computational results show that this approach is promising, as it utilizes the strengths of quantum(-inspired) computing to potentially provide better scalability than state-of-the-art methods such as GORT and BCP while yielding near-optimal solutions for larger instances.

Future research could focus on improving our approach in various ways. First, optimizing the computation time could involve implementing the heuristic more efficiently in C++ instead of using Python. However, optimizing the calculation time on the DA itself remains challenging, as no convenient termination criterion can be set for the DA. Secondly, the structure of our approach allows each phase to be approached using different methods. For instance, since the aim of this approach was to test to what extent the DA can be involved in solving the CVRPTW, no alternatives for finding the clusters in step 1 were tested. Future research could experiment to see how other cluster algorithms perform in this contextone could compare different approaches for clustering the set of vertices in phase 1 as well as employing different hardware and software for solving QUBOs.

Our work demonstrates the potential of leveraging quantum-inspired computing in conjunction with deep learning for complex combinatorial optimization tasks. As quantum and quantum-inspired technologies continue to evolve, the integration of diverse computational techniques and hardware has the potential to provide even more efficient and versatile solutions for complex combinatorial optimization problems.

Disclosure of Interests. Shaglel and Kliesch are co-financed by the ERDF of the European Union and by Fonds of the Hamburg Ministry of Science, Research, Equalities and Districts (BWFGB). Kliesch is also funded by the Fujitsu Germany GmbH as part of the endowed professorship "Quantum Inspired and Quantum Optimization."

References

1. Bent, R., Van Hentenryck, P.: Spatial, temporal, and hybrid decompositions for large-scale vehicle routing with time windows. In: Cohen, D. (ed.) CP 2010. LNCS, vol. 6308, pp. 99–113. Springer, Heidelberg (2010). https://doi.org/10.1007/978-3-642-15396-9_11
2. Bertsimas, D., Dunn, J.: Optimal classification trees. Mach. Learn. **106**(7), 1039–1082 (2017). https://doi.org/10.1007/s10994-017-5633-9
3. Bresson, X., Laurent, T.: Residual gated graph convnets. Preprint arXiv:1711.07553 (2017). https://doi.org/10.48550/arXiv.1711.07553
4. Desaulniers, G., Madsen, O.B., Ropke, S.: The vehicle routing problem with time windows. In: Toth, P., Vigo, D. (eds.) Vehicle Routing: Problems, Methods, and Applications, vol. 2, pp. 119–159 (2014). https://doi.org/10.1137/1.9781611973594.ch5
5. Dondo, R., Cerdá, J.: A cluster-based optimization approach for the multi-depot heterogeneous fleet vehicle routing problem with time windows. Eur. J. Oper. Res. **176**(3), 1478–1507 (2007). https://doi.org/10.1016/j.ejor.2004.07.077
6. Dornemann, J.: Solving the capacitated vehicle routing problem with time windows via graph convolutional network assisted tree search and quantum-inspired computing. Front. Appl. Math. Stat. **9** (2023). https://doi.org/10.3389/fams.2023.1155356
7. Errami, N., Queiroga, E., Sadykov, R., Uchoa, E.: VrpSolverEasy: a python library for the exact solution of a rich vehicle routing problem. INFORMS J. Comput. (2023). https://doi.org/10.1287/ijoc.2023.0103
8. Feld, S., et al.: A hybrid solution method for the capacitated vehicle routing problem using a quantum Annealer. Front. ICT **6** (2019). https://doi.org/10.3389/fict.2019.00013
9. Gillett, B.E., Miller, L.R.: A heuristic algorithm for the vehicle-dispatch problem. Oper. Res. **22**(2), 340–349 (1974). https://doi.org/10.1287/opre.22.2.340

10. Gurobi Optimization, L.: Gurobi optimizer reference manual (2022). https://www.gurobi.com
11. Helsgaun, K.: An extension of the Lin-Kernighan-Helsgaun TSP solver for constrained traveling salesman and vehicle routing problems: Technical report. Roskilde Universitet (2017). http://www.akira.ruc.dk/~keld/research/LKH-3/LKH-3_REPORT.pdf
12. Irie, H., Wongpaisarnsin, G., Terabe, M., Miki, A., Taguchi, S.: Quantum annealing of vehicle routing problem with time, state and capacity. In: Feld, S., Linnhoff-Popien, C. (eds.) QTOP 2019. LNCS, vol. 11413, pp. 145–156. Springer, Cham (2019). https://doi.org/10.1007/978-3-030-14082-3_13
13. Joshi, C.K., Laurent, T., Bresson, X.: An efficient graph convolutional network technique for the travelling salesman problem. Preprint arXiv:1906.01227 (2019). https://doi.org/10.48550/arXiv.1906.01227
14. Kingma, D.P., Ba, J.: Adam: A method for stochastic optimization. Preprint arXiv:1412.6980v9 (2014). https://doi.org/10.48550/arXiv.1412.6980
15. Kowalsky, M., Albash, T., Hen, I., Lidar, D.A.: 3-regular three-XORSAT planted solutions benchmark of classical and quantum heuristic optimizers. Quantum Sci. Technol. **7**(2), 025008 (2022). https://doi.org/10.1088/2058-9565/ac4d1b
16. Laporte, G., Semet, F.: Classical heuristics for the capacitated VRP. In: Toth, P., Vigo, D. (eds.) The Vehicle Routing Problem, pp. 109–128 (2002). https://doi.org/10.1137/1.9780898718515.ch5
17. Lucas, A.: Ising formulations of many np problems. Front. Phys. **2** (2014) https://doi.org/10.3389/fphy.2014.00005
18. Matsubara, S., et al.: Digital Annealer for high-speed solving of combinatorial optimization problems and its applications. In: Yang, H., Cheng, K.T.T. (eds.) Proceedings of the 25th Asia and South Pacific Design Automation Conference. ASP-DAC 2020, pp. 667–672 (2020). https://doi.org/10.1109/ASP-DAC47756.2020.9045100
19. Ouyang, Y.: Design of vehicle routing zones for large-scale distribution systems. Transp. Res. Part B: Methodol. **41**(10), 1079–1093 (2007). https://doi.org/10.1016/j.trb.2007.04.010
20. Perron, L., Furnon, V.: Or-tools routing library (2022). https://developers.google.com/optimization/
21. Poullet, J.: Leveraging machine learning to solve the vehicle routing problem with time windows. Master Thesis, Massachusetts Institute of Technology, Sloan School of Management, Operations Research Center (2020). https://hdl.handle.net/1721.1/127285
22. Qi, M., Lin, W.H., Li, N., Miao, L.: A spatiotemporal partitioning approach for large-scale vehicle routing problems with time windows. Transp. Res. Part E: Logist. Transp. Rev. **48**(1), 248–257 (2012). https://doi.org/10.1016/j.tre.2011.07.001
23. Rieffel, E.G., Venturelli, D., O'Gorman, B., Do, M.B., Prystay, E.M., Smelyanskiy, V.N.: A case study in programming a quantum Annealer for hard operational planning problems. Quantum Inf. Process. **14**(1), 1–36 (2014). https://doi.org/10.1007/s11128-014-0892-x
24. Sadykov, R., Uchoa, E., Pessoa, A.: A bucket graph–based labeling algorithm with application to vehicle routing. Transp. Sci. **55**(1), 4–28 (2021). https://doi.org/10.1287/trsc.2020.0985
25. Solomon, M.M.: Algorithms for the vehicle routing and scheduling problems with time window constraints. Oper. Res. **35**(2), 254–265 (1987). https://doi.org/10.1287/opre.35.2.254

26. Tran, T., et al.: A hybrid quantum-classical approach to solving scheduling problems. In: Baier, J.A., Botea, A. (eds.) Proceedings of the 9th International Symposium on Combinatorial Search. vol. 7, pp. 98–106 (2021). https://doi.org/10.1609/socs.v7i1.18390
27. Vinyals, O., Fortunato, M., Jaitly, N.: Pointer networks. In: Cortes, C., Lawrence, N., Lee, D., Sugiyama, M., Garnett, R. (eds.) Advances in Neural Information Processing Systems. vol. 28, pp. 2692–2700 (2015). https://proceedings.neurips.cc/paper/2015/file/29921001f2f04bd3baee84a12e98098f-Paper.pdf

Multi-action Sampling with Deep Reinforcement Learning for Traveling Salesman Problem

Wei Liu$^{(\boxtimes)}$, Thomas Bäck, and Yingjie Fan

LIACS, Leiden University, Leiden, The Netherlands
`w.liu@liacs.leidenuniv.nl`

Abstract. The Traveling Salesman Problem (TSP) is a classic combinatorial optimization challenge with broad applications in logistics and transportation. While traditional heuristics remain dominant due to their efficiency and reliability, recent developments in Deep Reinforcement Learning (DRL) have introduced promising new directions for data-driven approaches to solving the TSP. Although DRL-based methods show promise, they often do not yet match classical heuristics in terms of computational efficiency and solution quality. One promising direction to bridging this gap is the integration of classical heuristics with learning-based methods. The Learn-to-Improve (L2I) framework follows this hybrid paradigm, combining heuristics with reinforcement learning to iteratively refine solutions. In this paper, we propose a novel multi-action sampling strategy that further enhances the L2I framework for solving TSP. The core idea is to improve solution quality by averaging rewards over multiple actions, which reduces bias and encourages more effective exploration compared to single-action strategies. During inference, multi-action sampling is applied in the later stages to explore diverse solution paths in parallel, helping to prevent premature convergence. Experimental results demonstrate that the proposed method outperforms existing L2I approaches while maintaining competitive computational efficiency.

Keywords: Vehicle Routing · Deep Reinforcement Learning · Multi-action Sampling · Neural Networks · Learning-based Optimization

1 Introduction

Vehicle Routing Problems (VRPs) are critical yet challenging tasks in transportation and logistics due to their NP-hard nature. Efficient solution methods are essential to enable scalable and effective routing systems. Among these problems, the Traveling Salesman Problem (TSP) is a foundational case, with extensive research and wide-ranging applications in logistics, transportation, and scheduling domains [1]. Efficiently solving the TSP is a practical step toward addressing more complex VRPs.

Y. Zhang et al. (Eds.): LION 2025, LNCS 15745, pp. 129–141, 2026.
https://doi.org/10.1007/978-3-032-09192-5_9

Over the past few decades, numerous algorithms have been developed to tackle the TSP, enhancing both the quality of the solution and computational efficiency. Traditional approaches, including exact algorithms and heuristics algorithms [2], have been extensively studied and applied. Exact methods, such as branch-and-bound [3], ensure optimal solutions but are often limited by scalability due to their exponential time complexity. Moreover, heuristic algorithms [4–6], while capable of producing promising solutions in significantly less time, are more practical for large-scale routing problems. However, these methods often rely on manually designed decision rules that require extensive domain knowledge, which can limit their adaptability and performance on unfamiliar or complex problem instances.

The development of machine learning, particularly deep reinforcement learning (DRL) [7], has introduced promising alternatives for solving routing problems. DRL-based methods can outperform traditional heuristics by offering faster inference and removing the need for manually crafted rules. These models are more generalizable and adaptive, as they learn decision-making policies through environment interaction. DRL for VRPs can be categorized into two main approaches: Learn-to-Construct (L2C) and Learn-to-Improve (L2I) [8]. In L2C, the trained model represents an approximate implicit mapping function between the input and the final solution. During inference, solutions are generated sequentially by predicting one decision at a time, incrementally constructing a complete solution by adding a node to a partial solution at each step. In contrast, L2I uses trained models as auxiliary components for improvement operators. For instance, they can be used to select node pairs for local search operators or to guide destroy-repair operators. Unlike L2C approaches, which construct solutions from scratch, L2I methods start with an initial complete solution and iteratively refine it to enhance the quality of solutions.

In this paper, we propose a novel training and inference strategy to enhance L2I methods for solving the TSP. Unlike existing L2I approaches that rely on fixed or heuristic-based improvement operations, limiting their adaptability. Our method introduces multi-action sampling in both training and inference phases to support broader exploration and more effective learning. During training, the multi-action sampling method facilitates more stable policy training by averaging the rewards of multiple sampled actions. In addition, we employ multi-action parallel sampling method during the later stage of inference to systematically explore the unknown region of the optimal solutions. Thus, our proposed method enables better exploration of the solution space and reduces the risk of premature convergence.

Extensive experimental evaluations demonstrate that our proposed method improves solution quality and training stability, delivering superior performance compared to existing L2I baselines while maintaining competitive computational efficiency. It is important to note that the objective of this paper is not to develop a learning-based optimization method that competes directly with exact solvers such as Concorde, or classical heuristics like evolutionary algorithms. Rather, our goal is to investigate the extent to which learning-based methods can perform in terms of computational efficiency and solution quality within their own paradigm. Accordingly, we primarily compare our approach against other learning-based methods. To provide a clearer picture of the optimality gap between learning-based and classical approaches, we also include results from representative classical heuristics in our experimental evaluation.

2   Related Work

Bello et al. [9] introduced a DRL-based Pointer Network that replaced manually designed local search heuristics and outperformed its supervised counterpart on TSP50 and TSP100. However, it was not suitable for the capacitated vehicle routing problem (CVRP). Nazari et al. [10] extended this approach by incorporating dynamic embeddings, enabling adaptation to evolving customer demands and reducing the inference time by 60% compared to OR-Tools [13].

With the development of Graph Neural Networks (GNNs) [12], researchers observed that Pointer Networks struggle to effectively capture the complex topological structures in routing problems. As a result, instead of using the sequence-to-sequence paradigm, Khalil et al. [11] tackled the TSP by integrating reinforcement learning with a GNN-based approach called Structure2Vec. Their model achieved performance comparable to the model in [9].

Inspired by the success of Transformers [14] in sequential decision problems, Deudon et al. [15] were the first to apply the transformer architecture, using multi-head attention and a feed-forward encoder-decoder with a pointing mechanism in the decoding process to solve the TSP. Joshi et al. [16] introduced a graph convolutional network (GCN) trained under a supervised learning framework for TSP, while Kool et al. [17] retained the multi-head attention-based encoder and incorporated self-attention along with a modified context node vector during decoding. Their work presented a general framework for solving VRPs and remains one of the most widely adopted end-to-end approaches.

Ma et al. [18] proposed graph pointer networks with DRL for TSP, achieving better generalization on instances up to 1,000 nodes than previous learning-based methods. Peng et al. [19] built on [17] by introducing a dynamic attention model with an adaptive encoder-decoder architecture for CVRP, effectively incorporating graph structure information throughout the optimization process.

Kwon et al. [20] addressed symmetry issues in DRL-based TSP solutions by introducing parallel rollouts using the POMO framwork. In addition, they applied data augmentation to increase training diversity and overcome the limited number of solution trajectories.

Xin et al. [21] improved the model from [17] by using multiple decoders with distinct parameters, selecting the best decoder output to determine the next node. Fu et al. [22] successfully generalized a small pre-trained model to large TSP instances. Jiang et al. [23] exploited group distributionally robust optimization and designed a convolutional embedding layer to improve the generalization of solutions cross different data distributions.

Chen and Tian [24] proposed the first DRL-based L2I model, selecting solution fragments for re-optimization through a dual-policy framework. Lu et al. [25] enhanced this by incorporating a richer set of improvement operators and a threshold-based mechanism to decide between further refinement or perturbation.

Wu et al. [27] introduced Learning Improvement Heuristic (LIH), a DRL-based model that leverages a Transformer-like architecture to select node pairs for 2-opt operations in TSP and CVRP, which Ma et al. [26] later improved by incorporating cyclic positional encoding to better capture solution symmetry.

Hottung and Tierney [28] applied DRL to Large Neighborhood Search (LNS) for CVRP, training a pointer network to guide repair operations while relying on manually designed heuristics. Gao et al. [29] addressed this limitation by developing EGATE, a graph-attention network that used edge embeddings to enhance neighborhood search efficiency.

Xin et al. [30] integrated DRL-based learning with the powerful LKH heuristic, training a sparse graph network to guide edge selection, significantly improving TSP scalability. Li et al. [32] developed a divide-and-conquer strategy for large-scale CVRP, using a Transformer model to predict sub-problem costs before solving and merging them, achieving an order of magnitude speedup over LKH3 [31] while maintaining solution quality.

L2I methods iteratively refine complete solutions, gradually improving their quality in multiple steps. Although these approaches are effective in finding high-quality solutions, they typically entail higher computational costs during both training and inference compared to L2C methods, as they require exploring a significantly larger search space.

3   Preliminaries

3.1   Traveling Salesman Problem

The goal of the Traveling Salesman Problem (TSP) is to determine the shortest possible route that allows a salesman to start and end at the same city while visit each city in a given set exactly once. Let $C = \{c_1, c_2, \ldots, c_n\}$ denote the set of cities, where n is the total number of cities. The distance between any two cities i and j is denoted by $d(i, j)$. The objective is to find a permutation $x = (x_1, x_2, \ldots, x_n)$ of the cities that minimizes the total length of the tour. This can be formulated as the following minimization problem:

$$f_{\text{TSP}}(x) = \sum_{i=1}^{n-1} d(c_{x_i}, c_{x_{i+1}}) + d(c_{x_n}, c_{x_1}). \tag{1}$$

This expression represents the total distance traveled by visiting the cities in the order defined by x, including the final return to the starting city.

3.2   Deep Reinforcement Learning

Deep Reinforcement Learning (DRL) differs from traditional Deep Learning (DL) by learning decision-making policies through interactions with dynamic environments. At each time step t, an agent observes the current state s_t of the environment and selects an action a_t. The environment then transitions to a new state s_{t+1} and provides the agent with a reward r_t as feedback. This iterative process enables the agent to learn optimal strategies without the need for labeled data, an inherent limitation in supervised learning (SL).

The ability of DRL to explore and interact with the environment grants it superior generalization capabilities compared to SL methods. Therefore, DRL has been extensively applied to optimization problems in routing and logistics.

To formalize this, we define the agent's objective as maximizing the expected return (i.e., negative tour length) under policy p_θ for a given instance S as:

$$J(\theta \mid S) = -\mathbb{E}_{\tau \sim p_\theta(\cdot \mid S)}[L(\tau \mid S)], \tag{2}$$

where S represents a VRP instance, τ is a solution trajectory, and $L(\tau \mid S)$ is the tour length (negative reward). The policy $p_\theta(\tau \mid S)$ denotes the probability distribution of selecting solution τ based on parameters θ.

To optimize θ, the REINFORCE algorithm is commonly employed. Its gradient is given by:

$$\nabla_\theta J(\theta \mid S) = \mathbb{E}_{\tau \sim p_\theta(\cdot \mid S)}\left[(L(\tau \mid S) - b(S)) \cdot \nabla_\theta \log p_\theta(\tau \mid S)\right], \tag{3}$$

where $b(S)$ is a baseline function to reduce variance and stabilize training.

Recent advancements, such as the POMO method [20], have further improved the efficiency of DRL in VRPs by incorporating shared baselines and multiple starting points, enhancing both solution quality and training stability.

4   Methodology

4.1   Multi-action Sampling in Training

Each TSP instance consists of a set of N nodes, where each node i is represented by its 2D coordinates and other problem-specific features. In conventional DRL approaches, the positional features and coordinates of the nodes are fed into a neural network as input. The policy network then generates a probability matrix, from which a single action is selected at each time step based on the calculated probabilities. The selected action is subsequently applied to modify the original solution through state transitions in the environment. The path length $f(s_{t+1})$ of the updated state s_{t+1} is used as a reward to update the network, as illustrated in the left part of Fig. 1.

In this paper, we extend the standard framework by performing multiple action selections at each time step (see MS-improve in Fig. 1), enhancing exploration and improving solution quality. The policy network follows an encoder-decoder architecture with multi-layer self-attention, aligning with the design proposed in [27]. This architecture, inspired by Transformer models [14], enables the network to capture spatial dependencies among nodes by encoding the problem instance into contextual representations and decoding actions based on learned attention weights.

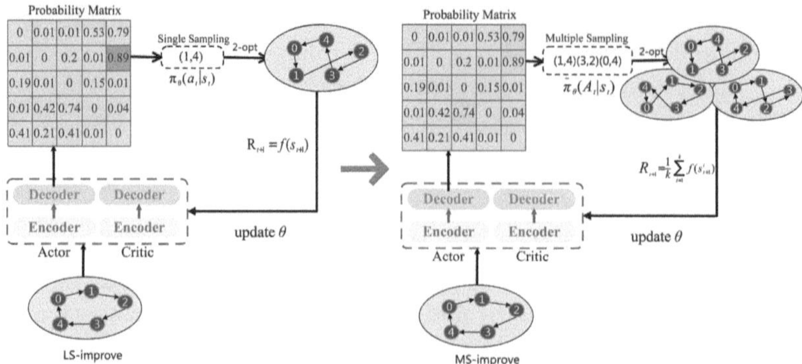

Fig. 1. Actor-Critic Learning Framework for TSP with 2-Opt Improvement Heuristic.

During the training phase, we propose a *multi-action sampling strategy* to improve the efficiency of policy learning. Instead of selecting a single action per time step from the policy network, we perform multiple samplings based on the probability distribution to obtain multiple node pairs. For each sampled action, we compute the corresponding solution length and use the average of these lengths to calculate the reward. In addition, we update the policy network using the average probability of the selected actions.

Action Set Generation: The policy network encodes the solution information using a neural network (encoder-decoder) to generate a probability matrix for node pairs. At time step t, a set of k actions $A_t = \{a_t^1, a_t^2, \dots, a_t^k\}$ is sampled from this matrix. The 2-opt heuristic is then applied to each selected node pair to modify the original path, resulting in a new state and an improved solution.

Reward Computation: For each action a_t^i in the sampled action set, the corresponding new state s_{t+1}^i is obtained by applying the 2-opt operator. The solution length $f(s_{t+1}^i)$ is then computed. The reward at step t is defined as the average tour length across the k generated solutions:

$$R_t = \frac{1}{k} \sum_{i=1}^{k} f(s_{t+1}^i). \tag{4}$$

This reward averaging reduce the variance introduced by individual actions, providing a more stable learning process and encouraging more balanced policy updates.

Policy Update: To update the policy network, we compute the average probability of the sampled actions in the action set A_t:

$$\bar{\pi}_\theta(A_t|s_t) = \frac{1}{k} \sum_{i=1}^{k} \pi_\theta(a_t^i|s_t). \tag{5}$$

The policy parameters are then updated using the REINFORCE gradient estimator:

$$\nabla_\theta J(\theta) = \mathbb{E}\left[\sum_{t=0}^{T-1} \nabla_\theta \log \bar{\pi}_\theta(A_t|s_t)G_t\right], \tag{6}$$

where G_t represents the cumulative reward from time step t onward. By using the average probability of multiple sampled actions, this update strategy ensures stable policy learning and prevents fluctuations caused by high-probability or high-reward individual actions.

The proposed multi-action sampling strategy enhances exploration by evaluating multiple actions simultaneously, thereby reducing the risk of premature convergence to suboptimal solutions. It mitigates policy learning bias by leveraging averaged rewards and log-probabilities, ensuring a more balanced update process. In addition, it improves generalization by learning from a wider variety of solution trajectories, enhancing adaptability across various problem instances. Lastly, it improves the stability of training by smoothing the gradient updates, reducing fluctuations, and facilitating a more stable learning process.

4.2   Multi-action Sampling in Inference

We also employ the multi-action sampling strategy during the inference stage to enhance solution exploration. After training, the model performs inference based on the learned policy, iteratively refining the initial path through a sequence of 2-opt operations at each time step until reaching the predefined time step threshold. Here, we introduce a multi-action sampling mechanism in the later stages of inference to expand the search space. Specifically, at each time step n, we sample k candidate actions from the probability distribution, applying k distinct 2-opt transformations to the solutions obtained at step $n - 1$. This process generates k candidate paths, from which we select the shortest as the updated solution for the next time step. These k transformations are executed in parallel in our experiment with minimal computational overhead, ensuring efficient search expansion.

5   Experiments

In this section, we evaluate the performance of the proposed multi-action sampling approach on the TSP. We generate synthetic datasets for TSP with varying numbers of nodes (20, 50, and 100). The coordinates of the nodes were sampled uniformly from the unit square [0,1]x[0,1]. All experiments were carried out on a system with an Intel i7-9700K CPU, 32 GB RAM, and an NVIDIA RTX 3090 GPU. The code was implemented in Python using Pytoch 2.3.0. The policy and value networks were trained using the Adam optimizer with a learning rate of 10^{-4}. We evaluate our method with a variety of baselines, including: 1) concorde [33], an efficient exact solver specialized for TSP; 2) LKH3 [31], a well-known heuristic solver that achieves state-of-the-art performance on various routing problems; 3) OR-Tools [13], a mature and widely used solver for routing problems based on metaheuristics; and 4) AM [17], the pioneering application of the transformer model to solving VRPs, which has yielded outstanding results. 5) GNNGLS [34], a hybrid data-driven approach based on graph neural networks and guided local search to solve the TSP. 6) LS-Improve [27], a classic learning-to-improve technique that iteratively applies 2-opt swaps to refine the initial solution, ultimately achieving an optimal route.

5.1   Results and Discussion

Table 1. Performance comparison across various methods for TSP20, TSP50, and TSP100. The table presents the objective value (Obj.), optimality gap (Gap), and computation time (Time) for each method. *MS-Improve (RO)* refers to our method with the modified reward mechanism only, while *MS-Improve (RO + MA20)* and *MS-Improve (RO + MA50)* incorporate both the modified reward mechanism and multi-action sampling with 20 and 50 actions, respectively, during inference.

Methods	TSP20			TSP50			TSP100		
	Obj.	Gap	Time	Obj.	Gap	Time	Obj.	Gap	Time
Concorde	3.83	0.00%	<1 m	5.69	0.00%	2 m	7.76	0.00%	20 m
LKH3	3.83	0.00%	<1 m	5.69	0.00%	1.5 m	7.76	0.00%	14 m
OR-Tools	3.86	0.94%	1.2 m	5.85	2.87%	5 m	8.06	3.86%	23 m
AM (N = 1,280)	3.84	0.30%	1.8 m	5.72	0.48%	7.4 m	7.95	2.45%	1.7 h
AM (N = 5,000)	3.83	0.04%	16 m	5.72	0.47%	45 m	7.93	2.18%	4.3 h
GNNGLS	3.83	0.00%	–	5.69	0.01%	–	7.81	0.70%	–
LS-Improve (T = 3,000)	3.83	0.00%	39 m	5.71	0.34%	45 m	7.91	1.85%	1.5 h
MS-Improve (RO)	3.83	0.00%	39 m	5.70	0.20%	45 m	7.85	1.10%	1.5 h
MS-Improve (RO + MA20)	3.83	0.00%	42 m	5.69	0.00%	1.1 h	7.81	0.64%	1.7 h
MS-Improve (RO + MA50)	3.83	0.00%	47 m	5.69	0.00%	1.3 h	7.78	0.24%	2.2 h

We evaluate all algorithms on 10,000 randomly generated TSP instances and report their average solution quality and runtime. For exact solvers such as Concorde and LKH3, the runtime per instance is extremely short. Since we batch 10,000 instances during evaluation, the reported total runtimes are approximate and may slightly overestimate due to I/O and orchestration overhead. The comparative performance of various methods across TSP20, TSP50, and TSP100 is summarized in Table 1.

Our proposed MS-Improve algorithm demonstrates strong performance, particularly when multi-action sampling is used during inference, achieving near-optimal solutions across all problem sizes. For TSP20, we successfully achieved the optimal objective value 3.83 even without multi-action sampling during inference. For TSP50 and TSP100, our enhanced reward mechanism yielded superior solutions compared to the LS-Improve method. Furthermore, the implementation of multiple sampling techniques during the inference phase resulted in additional performance improvements.

Notably, our approach ($RO + MA50$) achieves the best performance among learning-based methods, attaining an objective value of 7.78 with a minimal optimality gap of 0.24%. While the computation time of $RO + MA50$ (2.2 h) exceeds that of LKH3 (23 min), it remains competitive with other learning-based methods such as LS-Improve (1.5 h).

Although our algorithm increases the execution time compared to the original LS-Improve, it supports parallel execution of path exchanges. In other words, the training time required to perform 20 actions (samples) is almost equivalent to that for 50 actions (samples), excluding the time spent sequentially selecting actions.

In summary, our algorithm, MS-Improve, achieves near-optimal solutions with competitive computational time, demonstrating its effectiveness and scalability for solving TSP across varying instance sizes.

5.2 Ablation Study

To better understand the contributions of individual components in MS-Improve, we perform a series of ablation studies. Specifically, we analyze the effects of the proposed multi-action reward computation during training and the multi-action sampling strategy during inference.

Impact of Multi-action Reward Computation on Training

To examine the effect of the proposed multi-action reward computation method on the training process, we compare the training dynamics of LS-Improve with those of our method. As shown in Fig. 2, the multi-action reward computation significantly improves training stability compared to LS-Improve. Specifically, the objective value curves for our method exhibit notably smaller fluctuations, indicating a more consistent and stable convergence trajectory.

Moreover, the proposed method effectively accelerates the reduction of the objective value during early epochs, achieving competitive performance within fewer iterations. This improvement is primarily attributed to the averaging of rewards across multiple sampled actions, which reduces the variance in the policy gradient updates and improve learning efficiency.

Impact of Multi-action Sampling on Inference

We further investigate the impact of the number of sampled actions on inference performance by conducting a controlled study on multi-action sampling.

Figure 3 illustrates the performance variations across different values of k and corresponding inference time steps. Here, the time step represents the point at which multi-action sampling is activated, and k denotes the number of actions

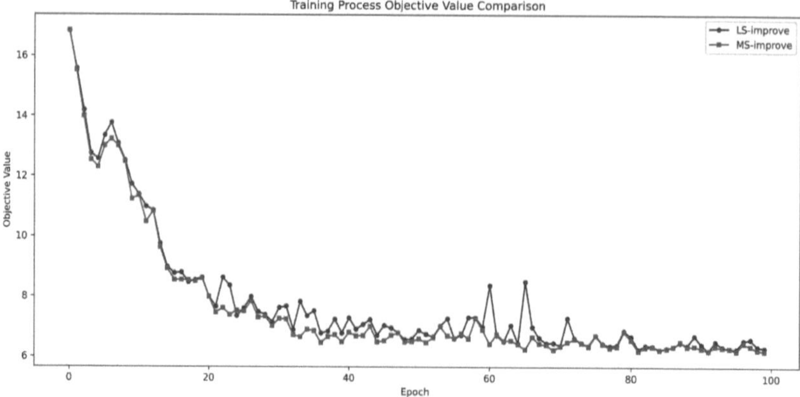

Fig. 2. Performance Variations Across Different Reward Computation Methods in the Training Phase of TSP50.

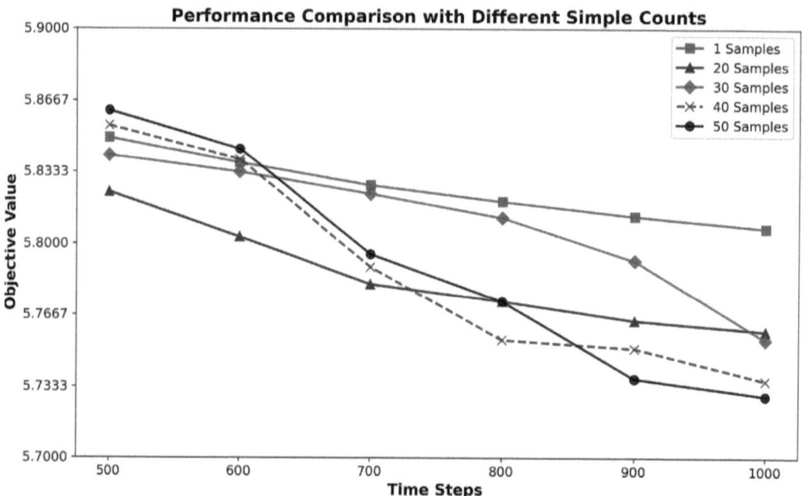

Fig. 3. Performance Variations Across Different k Values and Time Steps in the Inference Phase of TSP50.

sampled per step. Notably, $k = 1$ serves as the baseline, where only one action is selected at each time step to update the solution.

The results demonstrate that applying larger k values at earlier time steps (500–600) may cause instability in the convergence process, resulting in higher objective values. In contrast, introducing multi-action sampling at later time steps (900–1000) significantly improves solution quality. During these later stages, larger k values exhibit clear advantages by achieving lower objective values and further optimizing path quality. This suggests that while fewer actions are preferable in the early inference phase to maintain stability, leveraging a greater number of actions in the later phases expands the search space for potential optimal solutions, thereby improving the overall performance.

For instance, at the 500 time step, sampling fewer actions ($k = 20$) results in better objective values compared to higher k values, such as $k = 40$ or $k = 50$. However, as the time step progresses to 1000, the performance of larger k values overtakes that of smaller ones, indicating that multi-action sampling at later stages effectively enhances exploration and solution quality. This analysis underscores the importance of adapting the sampling strategy to different phases of inference, balancing exploration and convergence to achieve optimal results.

6   Conclusion

In this paper, we proposed a novel multi-action sampling strategy to enhance the training efficiency of Learn-to-Improve (L2I) methods for VRPs. Our approach integrates multi-action sampling in both the training and inference stages. By averaging rewards over multiple sampled actions during training, our method mitigates bias and stabilizes policy updates. In inference, we employ multi-action sampling to explore more possibilities of optimal solutions in parallel, reducing the risk of premature convergence and enhancing final solution quality.

Extensive experiments on the TSP demonstrate the effectiveness of our approach. Compared to state-of-the-art DRL baselines, our method achieves superior or comparable solution quality while maintaining competitive computational efficiency. The ablation study further validates the impact of multi-action sampling, showing that increasing the number of sampled actions improves optimization performance without significantly increasing computational overhead. From this perspective, dynamically adjusting the number of actions during the inference stage may further improve both solution quality and inference efficiency.

As a next step, we aim to further explore the potential of combining classical heuristics with learning-based methods, with the goal of developing more efficient optimization approaches for large-scale and real-world problems that are difficult to model explicitly.

References

1. Laporte, G., Asef-Vaziri, A., Sriskandarajah, C.: Some applications of the generalized travelling salesman problem. J. Oper. Res. Soc. **47**(12), 1461–1467 (1996)
2. Festa, P.: A brief introduction to exact, approximation, and heuristic algorithms for solving hard combinatorial optimization problems. In: Proceedings of the 16th International Conference on Transparent Optical Networks (2014)
3. Lawler, E.L., Wood, D.E.: Branch-and-bound methods: a survey. Oper. Res. **14**(4), 699–719 (1966)
4. Ghanem, N., et al.: Computed tomography in gastrointestinal stromal tumors. Eur. Radiol. **13**(7), 1669–1678 (2003). https://doi.org/10.1007/s00330-002-1803-6
5. Xiang, X., Tian, Y., Xiao, J., Zhang, X.: A clustering-based surrogate-assisted multiobjective evolutionary algorithm for shelter location under uncertainty of road networks. IEEE Trans. Industr. Inf. **16**(12), 7544–7555 (2020)

6. Su, Y., Guo, N., Tian, Y., Zhang, X.: A non-revisiting genetic algorithm based on a novel binary space partition tree. Inf. Sci. **512**, 661–674 (2020)
7. Silver, D., Schrittwieser, J., Simonyan, K., Antonoglou, I., Hassabis, D.: Mastering the game of Go without human knowledge. Nature **550**(7676), 354–359 (2017)
8. Shi, R., Niu, L.: A brief survey on learning based methods for vehicle routing problems (2025)
9. Bello, I., Pham, H., Le, Q.V., Norouzi, M., Bengio, S.: Neural combinatorial optimization with reinforcement learning. In: International Conference on Learning Representations (ICLR) (2017)
10. Nazari, M., Oroojlooy, A., Snyder, L., Takác, M.: Reinforcement learning for solving the vehicle routing problem. Adv. Neural Inf. Process. Syst. 31 (2018)
11. Khalil, E., Dai, H., Zhang, Y., Dilkina, B., Song, L.: Learning combinatorial optimization algorithms over graphs. Adv. Neural Inf. Process. Syst. 30 (2017)
12. Wu, Z., Pan, S., Chen, F., Long, G., Zhang, C., Yu, S.P.: A comprehensive survey on graph neural networks. IEEE Trans. Neural Netw. Learn. Syst. **32**(1), 4–24 (2020)
13. Google Inc.: OR-Tools: Google optimization tools. https://developers.google.com/optimization/
14. Vaswani, A., et al.: Attention is all you need. In: Proceedings of Neural Information Processing Systems, pp. 6000–6010 (2017)
15. Deudon, M., Cournut, P., Lacoste, A., Adulyasak, Y., Rousseau, L.-M.: Learning heuristics for the TSP by policy gradient. In: Integration of Constraint Programming, Artificial Intelligence, and Operations Research, pp. 170–181 (2018)
16. Joshi, C. K., Laurent, T., Bresson, X.: An efficient graph convolutional network technique for the Traveling Salesman Problem. arXiv preprint arXiv:1906.01227
17. Kool, W., Van Hoof, H., Welling, M.: Attention, learn to solve routing problems! In: International Conference on Learning Representations (ICLR) (2019)
18. Ma, Q., Ge, S., He, D., Thaker, D., Drori, I.: Combinatorial optimization by graph pointer networks and hierarchical reinforcement learning. In: AAAI Workshop on Deep Learning on Graphs: Methodologies and Applications (2020)
19. Peng, B., Wang, J., Zhang, Z.: A deep reinforcement learning algorithm using dynamic attention model for vehicle routing problems. In: Artificial Intelligence Algorithms and Applications, pp. 636–650 (2020)
20. Kwon, Y., Choo, J., Kim, B., Yoon, I., Gwon, Y., Min, S.: POMO: policy optimization with multiple optima for reinforcement learning. Adv. Neural. Inf. Process. Syst. **33**, 21188–21198 (2020)
21. Xin, L., Song, W., Cao, Z., Zhang, J.: Multi-decoder attention model with embedding glimpse for solving vehicle routing problems. In: Proceedings of the AAAI Conference on Artificial Intelligence, vol. 35, pp. 12042–12049 (2021)
22. Fu, Z., Qiu, K., Zha, H.: Generalize a small pre-trained model to arbitrarily large TSP instances. In: Proceedings of the AAAI Conference on Artificial Intelligence, vol. 35, pp. 7474–7482 (2021)
23. Jiang, Y., Wu, Y., Cao, Z., Zhang, J.: Learning to solve routing problems via distributionally robust optimization. In: Proceedings of the AAAI Conference on Artificial Intelligence, vol. 36, pp. 9786–9794 (2022)
24. Chen, X., Tian, Y.: Learning to perform local rewriting for combinatorial optimization. In: Proceedings of Neural Information Processing Systems, pp. 6278–6289 (2019)
25. Lu, H., Zhang, X., Yang, S.: A learning-based iterative method for solving vehicle routing problems. In: International Conference on Learning Representations (ICLR) (2020)

26. Ma, Y., et al.: Learning to iteratively solve routing problems with dual-aspect collaborative transformer. Adv. Neural. Inf. Process. Syst. **34**, 11096–11107 (2021)
27. Wu, Y., Song, W., Cao, Z., Zhang, J., Lim, A.: Learning improvement heuristics for solving routing problems. IEEE Trans. Neural Netw. Learn. Syst. 1–13 (2021)
28. Hottung, A., Tierney, K.: Neural large neighborhood search for the capacitated vehicle routing problem. In: ECAI 2020, pp. 443–450 (2020)
29. Gao, L., Chen, M., Chen, Q., Luo, G., Zhu, N., Liu, Z.: Learn to design the heuristics for vehicle routing problem. arXiv preprint arXiv:2002.08539
30. Xin, L., Song, W., Cao, Z., Zhang, J.: NeuroLKH: combining deep learning model with Lin-Kernighan-Helsgaun heuristic for solving the traveling salesman problem. Adv. Neural. Inf. Process. Syst. **34**, 7472–7483 (2021)
31. Helsgaun, K.: An effective implementation of the Lin-Kernighan traveling salesman heuristic. Eur. J. Oper. Res. **126**(1), 106–130 (2000)
32. Li, S., Yan, Z., Wu, C.: Learning to delegate for large-scale vehicle routing. Adv. Neural. Inf. Process. Syst. **34**, 26198–26211 (2021)
33. Applegate, D., Bixby, R., Chvátal, V., Cook, W.: Concorde TSP solver. http://www.math.uwaterloo.ca/tsp/concorde (2006)
34. Hudson, B., Li, Q., Malencia, M., Prorok, A.: Graph neural network guided local search for the traveling salesperson problem. In: Proceedings of the International Conference on Learning Representations (2021)

Adaptive Bias Generalized Rollout Policy Adaptation on the Flexible Job-Shop Scheduling Problem

Lotfi Kobrosly[1,2], Marc-Emmanuel Coupvent des Graviers[1], Christophe Guettier[1], and Tristan Cazenave[2(✉)]

[1] Safran Electronics and Defense, Paris, France
[2] LAMSADE, Université Paris Dauphine–PSL, Place du Maréchal de Lattre de Tassigny, Paris, France
`tristan.cazenave@dauphine.fr`

Abstract. The Flexible Job-Shop Scheduling Problem (FJSSP) is an NP-hard combinatorial optimization problem, with several application domains, especially for manufacturing purposes. The objective is to efficiently schedule multiple operations on dissimilar machines. These operations are gathered into jobs, and operations pertaining to the same job need to be scheduled sequentially. Different methods have been previously tested to solve this problem, such as Constraint Solving, Tabu Search, Genetic Algorithms, or Monte Carlo Tree Search (MCTS). We propose a novel algorithm derived from the Generalized Nested Rollout Policy Adaptation, developed to solve the FJSSP. We report encouraging experimental results, as our algorithm performs better than other MCTS-based approaches, even if makespans obtained on large instances are still far from known upper bounds.

Keywords: Monte Carlo Tree Search · Generalized Nested Rollout Policy Adaptation · Policy Gradient Descent · Markov Decision Process · Flexible Job-Shop Scheduling Problem

1 Introduction

The Flexible Job-Shop Scheduling Problem (FJSSP), as an extension of the more widely known Job-Shop Scheduling Problem (JSSP), has been studied for years in the research community. This is due to its proximity to manufacturing, aeronautics, and medicine applications [19,44], but also because of the computational challenge it poses, since it qualifies as an NP-hard problem [35,41,51,53]. The usual branch-and-bound methods that work for the JSSP do not generalize well to the FJSSP [19], suggesting the need for alternative methods to solve this problem.

Supplementary Information The online version contains supplementary material available at https://doi.org/10.1007/978-3-032-09192-5_10.

Y. Zhang et al. (Eds.): LION 2025, LNCS 15745, pp. 142–156, 2026.
https://doi.org/10.1007/978-3-032-09192-5_10

The goal of the FJSSP is to assign ordered operations of different jobs sequentially on the available machines, while minimizing a certain objective function [21]. To address the problem, Monte Carlo Tree Search (MCTS), which can be considered as a form of Reinforcement Learning [49] that quantifies the tradeoff between exploration and exploitation through probing [7], qualifies as a promising solving method. Indeed, we can consider the FJSSP as a sequential decision problem. MCTS methods proved successful in solving this type of problems, such as one-player and two-player games [10,47] or in Operational Research (OR) problems like the Vehicle Routing Problem [13].

In this paper, we present a new approach called the *Adaptive Bias Generalized Nested Rollout Policy Adaptation (ABGNRPA)* on the FJSSP and compare it to the Generalized Nested Rollout Policy Adaptation (GNRPA) from which it stems, as well as to other variations of the MCTS. Section 2 describes the FJSSP problem and the GNRPA while presenting some of the previous work on these subjects. Section 3 presents the ABGNRPA and finally Sect. 4 showcases our experimental procedure and results.

2 Background and Motivation

2.1 The Job-Shop Scheduling Problem

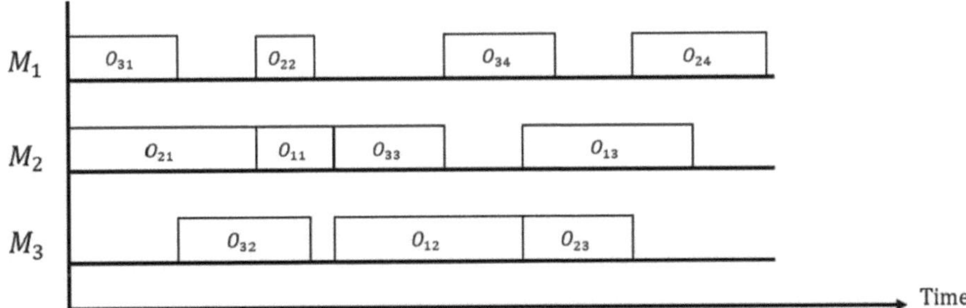

Fig. 1. Gantt's diagram of an example of a JSSP instance. The operations denoted by O_{ij} are assigned to machines M_l while respecting the constraints of order of operations in a job, of compatibility with machines and no interruption of the processing of operations. Blank spaces in a machine's line of processing represent inactivity.

The JSSP is a combinatorial optimization and scheduling problem. It is a theoretical representation of several industrial use cases where we schedule the processing of several operations in an orderly fashion to optimize a cost function. The formalized setting of the JSSP can be defined as follows, with a visual representation available in Fig. 1:

- A set of n jobs $\mathcal{J} = \{\mathcal{J}_i \mid i \in \{1..n\}\}$.
- A set of m machines, $\mathcal{M} = \{m_l, l \in \{1..m\}\}$.
- For every job $\mathcal{J}_i$, $i \in \{1..n\}$, there are n_i operations $\mathcal{J}_i = \{o_{ij}, j \in \{1..n_i\}\}$.

- The operations in each job are ordered, which means that for operations $o_{ij}, o_{ik} \in \mathcal{J}_i$, $j, k \in \{1..n_i\}$ and $j < k$, o_{ij} must be processed before o_{ik}.
- Every operation o_{ij} is defined in the problem's instance to be compatible with only one machine m_l, $l \in \{1..m\}$ with a fixed processing time d_{ij}.
- Once the processing of an operation o_{ij} starts on a machine m_l, it is done until its completion, *i.e.* it cannot be interrupted.
- The objective is to optimize a cost function.

Several approaches have been adopted to address JSSP. Some of them model it as a Mixed Integer Linear Program [37], while others considered solving it with Constraint Programming (CP) [17,57], Tabu Search [39] or other local search techniques [3]. More recently, Deep Reinforcement Learning (DRL) has been used as one of the most prominent approaches considering the JSSP as a sequential decision problem [32,58]. Graph Neural Networks have gained a good amount of consideration, whether used individually or combined with DRL [23] or a randomized ϵ-greedy that can select the second best action [1].

2.2   The Flexible Job-Shop Scheduling Problem

The FJSSP is an extension of the JSSP, where in addition to the aforementioned setting, the assignment of operations on machines becomes an extra decision. Indeed, each operation can be processed by more than one machine instead of being statically allocated on a unique one as described in Sect. 2.1. Every couple of operation and machine (o_{ij}, m_l) is associated with a processing time d_{ijl} if they are compatible. These additions enlarge the search space as the choice of the assignment of an operation on a machine changes the duration, remove homogeneity within the problem structure, for instance where symmetry breaking is harder to apply, and therefore make the problem more difficult to approach. Solving it requires assigning operations to machines, on top of ordering them and setting their start times. The objective can vary from the most studied criterion of the minimization of the makespan *i.e.*, the total execution time [19] to other ones *e.g.* the minimization of the workload at all instants t [15] or the completion times of jobs [20]. In our case, we focus on makespan minimization, which is more documented, providing us with more benchmarks to compare with our results.

The FJSSP can be represented with different approaches in addition to some of those applicable to the JSSP [19,28,38]. Several solving methods have been developed ranging from using Representation Learning via Graph Neural Networks [40,48] to Deep Reinforcement Learning [55], Memetic Algorithm [34] and Genetic Algorithm [20], considering the problem as a Markov Decision Process (MDP) [5,40]. For the Dynamic FJSSP, another variant of the FJSSP, a time-based approach is adopted as operations and jobs can be added during the processing of already existing operations [56,58].

2.3   Generalized Nested Rollout Policy Adaptation

Stochastic methods are useful for searching for good solutions, typically on an NP-hard problem such as the FJSSP. We can consider the FJSSP as a sequential decision making problem modeled by a search tree (see Sect. 4.1 for the modeling details). Thus, a MCTS approach is a good candidate, in particular, the Generalized Nested Rollout Policy Adaptation (GNRPA) [9]. In the search tree, a node represents a state of the advancement of a solution, and an arc from a parent node to a child represents an action.

In the general setting of a GNRPA, a rollout, or a playout, is one path through the search tree, from its root to a leaf. The choice of actions to generate this path is operated using a policy [31] that assigns a weight to every action independently from the state. The weight defines the probability of selecting the corresponding action from available ones in the current state, using a *softmax with temperature* function. The state-independent feature means that even if the set of possible actions differs from one state to another, an action has the same weight in all action sets in which it appears. Changes that occur to this weight are propagated to other action sets. The informal description of a *level 1* GNRPA is as follows [9]:

1. The policy is initialized with the same value for all actions.
2. A rollout is conducted through the search tree using the policy as a weighted random choice on the actions' set to select its next move. For every action, we compute a *bias* β_{action} statically via a *heuristic* [45]. For the FJSSP, we would refer to them as β_{ijl} for job i, operation j and machine l. This bias helps to initialize the probabilities of the actions.
3. At every state, we compute the probabilities of the possible actions, according to the formula in Eq. 1, and the weighted randomized choice is performed on a softmax computation with τ a *temperature*.
4. At the end of this playout, we compare to the last best sequence and take the best out of the two based on their respective scores. Through this sequence, the weights of the encountered possible actions are updated, as shown in Eq. 2, where α is a learning rate. $\delta_{aa'} = 1$ for the chosen action in the trajectory $a = a'$ (assigning operation j of job i on machine l), thus increases its weight, and $\delta_{aa'} = 0$ otherwise decreasing those weights. This changes the probability distribution exploiting the results of the playouts.
5. The new policy is then used for the following playout, as well as the biases.
6. This process is repeated for $n_{policies}$, through which the policy is updated at each iteration and used in the next one, whether a better solution is found, in which case, its actions' weights are increased, or the actions' weights of the last best solution are increased.

$$p_{a,s} = \frac{e^{\frac{w_a}{\tau}+\beta_a}}{\sum_{a'} e^{\frac{w_{a'}}{\tau}+\beta_{a'}}} \tag{1}$$

$$w_a \leftarrow w_a - \alpha \frac{p_{a,s} - \delta_{aa'}}{\tau} \tag{2}$$

If we consider a GNRPA with bias-free policy, we get a simple Nested Rollout Policy Adaptation (NRPA) algorithm [9,43] which follows the same procedure.

For a level 2 GNRPA, we consider *two levels of nesting*. At the first level, a policy is initialized with the biases. Then, at the second level, a GNRPA is conducted as described above for $n_{nested_policies}$ iterations using a copy of the initial policy. At the end of these iterations, the best sequence is used to update the weights of the first level policy, regardless of the weights of the nested policy resulting at the second level. We then use this recently updated policy of level 1 for a second run of a lower level GNRPA. From this run, we also take the best sequence and update the higher level policy, and so on. Using more levels of nesting increases the computation time exponentially, with GNRPA depending on the depth of the search tree and the number of nested policies.

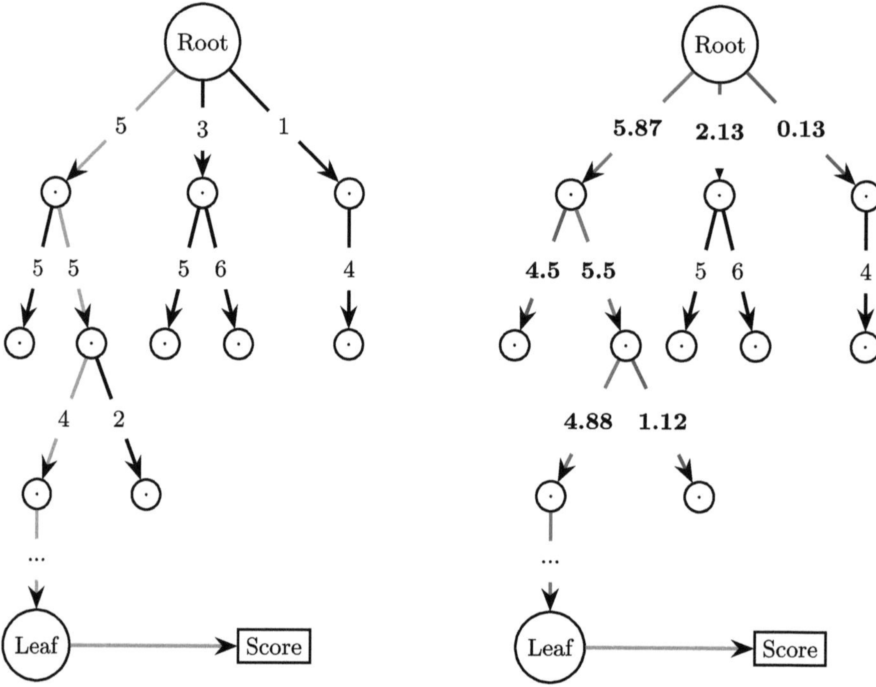

Fig. 2. Representation of a Nested Rollout Policy Adaptation, also compatible with GNRPA. The figure on the left represents a rollout, and arcs in green represent the chosen actions at each step. In this figure, the sequence obtained is the best one found so far. We use it to adapt the policy, as described in Sect. 2.3. Otherwise, we use the last best sequence for the policy update. Then we run the next rollout shown on the right. The values in bold are the ones that were updated, the incremented ones have their arcs in red, belonging to the best sequence and the decremented ones are in blue. (Color figure online)

2.4   Motivation

Improving the total makespan of an industrial assembly line has a clear impact on the productivity and cost-efficiency, which highlights the importance of FJSSP. Exact methods are effective in providing optimal or near-optimal results [17,37,57]. However, they can struggle in some extensions of the FJSSP where operations can be added dynamically [14,42], transportation times between machines are considered [27] or when uncertainties arise regarding processing times [25], machine capacity [2], machine failures or cost fluctuation [36].

MCTS-based methods have proven their efficacy in dealing with several problems and provide state-of-the-art results on games like Go [47] and Chess [16] or biology applications like inverse RNA folding [54] as well as energy systems, graph navigation, and, most importantly for us, combinatorial optimization problems [29]. These methods require little computational and memory resources, rendering them compatible with embedded applications.

To the best of our knowledge, only a few attempts use MCTS to solve the FJSSP or its variants [30,33,44,52] and they show promise. Yet, they only rely on classic MCTS and do not leverage the more effective potential of its more recent variants such as GNRPA shown in other works [8,43,45] where the information is mostly held by the choice of action independently from the state. We do highlight that this characteristic does not generalize to all problems such as navigation and some games where states contain most of the information with a limited set of possible actions in every state. In addition, previous research only focused on generated or private instances related to industrial needs or on a small number of public benchmarks. Thus, we investigate the performance of these methods in this particular setting on more public benchmark datasets.

We have initially conducted a relevance study, available in the appendix, to assess the usefulness of using such algorithms on this particular problem. Furthermore, we propose a novel variant that relies on the heuristic used to compute the biases. Our intuition is that adapting the biases throughout the playouts will provide a more adequate prior for the considered instance. Although it can still converge to a local minimum, this minimum might prove to be better than what previous variations of the MCTS can provide.

2.5   Disclaimer

Our objective is not necessarily to outperform state-of-the-art methods, given that CP-models are more effective than search-based ones here. Indeed, the results available in the literature and to which we are comparing ours are given mostly by CP-based models [19]. However, we aim to improve the performance of existing MCTS algorithms and showcase it on an academic problem with several benchmarks.

3   The Adaptive Bias Generalized Nested Rollout Policy Adaptation (ABGNRPA)

Our main contribution is an enhancement of GNRPA, called ABGNRPA. Compared to GNRPA, after initializing the biases in the same way, we **update them at each step of the playout**. This forces the bias to be dynamic and to rely on the results of the exploration for a better estimation of the potential of the trajectory. This differs from the update of the weights of the policy which happens at the end of the trajectory using a reward, *i.e.* the makespan, whereas biases are updated at every step of every playout.

Application to FJSSP. The GNRPA and ABGNRPA biases are initialized using the same heuristic called *Earliest Ending Time*. For every action, we compute the minimal starting time possible of processing the operation, given the current state, and add the duration corresponding to the couple (operation, machine). The operations finishing earlier are given a higher value as in Eq. 3. We also normalize the biases taking into account all possible actions, see Eq. 4. The choice of this heuristic is based on experimental results as it gives a better overall performance compared to other ones, as per the appendix. The superiority of this heuristic appears to come from the facts that it is optimistic and captures some information from the current state.

The update of the biases, which is specific to ABGNRPA, is based on a FJSSP makespan lower bound ($\mathcal{LB}$). It is shown in Eq. 5 where i and j refer respectively to the job and operation, l to the machine, $\mathcal{LB}_{ijl}$ is the makespan lower bound value in the current state, γ is a learning rate, and $\delta_{ijl} = 1$ if assigning operation j of job i on machine l is chosen or $\delta_{ijl} = -\frac{1}{n_{actions}}$ otherwise.

The $\mathcal{LB}$ is the maximum remaining time on all jobs $i \in \{1..n\}$, added to the current makespan at time t, $\mathcal{M}_t$. For every job, we compute the sum of the minimal durations of every operation that remains, which are represented by the set $\mathcal{U} = \{o_{i,j}, i \in \{1..n\}$ where $o_{i,j}$ has not been assigned to a machine yet$\}$ and take the maximum value obtained, as detailed in Eq. 6. These updated biases are then used for the following playouts, as shown in algorithm 3. We believe we can enlarge the application of this algorithm to other problems given a relevant lower bound function.

$$EET_{ijl} \leftarrow -(earliest_start_time(action_{ijl}, state) + d_{ijl}) \tag{3}$$

$$\beta_{ijl} \leftarrow \frac{EET_{ijl}}{\sum_{action} EET_{action}} \tag{4}$$

$$\beta_{ijl} \leftarrow \beta_{ijl} + \gamma * \delta_{ijl} * \mathcal{LB}_{ijl} \tag{5}$$

$$\mathcal{LB}_{ijl} = \mathcal{M}_t + \max_{i' \in \{1..n\}} \left[\sum_{o_{i'k} \in \mathcal{U}} \min_{l' \in \{1..m\}} (d_{i'kl'}) \right] \tag{6}$$

Algorithm 1. Biased Policy Playout for ABGNRPA

Input: state, policy, heuristic_values, γ, τ
Begin
list_of_moves $\leftarrow$ state.possible_moves() $\triangleright$ *Moves are tuples (job, operation, machine)*
while list_of_moves is not empty **do**
    for move $\in$ list_of_moves **do**
        probabilities[move] $\leftarrow e^{\frac{\alpha_{move}}{\tau} + \beta_{move}}$ $\triangleright$ *table of p_{move} values of size n_{moves}*
    end for
    next_move $\leftarrow$ random_weighted_choice(list_of_moves, probabilities)
    $\mathcal{LB} \leftarrow$ LB_function(next_move)
    state.play(next_move)
    for move $\in$ list_of_moves **do**
      if move = next_move **then**
        $\delta \leftarrow 1$
      else
        $\delta \leftarrow \frac{-1}{\text{size(list_of_moves)}}$
      end if
      $\beta_{move} \leftarrow \beta_{move} + \gamma * \delta * \mathcal{LB}$
    end for
    list_of_moves $\leftarrow$ state.possible_moves()
end while
End

We highlight that our contribution differs from [45]. Indeed, for the Capacitated Vehicle Routing Problem with Time Windows (CVRPTW), at each state they take into account the distance to the arrival point, the wait time and the arrival time, making the bias *dynamic*. In our case, we initialize the biases with the heuristic value and we update at each step them using $\mathcal{LB}$ which is computed from the current state instead of recomputing the bias. This means that the bias accumulates modifications every time the action is encountered.

4   Experiments and Results

4.1   Model Description and Implementation

We construct our model considering the problem as a sequential decision making process by ordering the different operations on the machines using the inherent partial order given by the jobs, and compute the starting time of each of these operations after the allocation. The idea is to visualize the problem as a tree search problem and implement MCTS methods to solve it as explained in the following:

1. Since operations in a job have a specific order, we will allocate them following that order and we start by the first operation. The choices for the first action are then the couples containing the first operations of all jobs, *i.e.*, the subset $\mathcal{RO} = \{o_{i,0} , i \in \{1..n\}\}$, and their compatible machines $\mathcal{A} = \{(o_{i,0}, m_l), i \in \{1..n\}, l \in \{1..m\}$, if $o_{i,0}$ and m_l are compatible$\}$.

2. An operation is chosen, using the policy, and allocated on a compatible machine m_{l_0}. The starting time is then $t_{start,i_0,l_0} = 0$ and the ending time is $t_{end,i_0,0} = d_{i_0,0,m_{l_0}}$.
3. Given that we decided to process the operations following the job order, the only operation from job i_0 that can be allocated at this point is $o_{i_0,1}$. It then replaces $o_{i_0,0}$ in $\mathcal{RO}$.
4. We repeat the previous steps, each time removing the chosen operation and replacing it by the following one in the same job until there are no operations left in $\mathcal{RO}$.

We implemented the model using `Python`, version 3.9.19. All compared algorithms have been implemented from scratch and use native `Python` functions and class types, which enables a fair comparison of all baseline algorithms described in Sect. 4.3.

4.2   Datasets

Several benchmarking instances are available in the literature [19][1]. We choose to consider a subset of these instances, with $|\mathcal{M}_{ij}|$ representing the flexibility, *i.e.* the number of compatible machines with operation o_{ij}:

- 10 instances from Brandimarte's [6]: $n = 10..20$, $m = 4..15$, $|\mathcal{J}_i| = 3..15$, $d_{ijl} = 1..20$, $|\mathcal{M}_{ij}| = 2..6$.
- 4 instances from Kacem's [26]: $n = 4..10$, $m = 5..10$, $\sum_i |\mathcal{J}_i| = 12..56$, $|\mathcal{M}_{ij}| = m$, $d_{ijl} = 1..10$ with some outliers such as 54 in *k1*
- 66×3 instances from Hurink's [24] (*edata, rdata, vdata*) derived from *sdata-instances*. With $n = 6..30$, $m = 5..15$, flexibility varies as:
  - *edata*: average is 1.15, max is 3 (few operations have flexibility)
  - *rdata*: average is 2, max is 3 (most operations have flexibility)
  - *vdata*: average is $\frac{1}{2}m$, max is $\frac{4}{5}m$ (all operations have flexibility).

All these descriptions are taken from [4]. These instances vary in size and flexibility, giving a variety of instance types to benchmark our approach. In addition, optimal values were found for all instances, except *mk10* from [6].

4.3   Baseline Algorithms

We compare our algorithm to some variations of MCTS algorithms: classic MCTS, Nested Monte Carlo Tree Search (NMCTS) [8], Nested Rollout Policy Adaptation (NRPA) [43], Generalized Nested Rollout Policy Adaptation (GNRPA) [9] as well as other devised ones:

- *Heuristic-Based NMCTS* (**HBNMCTS**): the playouts are biased instead of uniformly random.

---

[1] https://github.com/SchedulingLab/fjsp-instances.

- *Bias Initialized NRPA* (**BINRPA**): the weights of actions are initialized using the biases provided by the heuristic instead of a uniform distribution.
- *Randomized Greedy* (**RandGreedy**): no exploration is done, only a weighted random playout following the values of a heuristic.

All these algorithms are run for level 1, as explained in Sect. 2.3 for GNRPA-based approaches and as described in [8] for NMCTS-based approaches. *Randomized Greedy* is the only exception as it only plays one rollout and does not conduct an exploration in the tree. We allow a budget of 60 s for every algorithm, running on 100 independent iterations.

4.4   Experimental Results

Table 1. The table on the left showcases the number of times each algorithm obtains the best makespan compared to the others, and the number of times it finds the known Upper Bounds (UB) from literature [19,22]. The higher the value the better. The table on the right shows the average STandard Deviation (STD) of the makespan for each algorithm and the average gap on all instances between the upper bound and minimal and mean makespan values found. In this table, the lower the better.

Algorithm	Best value frequency	UB reached count
NRPA	1	1
MCTS	2	2
NMCTS	0	0
HBNMCTS	32	3
BINRPA	1	1
RandGreedy	1	1
GNRPA	72	4
ABGNRPA	**126**	**8**

Average STD	Gap to UB	
	Min makespan	Mean makespan
7.59%	54.21%	83.16%
2.88%	52.93%	66.66%
5.01%	61.00%	83.00%
3.05%	15.52%	23.91%
2.94%	52.71%	66.88%
9.61%	53.33%	110.45%
2.71%	14.27%	18.79%
1.43%	**12.08%**	**17.84%**

Table 2. Bias deviation between the start and the end of an ABGNRPA run in percentages, aggregated on instances' datasets from [6,24,26]

Instanes groups	*edata*	*rdata*	*vdata*	Kacem	Brandimarte
Max deviation	31.51%	32.14 %	34.14%	28.06%	30.07%
Min deviation	7.59%	7.20%	5.30%	13.38%	10.19%
Average deviation	15.27%	14.59%	13.28%	20.86%	17.31%

An aggregation of the results obtained are available in Table 1. Exhaustive results are shown in the appendix as per-instance results (minimal and average makespans and standard deviation) with averages ranking. We also provide further details on nesting levels' impact in the appendix.

4.5   Observations and Discussion

In Table 1, we observe that, on **126 out of 212** instances, ABGNRPA performs better than the other algorithms. It gets closest to the optimal and even reaches it on **8 instances**, and GNRPA and HBNMCTS are right behind it. This shows that the addition of biases helps to improve the initial performance of NRPA. However, GNRPA does not appear to sufficiently exploit the information that can be extracted from the problem structure. The better performance of ABGNRPA implies that the adaptive bias feature abstracts additional information helping it get a better bias estimation. This is also highlighted by its average performance and stability. We can conjecture that the biases are *evolving* to become more compatible with the problem at hand and that this evolution, which occurs during the playouts, helps capture information before its end and capitalize on it to improve the search.

We compute the bias deviation between the start and the end of a ABGNRPA run. Aggregated results are in Table 2, and per-instance results are available in the appendix. Varying between instances, this behavior can explain the better performance of ABGNRPA and indicates that the adaptation of the biases leads to a better representation of the instance's characteristics. However, contrary to other existing applications of GNRPA [12,18], increasing the nesting level does not improve performance for GNRPA variants. Our hypothesis is that these algorithms, with this specific model, stagnate around a local minimum.

The last point suggests that further training is rather useless as it would make the policy more deterministic. This hinders the possibility of further exploration beyond the visited branches. To counter this, we can consider using Limited Repetitions [11], a combination of different heuristics with a learning scheme for the weights on these heuristics from which the bias is constructed [46], or a nested bias system. Another possible improvement is the consideration of a state-action-based policy, and not only an action-based policy, which may resemble a Q-learning scheme [50]. For the $\mathcal{LB}$ function, we have tested the formulation mentioned in Eq. 6 but we can also try others that may be more expressive.

5   Conclusion

Through this work, we compared the performance of our algorithm, the ABGN-RPA, to that of other MCTS variations on the FJSSP and concluded that we have indeed improved the results. However, many parameters are to be considered, such as the heuristic function, the $\mathcal{LB}$ estimation, and the hyperparameters of the algorithm. The bias deviation, presented in Table 2, seems to characterize the GNRPA family adaptability. This suggests the need for further investigation to fully explore its potential to handle larger and different problems. Improving the scope and extent of ABGNRPA may prove useful when tackling other planning and scheduling applications. Taking into account different factors that hinder the prowess of CP-based approaches, we can foresee the relevance of our approach in bypassing these obstacles with its stochastic nature.

References

1. Abgaryan, H., Harutyunyan, A., Cazenave, T.: Randomized greedy sampling for jssp. In: Proceedings of LION 18, the 18th Learning and Intelligent Optimization Conference, pp. 9–19 (2024)
2. Amiri, F., Shirazi, B., Tajdin, A.: Multi-objective simulation optimization for uncertain resource assignment and job sequence in automated flexible job shop. Appl. Soft Comput. **75**, 190–202 (2019). https://doi.org/10.1016/j.asoc.2018.11.015
3. Beck, J.C., Feng, T., Watson, J.P.: Combining constraint programming and local search for job-shop scheduling. INFORMS J. Comput. **23**(1), 1–14 (2011)
4. Behnke, D., Geiger, M.J.: Test instances for the flexible job shop scheduling problem with work centers. Arbeitspapier/Research Paper/Helmut-Schmidt-Universität, Lehrstuhl für Betriebswirtschaftslehre, insbes. Logistik-Management (2012)
5. Bellman, R.: A markovian decision process. J. Math. Mech. pp. 679–684 (1957)
6. Brandimarte, P.: Routing and scheduling in a flexible job shop by tabu search. Ann. Oper. Res. **41**(3), 157–183 (1993)
7. Browne, C.B., et al.: A survey of monte carlo tree search methods. IEEE Trans. Comput. Intell. AI in games **4**(1), 1–43 (2012)
8. Cazenave, T.: Nested monte-carlo search. In: Twenty-First International Joint Conference on Artificial Intelligence (2009)
9. Cazenave, T.: Generalized nested rollout policy adaptation. In: Monte Carlo Search: First Workshop, MCS 2020, Held in Conjunction with IJCAI 2020, Virtual Event, January 7, 2021, Proceedings 1, pp. 71–83. Springer (2021)
10. Cazenave, T.: Monte carlo game solver. In: Monte Carlo Search: First Workshop, MCS 2020, Held in Conjunction with IJCAI 2020, Virtual Event, January 7, 2021, Proceedings 1, pp. 56–70. Springer (2021)
11. Cazenave, T.: Generalized nested rollout policy adaptation with limited repetitions. arXiv preprint arXiv:2401.10420 (2024). https://doi.org/10.48550/arXiv.2401.10420
12. Cazenave, T.: Learning a prior for monte CARLO search by replaying solutions to combinatorial problems. In: Parallel Problem Solving from Nature, PPSN 2024. Lecture Notes in Computer Science, vol. 15148, pp. 85–99. Springer (2024), https://doi.org/10.1007/978-3-031-70055-2_6
13. Cazenave, T., Lucas, J.Y., Kim, H., Triboulet, T.: Monte CARLO vehicle routing. In: ATT at ECAI 2020 (2020)
14. Chang, J., Yu, D., Hu, Y., He, W., Yu, H.: Deep reinforcement learning for dynamic flexible job shop scheduling with random job arrival. Processes **10**(4), 760 (2022). https://doi.org/10.3390/pr10040760
15. Chaudhry, I.A., Khan, A.A.: A research survey: review of flexible job shop scheduling techniques. Int. Trans. Oper. Res. **23**(3), 551–591 (2016)

16. Clark, G.: Deep synoptic monte-carlo planning in reconnaissance blind chess. In: Ranzato, M., Beygelzimer, A., Dauphin, Y., Liang, P., Vaughan, J.W. (eds.) Advances in Neural Information Processing Systems. vol. 34, pp. 4106–4119. Curran Associates, Inc. (2021). https://proceedings.neurips.cc/paper_files/paper/2021/file/215a71a12769b056c3c32e7299f1c5ed-Paper.pdf
17. Da Col, G., Teppan, E.C.: Industrial-size job shop scheduling with constraint programming. Oper. Res. Perspect. **9**, 100249 (2022)
18. Dang, C., Bazgan, C., Cazenave, T., Chopin, M., Wuillemin, P.: Warm-starting nested rollout policy adaptation with optimal stopping. In: Thirty-Seventh AAAI Conference on Artificial Intelligence, AAAI 2023, pp. 12381–12389. AAAI Press (2023). https://doi.org/10.1609/aaai.v37i10.26459
19. Dauzère-Pérès, S., Ding, J., Shen, L., Tamssaouet, K.: The flexible job shop scheduling problem: a review. Eur. J. Oper. Res. **314**(2), 409–432 (2024)
20. De Giovanni, L., Pezzella, F.: An improved genetic algorithm for the distributed and flexible job-shop scheduling problem. Eur. J. Oper. Res. **200**(2), 395–408 (2010)
21. Fera, M., et al.: Production scheduling approaches for operations management. InTech (2013)
22. Graviers, M.E.C.d., Kobrosly, L., Guettier, C., Cazenave, T.: Updating lower and upper bounds for the job-shop scheduling problem test instances. arXiv preprint arXiv:2504.16106 (2025). https://doi.org/10.48550/arXiv.2504.16106
23. Hameed, M.S.A., Schwung, A.: Reinforcement learning on job shop scheduling problems using graph networks. arXiv preprint arXiv:2009.03836 (2020)
24. Hurink, J., Jurisch, B., Thole, M.: Tabu search for the job-shop scheduling problem with multi-purpose machines. Oper. Res. Spektrum **15**, 205–215 (1994)
25. Jamrus, T., Chien, C.F., Gen, M., Sethanan, K.: Hybrid particle swarm optimization combined with genetic operators for flexible job-shop scheduling under uncertain processing time for semiconductor manufacturing. IEEE Trans. Semicond. Manuf. **31**(1), 32–41 (2017). https://doi.org/10.1109/TSM.2017.2758380
26. Kacem, I., Hammadi, S., Borne, P.: Pareto-optimality approach for flexible job-shop scheduling problems: hybridization of evolutionary algorithms and fuzzy logic. Math. Comput. Simul. **60**(3–5), 245–276 (2002)
27. Karimi, S., Ardalan, Z., Naderi, B., Mohammadi, M.: Scheduling flexible job-shops with transportation times: mathematical models and a hybrid imperialist competitive algorithm. Appl. math. modellíng **41**, 667–682 (2017). https://doi.org/10.1016/j.apm.2016.09.022
28. Kasapidis, G.A., Paraskevopoulos, D.C., Repoussis, P.P., Tarantilis, C.D.: Flexible job shop scheduling problems with arbitrary precedence graphs. Prod. Oper. Manag. **30**(11), 4044–4068 (2021)
29. Kemmerling, M., Lütticke, D., Schmitt, R.H.: Beyond games: a systematic review of neural monte CARLO tree search applications. Appl. Intell. **54**(1), 1020–1046 (2024). https://doi.org/10.1007/s10489-023-05240-w
30. Li, K., Deng, Q., Zhang, L., Fan, Q., Gong, G., Ding, S.: An effective MCTS-based algorithm for minimizing makespan in dynamic flexible job shop scheduling problem. Comput. Ind. Eng. **155**, 107211 (2021)
31. Li, Y.: Deep reinforcement learning: an overview. arXiv preprint arXiv:1701.07274 (2017)
32. Liu, C.L., Chang, C.C., Tseng, C.J.: Actor-critic deep reinforcement learning for solving job shop scheduling problems. IEEE Access **8**, 71752–71762 (2020)

33. Lu, C.L., Chiu, S.Y., Wu, J., Chao, L.P.: Dynamic monte-CARLO tree search algorithm for multi-objective flexible job-shop scheduling problem. Appl. Math. **10**(4), 1531–1539 (2016)
34. Luo, Q., Deng, Q., Gong, G., Zhang, L., Han, W., Li, K.: An efficient memetic algorithm for distributed flexible job shop scheduling problem with transfers. Expert Syst. Appl. **160**, 113721 (2020)
35. Mastrolilli, M., Svensson, O.: Hardness of approximating flow and job shop scheduling problems. J. ACM **58**(5) (2011). https://doi.org/10.1145/2027216.2027218
36. Meilanitasari, P., Shin, S.J.: A review of prediction and optimization for sequence-driven scheduling in job shop flexible manufacturing systems. Processes **9**(8), 1391 (2021). https://doi.org/10.3390/pr9081391
37. Meng, L., Zhang, C., Ren, Y., Zhang, B., Lv, C.: Mixed-integer linear programming and constraint programming formulations for solving distributed flexible job shop scheduling problem. Comput. Ind. Eng. **142**, 106347 (2020)
38. Naderi, B., Roshanaei, V.: Critical-path-search logic-based benders decomposition approaches for flexible job shop scheduling. INFORMS J. Optim. **4**(1), 1–28 (2022)
39. Nowicki, E., Smutnicki, C.: An advanced tabu search algorithm for the job shop problem. J. Sched. **8**, 145–159 (2005)
40. Park, J., Chun, J., Kim, S.H., Kim, Y., Park, J.: Learning to schedule job-shop problems: representation and policy learning using graph neural network and reinforcement learning. Int. J. Prod. Res. **59**(11), 3360–3377 (2021)
41. Pezzella, F., Morganti, G., Ciaschetti, G.: A genetic algorithm for the flexible job-shop scheduling problem. Comput. Oper. Res. **35**(10), 3202–3212 (2008)
42. Rajabinasab, A., Mansour, S.: Dynamic flexible job shop scheduling with alternative process plans: an agent-based approach. Inter. J. Adv. Manufact. Technol. **54**, 1091–1107 (2011). https://doi.org/10.1007/s00170-010-2986-7
43. Rosin, C.D.: Nested rollout policy adaptation for monte CARLO tree search. In: Ijcai. vol. 2011, pp. 649–654 (2011)
44. Saqlain, M., Ali, S., Lee, J.: A monte-CARLO tree search algorithm for the flexible job-shop scheduling in manufacturing systems. Flex. Serv. Manuf. J. **35**(2), 548–571 (2023)
45. Sentuc, J., Cazenave, T., Lucas, J.Y.: Generalized nested rollout policy adaptation with dynamic bias for vehicle routing. arXiv preprint arXiv:2111.06928 (2021)
46. Sentuc, J., Ellouze, F., Lucas, J.Y., Cazenave, T.: Learning the bias weights for generalized nested rollout policy adaptation. In: International Conference on Learning and Intelligent Optimization, pp. 194–207. Springer (2023). https://doi.org/10.1007/978-3-031-44505-7_14
47. Silver, D., et al.: Mastering the game of go with deep neural networks and tree search. nature **529**(7587), 484–489 (2016)
48. Song, W., Chen, X., Li, Q., Cao, Z.: Flexible job-shop scheduling via graph neural network and deep reinforcement learning. IEEE Trans. Industr. Inf. **19**(2), 1600–1610 (2022)
49. Vodopivec, T., Samothrakis, S., Ster, B.: On monte CARLO tree search and reinforcement learning. J. Artif. Intell. Res. **60**, 881–936 (2017)
50. Watkins, C.J., Dayan, P.: Q-learning. Mach. Learn. **8**, 279–292 (1992). https://doi.org/10.1007/BF00992698
51. Watson, J.P., Beck, J., Howe, A.E., Whitley, L.: Problem difficulty for tabu search in job-shop scheduling. Artif. Intell. **143**(2), 189–217 (2003). https://doi.org/10.1016/S0004-3702(02)00363-6

52. Wu, T.Y., Wu, I.C., Liang, C.C.: Multi-objective flexible job shop scheduling problem based on monte-CARLO tree search. In: 2013 Conference on Technologies and Applications of Artificial Intelligence, pp. 73–78. IEEE (2013)
53. Xie, J., Gao, L., Peng, K., Li, X., Li, H.: Review on flexible job shop scheduling. IET collaborative Intell. Manufact. **1**(3), 67–77 (2019)
54. Yang, X., Yoshizoe, K., Taneda, A., Tsuda, K.: RNA inverse folding using monte CARLO tree search. BMC Bioinformatics **18**, 1–12 (2017). https://doi.org/10.1186/s12859-017-1882-7
55. Zhang, C., Song, W., Cao, Z., Zhang, J., Tan, P.S., Chi, X.: Learning to dispatch for job shop scheduling via deep reinforcement learning. Adv. Neural. Inf. Process. Syst. **33**, 1621–1632 (2020)
56. Zhang, F., Mei, Y., Nguyen, S., Zhang, M.: Evolving scheduling heuristics via genetic programming with feature selection in dynamic flexible job-shop scheduling. IEEE Trans. Cybern. **51**(4), 1797–1811 (2020)
57. Zhou, J.: A constraint program for solving the job-shop problem. In: Freuder, E.C. (ed.) CP 1996. LNCS, vol. 1118, pp. 510–524. Springer, Heidelberg (1996). https://doi.org/10.1007/3-540-61551-2_97
58. Zhou, L., Zhang, L., Horn, B.K.: Deep reinforcement learning-based dynamic scheduling in smart manufacturing. Procedia Cirp **93**, 383–388 (2020)

Codetector: A Framework for Zero-Shot Detection of AI-Generated Code

Nadim Adham[1(✉)] [iD], Henning Duwe[1] [iD], and Holger H. Hoos[1,2] [iD]

[1] AI Methodology, RWTH Aachen University, Aachen, Germany
nadim.adham@rwth-aachen.de, {duwe,hh}@aim.rwth-aachen.de
[2] LIACS, Leiden University, Leiden, The Netherlands

Abstract. Generative artificial intelligence (AI) models are advancing rapidly, and their ability to generate code has significant implications for software development. Their use in code generation raises concerns about plagiarism, malware generation and dataset contamination. Previous work proposed methods to address these issues, using developments from the field of natural language detection. However, due to the infancy of the field, there is a deficit in established benchmarks and datasets, making a comprehensive comparison between detection methods difficult.

In this work, we investigated the efficacy of five existing zero-shot detection methods on AI-generated code. To do so, we used 17 large language models to generate and analyse 113 776 code samples from six common datasets and sources. By collecting new, previously unseen data with their respective time-stamps, we address the issue of data leakage, i.e., the exposure of LLMs to testing data during pre-training. Our proposed framework enables easy integration of new LLMs and datasets, facilitating the generation, detection and analysis of AI-generated code. Following this, we examined the impact of different hyperparameters used during code generation (such as the temperature) on detection performance. Our code is available under https://github.com/ Progyo/Codetector.

Keywords: Zero-shot Detection · LLMs · Code Generation · Framework

1 Introduction

Development in AI has made significant strides within the last ten years, and AI techniques are now being integrated into many systems we interact with, ranging from search engines to weather pattern prediction and more. Additionally, in recent years, an increasing number of AI systems have been developed to not only process data but to also generate content. Large language models (LLMs), in

Supplementary Information The online version contains supplementary material available at https://doi.org/10.1007/978-3-032-09192-5_11.

Y. Zhang et al. (Eds.): LION 2025, LNCS 15745, pp. 157–173, 2026.
https://doi.org/10.1007/978-3-032-09192-5_11

particular, have shown a rapid rate of improvement in a multitude of tasks including several related to coding. Despite their popularity and success, LLMs come with risks: From malicious actors creating backdoors through dataset contamination [18], to careless users generating plagiarised or dangerous code [9,12,17]. Moreover, if left unchecked, such AI-generated content (AIGC) will bleed into the training sets of future models, possibly leading to phenomena such as model collapse, where models degrade due to training on their own synthetic outputs, causing them to converge into narrower modes and decrease in diversity [1]. These challenges underscore the urgent need for AI detection systems. However, AIGC detection, particularly for code, remains relatively immature, with the earliest work we found dating to 2023, allowing little time for the development of benchmarks or datasets [23,24].

Frequently, code detection performance is benchmarked using output from a single LLM [9,17,23], which can be problematic, especially for zero-shot methods that perform best when detecting text generated by the same base model [14]. This is also a common pitfall for fine-tuned detection methods that tend to specialise in the corpus or domain they were trained on, leading to weakened cross-domain performance [24]. Thus, focusing on the detection performance of code generated by a single model cannot accurately represent the performance of a given detection method. In addition to the models used to generate AIGC, the underlying data sources may also impact detection performance. All in all, we aimed to answer the following research questions in this work:

1. RQ1: Can zero-shot detection methods from other domains be effectively applied to code detection?
2. RQ2: What is the impact of data leakage on detection performance?
3. RQ3: How do generation parameters of LLMs, such as temperature, affect detection performance?

To answer these questions, we evaluated five existing zero-shot detection methods on AI-generated code from 17 LLMs. This code was generated based on samples scraped from different sources, such as LeetCode[1] and Stack Overflow.[2] After running the generated samples through our detection pipeline, individual detection values were assigned based on specific method-model combinations and stored as "detection samples". We also analysed the impact of these different data sources on detection performance and propose an end-to-end framework for generating and detecting AIGC. Overall, our contributions can be summarised as follows:

- We developed a framework called "Codetector" that enables the modular integration of new data sources and LLMs for generating and detecting AIGC.
- We evaluated the effectiveness of five existing zero-shot methods for detecting AI-generated code on 113 776 code samples, generated by 17 LLMs, and analysed 12 294 388 detection samples.

---

[1] https://leetcode.com/.
[2] https://stackoverflow.com/.

- We empirically demonstrate that the performance of existing zero-shot detection methods can be improved by using new, unseen samples for generation and detection.
- On our processed dataset, newer detection methods achieved AUROC scores of up to 97% with an average of around 83%, showing that some detectors can be directly applied to code detection.

The remainder of this paper is structured as follows: In Sect. 2, we introduce key concepts and existing work in AI-generated code detection. Following this, in Sect. 3, we briefly present the framework that we developed. In Sect. 4, we present the generated and detection datasets and the necessary steps we took to produce them. We then analyse the results in Sect. 5 and determine possible impact factors on detection scores. Finally, we summarise our findings in Sect. 6.

2   Background and Related Work

Detecting AIGC can be defined as a binary classification task where AIGC is typically assigned as the positive class, while human-generated content is assigned the negative class [24]. Therefore, the false positive rate (FPR) is the rate of incorrectly labelled human-generated samples; conversely, the true positive rate (TPR) is the rate of correctly labelled AIGC samples. Plotting the relation between the TPR and FPR for varying detection thresholds gives the receiver operating characteristic (ROC) curve of a given detector. The area under the ROC curve (AUROC) is a value ranging between 0 and 1 that indicates how well the detector can separate the two classes [17].

Contemporary work focuses on the task of detecting AI-generated natural language texts, leaving out a large pool of questions specific to the detectability of AI-generated code. A few studies have begun investigating this area, with Wang *et al.* [23] being the first to compare commercial and open-source detection methods for AI-generated code. They showed poor performance across the board, which raised the question of whether machine-generated code is more difficult to detect than machine-generated text in natural language, and entirely different techniques are needed. Several subsequent studies [9,20,25] have provided evidence that AIGC detection, in the context of code, may be more feasible than initially thought.

Much of the current research for code detection occurs in the context of education, usually with a focus on detecting submissions in beginner-level programming courses. Beginners often write inefficient code, a trait some detection methods inadvertently exploit [9]. It has been observed that ChatGPT [15,16] writes code utilising practices akin to those of a professional programmer, which stands out from the submissions of novices [9]. A lot of research focuses on zero-shot detection methods, because they do not require any fine-tuning, making them cheaper to iterate and develop. Ideally, a robust zero-shot method would also perform well across different model outputs, a challenge that current methods still face [14,24].

Some research investigated the performance of DetectGPT [14] in detecting modified Python code submissions with accuracies between 40–54%, reflecting the poor performance in previous literature [17,23]. DetectGPT4Code [25] investigates the detection capabilities of code generated by GPT-3.5 and GPT-4, reaching an AUROC of up to 86% when detecting Python samples generated by GPT-4. Meanwhile, CodeDetectGPT [20] replaces the costly LLM perturbation model of DetectGPT with a randomised algorithm that perturbs the sample by inserting spaces and newlines. This method achieved AUROC scores up to 99% in white-box settings at a temperature of 0.2, though performance dropped by 27% at 1.0 [20].[3] This raises an important question: Does detection performance consistently decrease with higher temperatures across all zero-shot detection methods? Overall, the current literature deviates from previous expectations, suggesting the plausibility of reliably detecting machine-generated code.

A common shared trait of the aforementioned studies is the near-exclusive focus on ChatGPT-generated code.[4] It is important to note that the tested versions of ChatGPT have been shown to have strong memorisation tendencies when completing code-oriented tasks [22]. This may have had an impact on the resulting AUROC scores reported in several studies by affecting their TPR and highlights the importance of choosing suitable datasets to generate and select code samples for detection from. While ChatGPT is among the most popular services used for tasks such as code completion, it is of utmost importance to evaluate the detection performance across a larger set of LLMs, to better gauge the state of the art in AIGC detection. In particular, as more services begin to build and adopt proprietary models, the need for strong cross-model detection methods will only increase. This fact, alongside the seemingly haphazard selection of code samples, is the motivation behind the extensive analysis and scrutiny of dataset sources and LLMs discussed in our work described in the following.

3   Framework

To generate the datasets used for code detection, we developed a framework called "Codetector". Our framework consists of two pipelines: A dataset generation pipeline and a detection pipeline. The dataset generation pipeline allows for the modular and easy integration of new data sources (Github, Stack Overflow, etc.) and LLMs used for code generation. The detection pipeline allows for the modular implementation of different detection methods and the LLMs they use. This design stands in contrast to many monolithic implementations of detection methods, where functionality is hard-coded and activated via command line arguments. The individual use cases and repositories utilised by our pipelines are unit-tested to ensure proper functionality. Another goal in designing both

---

[3] In our dataset, we utilise the temperatures 0.2 and 0.97. The latter was intended to be 1.0 to match the literature. This small difference was noticed too late in the generation process but arguably does not distort our findings.

[4] Or, more precisely, on code produced by models such as GPT-3.5 and GPT-4.

pipelines was for the implementation to be agnostic to the underlying frameworks used, such as PyTorch and TensorFlow, as well as higher-level frameworks, such as Huggingface Transformers.

While this model framework was initially developed for code generation and detection, it supports natural language samples as well. The impact of the underlying datasets, as well as other factors, are discussed in Sect. 5.

3.1 Dataset Generation Pipeline

The efficacy of various detection methods is strongly dependent on the quality and diversity of the samples contained in the datasets. Using our framework, new datasets and sources can easily be implemented by either extending existing dataset class formats like XML and Apache Parquet[5] or by implementing the required methods specified by the dataset abstract class. Additionally, datasets can be converted between different file formats out of the box, allowing data to be stored in both compressed and human-readable formats. Various filters can also be applied to datasets to prevent unwanted samples from being used for generation. Moreover, LLMs that inherit from the BaseModel class can be marked as generators by implementing the abstract GeneratorMixin. For further implementation details or examples regarding datasets and data sources, we refer to the IPython notebooks and the supplied source code.

3.2 Detection Pipeline

Once the dataset generation pipeline has produced a large corpus of human and machine-generated samples, these samples can be used for detection. Our framework currently supports single and two-model zero-shot detectors, allowing for flexibility in detection approaches. New detection methods can be implemented by extending either of the abstract classes, depending on the use case. To ensure compatibility within the detection pipeline, all LLMs used for detection must implement the DetectorMixin. LLMs that are used for both generating and detecting samples simply need to implement both mixins, reducing code redundancy while maintaining versatility.

4 Dataset Collection and Generation

A major contribution of our work is the collection and generation of a large corpus of human- and AI-generated code samples, using 17 models capable of code generation, of which the majority are open-source. It is important to note that we refrained from using more closed-source models for two reasons. Firstly, without access to the model weights, only a black-box analysis would have been possible, a known weakness of current zero-shot detection methods. Secondly, by using open-source models, we were able to reduce the research costs and

---

[5] https://parquet.apache.org/.

direct them to other purposes, such as generating output from a more expensive reasoning model (o1-mini). These models were chosen based on size, training date, coding capabilities and prior use in research.

Several factors were considered during the collection of this large corpus. Many earlier studies neglect the possible impact of data leakage in their detection datasets [20,23]. Evaluation benchmarks are often purged before pre-training, yet copies sometimes slip into the training data, affecting code quality scores [22]; thus, the impact of data leakage on detection performance must be considered as well. Evaluation benchmarks, such as APPS and CodeSearchNet, are often used as a source for generating AIGC [13,20,23,25]. When using samples seen in training, the chances of regurgitating fragments or entire snippets of code verbatim during generation increases [3,10]. This increase in false negatives would, in turn, reduce the TPR. It therefore makes sense to keep track of the inception dates of code samples contained in a dataset as well as the training cut-off date of the LLMs used to generate samples.

The length of the code samples also has a significant impact on the maximum achievable AUROC score [4]. Thus, the samples across different code detection datasets should also be similarly distributed in length for a fair comparison between them. How this was achieved in our work is explained in more detail in Subsect. 4.2. In addition to this, some related work sets a maximum output length for generated code samples [20]. This induces an upper bound for the AUROC score of the best detector and a pessimistic precedent for code detection capabilities [4,19].

4.1 Data Sources

To generate a diverse set of data that mimics real-life use cases, various data sources were compiled. These sources include common datasets used by several earlier studies on detection, such as APPS and CodeSearchNet, as well as code samples that were scraped from the web to further increase diversity. To investigate the impact of possible data leakage, two disjunct sets of data were generated with code samples from two different periods: One prior to the release of Chat-GPT in November 2022 (the "Pre" datasets) and the other from August 2023 to April 2024 (the "Post" datasets). August 2023 was chosen as the lower bound for the "Post" datasets by determining the most recent training date of the 11 LLMs that were initially chosen for analysis. As a result, the samples in the "Post" datasets should be novel to the LLMs, ensuring that they have not been seen during training. The generated samples of the six remaining LLMs, including o1 mini, were added later to the final dataset but excluded from the dataset analysis. The goal was to inspect whether older code samples have a noticeable impact on the AUROC scores of detection methods. Another factor to consider here is the prevalence of machine-generated code in some of these sources, which may be marked as human-generated, increasing the number of false positives. The chance of this occurring rises substantially with samples collected after the release of ChatGPT.

Stack Overflow. Stack Overflow is one of the largest repositories of publicly available code in the world. Its strength lies in the large number of code snippets where functionality is often explained in natural language in the accompanying post, providing a great starting point for generating prompts. Despite this advantage, Stack Overflow is filled with non-functional code, hence why the code is in a post on the website. It is therefore reasonable to only consider code from accepted solutions or the most upvoted ones. This narrows down the number of applicable code snippets but increases the average quality of the code. After the initial parsing and filtering stage, the posts can then be labelled. For labelling, GPT-3.5-turbo was instructed to return a JSON-formatted string describing the functionality of the code and the programming language. The so-called "Stack Overflow Pre" dataset consists of 2 587 804 posts between August 1st, 2015, and April 7th, 2016. The "Stack Overflow Post" dataset consists of 791 122 posts between August 1st, 2023, and April 7th, 2024. Labelling all of these would be costly and excessive, as the goal for each dataset is 1700 samples.[6] Moreover, since the code is unknown, an issue arises: The distribution of programming languages in the respective datasets is hidden and is only revealed after labelling has been completed. To address this, code samples from the parsed XML files were sampled to approximate a log-normal distribution with a total of 30 000 samples each. This gives enough headroom for the target programming language distributions to be reached. Similar precautions were taken for the following three data sources.

LeetCode. The performance of an LLM on LeetCode is often used as a metric to gauge a model's coding capabilities and understanding of diverse problem sets, given that it consists of short code samples ranging from easy to hard difficulties. Unlike Stack Overflow, LeetCode consists solely of functional code, making it a favourable candidate as a source dataset. Although many web-scraped Leet-Code datasets exist containing thousands of problem sets with solutions written in different programming languages, they cannot be used directly for generation for the following reasons. A key goal of this paper is to investigate the impact of possible data leakage on detection performance. This requires each problem to be timestamped to claim with high certainty that the problem is novel to the LLMs being used to generate new samples. Unfortunately, LeetCode problems are chronologically numbered but not timestamped. Using the Wayback Machine[7] to access previous versions of LeetCode's sitemap XML files and linear regression, an estimate for the problem numbers matching the November 2022 and August 2023 cut-off could be made. Problem set 2517 is estimated to be the upper bound for the pre-ChatGPT dataset and problem set 2883 is the lower bound for the August 2023 dataset.

---

[6] We chose this number based on computational estimates and budget constraints.

[7] https://web.archive.org/. The Wayback Machine archives versions of accessible websites on the web.

APPS and CodeSearchNet. Automated Programming Progress Standard (APPS) [8] and CodeSearchNet [11] are two popular datasets used to generate AIGC samples for detection [13,20,23,25]. APPS contains 10 000 Python problems at varying levels of complexity. These problems are used to test a model's natural language to code translation abilities [8]. In addition to the reference code, APPS supplies over 130 000 test cases for testing the generated code. These were not used in this paper but could be a way to filter out low-quality code. CodeSearchNet is comprised of 2 326 976 documented functions. These functions were sourced from non-fork GitHub repositories that were popular and had permissive licenses [11]. For comparison across multiple datasets, only the Python subset containing 503 502 samples was used.

4.2 Dataset Processing

It has been observed that some programming languages are more difficult to detect than others. For example, various detectors have been observed to report strictly higher AUROC scores on Python code samples from CodeContest than on Java samples [25]. Other work suggests that the detectability of samples written in different programming languages varies across detection methods [23]. This discrepancy further highlights the importance of closely inspecting the sources and distributions of the code samples. If one dataset contained more of a difficult-to-detect language, this could skew the detection score and lead to misleading results. Similarly, code length has an impact on the maximum achievable AUROC score of a detector [4,19], and by having two different code length distributions, the performance will be intrinsically different between the two datasets leading to inconclusive results. To investigate the impact of the underlying data sources on detection capabilities, the disjunct datasets must have similar distributions in both programming language and code length to ensure a fair comparison.

To address this issue, we developed a utility class called the "dataset helper" to generate a code length distribution based on the intersection of two separate data sources. This was calculated on a programming language basis to mitigate the aforementioned issue. The distribution was then sampled to fit a scaled log-normal distribution. A log-normal distribution was chosen, because it closely mirrors length distributions found in reality, and many unfiltered distributions exhibit similar characteristics. Additionally, this choice increases the presence of longer samples in the tail. A uniform distribution is infeasible, due to the large number of longer samples that would need to be generated, which would create an overly optimistic portrayal of code detection given its high mean code length. After limiting the distributions to a log-normal distribution, the Python code length histograms of the Stack Overflow, APPS, and CodeSearchNet datasets are identical.

Using the final distributions, a list of code sample hashes can be generated. These hashes, derived from the SHA256 hash of the UTF-8 string representation of each sample, serve as compact identifiers. They can be stored in a file and used as a filter during dataset loading to exclude other samples, ensuring adherence to the intended distribution. Notably, problem difficulty was not factored into sample selection in the generation pipeline, though it is reasonable to assume that problems requiring longer solutions tend to be more complex.

4.3   Generated Dataset

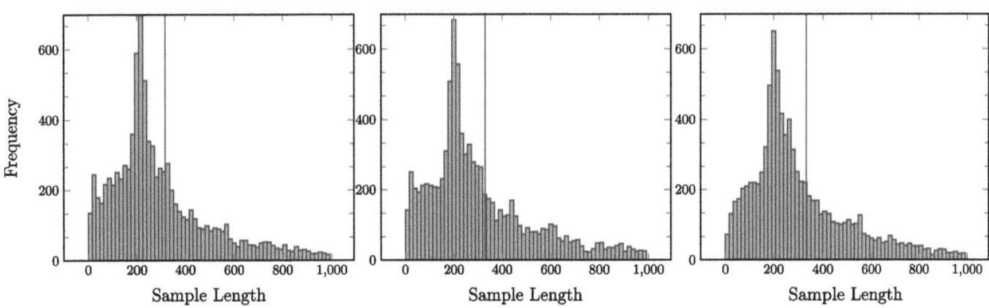

Fig. 1. Generated Python code length distribution of APPS (Left), CodeSearchNet (Middle), and Stack Overflow Post (Right) with a bin size of 16. Length in number of tokens using the cl100k base TikToken tokenizer.

In total, we generated 113 776 code samples across 6 datasets and 17 LLMs. It is important to note that samples with less than eight tokens were discarded, since they were either empty strings or filled with white space. Discarding samples from the dataset introduces the notion of imbalance between human and AIGC samples in the dataset. Comparing the human samples to samples generated by a single LLM, we can calculate an average human-to-generated ratio (see Table 1). To prevent class imbalance, it is important that the ratio stays close to 1 between the samples. Where possible, the parameters for temperature and top-p were set to 0.97 and 0.95, respectively. Temperature is a floating point value applied to the softmax function in the final layer of an LLM that flattens or sharpens the probability distribution of the next token.[8] Top-p sampling (or nucleus sampling) is a randomised sampling strategy that samples from the smallest subset of tokens with the largest probability whose sum exceeds the top-p value [5].

The temperature value and sampling strategy used by the LLM can have a large impact on the detectability of a code sample [20]. A lower temperature value sharpens the conditional probability distribution of the next token leading to a more deterministic generated output. This makes detection easier in a white box context because it increases the probability of a base model "recognizing" its own samples. Increasing the temperature has the opposite effect. Similarly, decreasing the top-p value increases determinism and vice versa. To verify the impact of these parameters, 26 636 additional CodeSearchNet samples were generated with varying temperature and top-p combinations. To compare results, some values

---

[8] Specifically LLMs utilising a decoder-only transformer architecture.

Table 1. List of the data sources, total human- and machine-generated samples, and average human-to-generated ratio (per LLM). E.g., For APPS, we have 561 human-generated samples. Each LLM has slightly fewer than 561, resulting in 9149 machine-generated samples and an average ratio of 1:0.96. *Machine-generated total contains additional samples generated at varying temperature and top-p values.

Data source	Human Total	Machine-Generated Total	Avg. Ratio
Stack Overflow (Pre + Post)	3400	25 544 + 27 243	1:0.94
APPS	561	9149	1:0.96
CodeSearchNet*	561	35 925	1:0.98
LeetCode (Pre + Post)	1000	8025 + 7890	1:0.94

were taken from previous work [20]. A total of four combinations of $T = 0.97$, $T = 0.2$, $p = 0.95$, and $p = 0.5$ were tested.

After generation, it is to be expected that some variance between the dataset distributions is introduced. Since models behave differently from one another and the code may be semantically similar but structurally different, the code lengths will vary as well. Although the distributions are not identical, the mean did not deviate significantly (see Fig. 1). The generated output was then analysed using various zero-shot detection methods within the framework developed in this paper. Both the generated data and the real values assigned by various zero-shot detection methods are published as separate datasets alongside this paper.

4.4   Computational Requirements

Throughout the creation of the datasets and results presented in this paper, we used a high performance cluster to run the LLMs during generation and detection. We ran the generation and detection phases sequentially across 10 parallel instances, each with their own H100 GPU, resulting in approximately 1042 compute hours being spent overall. Around 112 h were spent generating the samples, and another 930 running the detection methods across various model combinations. We determined the batch size, among other optimisation parameters, by running preliminary benchmarks to maximize token throughput. It should be noted that our framework automatically groups samples by size to reduce unnecessary padding added during batching that wastes memory. On average it took around 3.5 s to generate a sample and around 270 ms per detection sample. Although we had access to 96 GB of VRAM per instance, we used quantised models to allow for larger batch sizes to increase inference speed. The continuous progress of LLM quantisation will allow these models to be run on lower-end hardware, making zero-shot detection more viable in the future. The impact of quantisation on detection performance, however, is unknown and remains an open research question.

5   Analysis

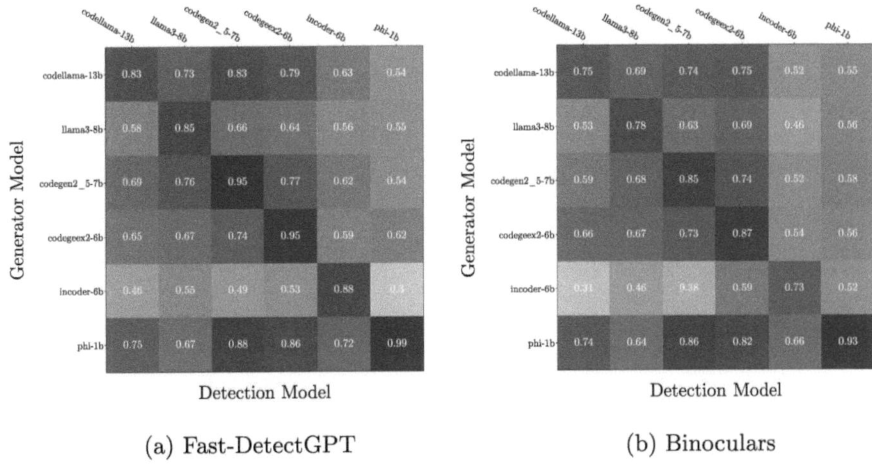

(a) Fast-DetectGPT

(b) Binoculars

Fig. 2. Cross-model performance on two datasets. (a) Fast-DetectGPT on Stack Overflow Post and (b) Binoculars detector on APPS.

It is crucial to investigate the impact of various factors pertaining to the properties of the samples themselves. Key factors include sample length, programming language, and generation parameters, like temperature and top-p value. This yielded 12 294 388 detection samples from five detection methods [2,6,7,21] and 16 LLMs.[9] To examine whether data leakage affects detection performance, we analysed a subset of detection models trained before our cut-off date, focusing on samples from 11 of the 17 LLMs. While initially working on constructing the source datasets and theorising the impact of data leakage, we hypothesised that introducing previously seen samples could have two opposing effects. Hypothesis A: Detection performance increases because the underlying LLM "recognises" its own generated samples better. Hypothesis B: Detection performance decreases because the generated samples closely match the original human samples, making differentiation more difficult. Our goal was to highlight the importance of dataset selection and generation, urging standardisation in the field through benchmarks and frameworks to allow for a more reliable comparison between detection methods and to answer to following research questions:

RQ1: Can Zero-Shot Detection Methods from Other Domains be Effectively Applied to Code Detection?

Heat maps are a useful way to visualise the cross-model performance of a detection method. It should be noted that, to maintain readability of the data con-

---

[9] o1-mini cannot be used as a detection model because OpenAI do not expose the logits produced by the LLM that are required for the investigated methods.

tained in the figures, we only present six of the LLMs in the following figures. We chose to represent only the largest model of a given family and remove instruct versions from the figures. After inspecting the heat maps of several detection methods and datasets, it becomes clear that current methods still generally perform better in a white box context compared to a black box one. This is evident from the darker diagonal in the heat maps (see Fig. 2). Alongside the code and datasets, we provide a large collection of complete figures, including various heat maps of detection method and dataset combinations, which can also be generated using the supplied IPython notebook files in the code repository. Overall, there is strong evidence supporting the applicability of existing detection methods for AI-generated code.

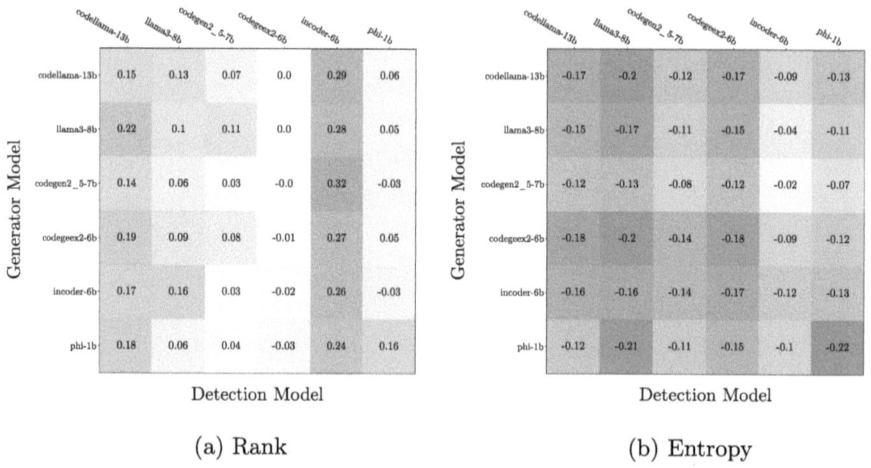

Fig. 3. (a) Python-only cross-model performance comparison between CodeSearchNet and Stack Overflow Post using rank (b) Cross-model performance comparison between LeetCode Pre and LeetCode Post using entropy. Green indicates an improvement in AUROC. Red indicates a decrease in AUROC. (Color figure online)

One can also see a large discrepancy in AUROC scores between detection methods. Directly comparing AUROC scores without considering the underlying ROC curve can be misleading, as AUROC alone does not reflect key detector characteristics, such as the FPR-TPR relationship at fixed thresholds. Filtering AIGC from future datasets requires a high TPR, as being overly cautious is preferred, while the FPR is less relevant. Conversely, detecting AIGC in student submissions requires a low FPR at the cost of letting some AIGC evade detection. Thus, we also investigate the ROC curves. Only curves that dominate others can be called superior detectors. Inspecting several ROC curves with varying detection methods, generator and base model combinations reveals a clear trend: Generally, the larger the difference between the AUROC scores, the greater the likelihood that the ROC curve dominates the other, meaning that the TPR lies above the other across all FPRs. This is notable because it allows us to define a heuristic that we call a "difference heat map" to better visualise differences in AUROC scores. The difference heat map visualises performance improvements of dataset B over A (A vs B) by calculating B - A per tile, highlighting positive differences in green and negative in red. In particular, the larger the absolute difference the more accurate the heuristic.

RQ2: What Is the Impact of Data Leakage on Detection Performance?

Figure 3 highlights the difference in AUROC between datasets before and after the cut-off date using the same detection methods. Although most detection methods showed an improvement, supporting hypothesis B, the entropy detector often decreased with novel samples (see Fig. 3).

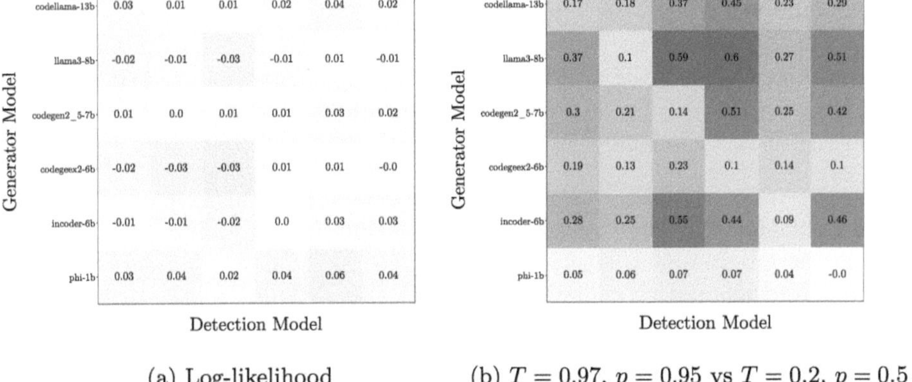

(a) Log-likelihood   (b) $T = 0.97$, $p = 0.95$ vs $T = 0.2$, $p = 0.5$

Fig. 4. (a) Cross-model performance comparison between Stack Overflow Pre and Stack Overflow Post using log-likelihood (b) Python-only cross-model performance comparison between CodeSearchNet at varying temperature and top-p using Fast-DetectGPT. Green indicates an improvement in AUROC. Red indicates a decrease in AUROC. (Color figure online)

Unlike when comparing Stack Overflow Post to APPS and CodeSearchNet, there was no significant change in detection performance between Stack Overflow Pre and Post (see Fig. 4a). While contradictory to both our hypotheses, this is not an inexplicable result. Firstly, less than 1% of samples were used in both of the 250-day periods of Stack Overflow Pre and Post. Additionally, the labelling process helped mitigate dataset contamination by encouraging novel outputs that better reflect the model's true coding abilities. This likely impacted how recognisable these samples were, even if seen during training. Since these steps were applied to both datasets, the generated samples should be of similar quality leading to the observed results. These findings may change if the natural language-to-code generation performance continues to increase or methods like retrieval augmented generation (RAG) are utilised during generation. Future work may look further into the impact of rephrasing problem descriptions from existing datasets on detection performance. The minimal performance difference between Stack Overflow Pre and Post contrasts with the significant variations observed between APPS, CodeSearchNet, and Stack Overflow Post, as well as LeetCode Pre and Post. This highlights a key finding: for current LLMs, collecting newly labelled data outside of benchmarks and common training datasets is just as crucial as the cut-off date.

RQ3: How Do Generation Parameters of LLMs, Such as Temperature, Affect Detection Performance?

Difference heatmaps can also be used to examine how detection methods perform under varying generation parameters. Since AUROC values across different top-p and temperature configurations are derived from the same human reference samples, comparing them is meaningful. The difference heat maps indicate that lower top-p and temperature values resemble the results of newly labelled data (see Fig. 4b). In most cases, the AUROC score of detection methods increased significantly across the board except for entropy detection. These results further underscore the need for standardised detection datasets and frameworks, as variations in generation parameters can further complicate the comparability of presented results.

Additionally, difference heat maps can also be used to investigate the performance difference of detection methods across different programming languages. While the programming language distributions also follow a log-normal distribution, it is advised to cautiously interpret these results, as the distributions differ far more across programming languages than across datasets. However, these significant differences in detection performance still raise the question of the cause underlying this discrepancy. Hoq *et al.* [9] suggest that syntactic differences are a sufficiently strong indicator for differentiating AI-generated code from student submissions. This is partially substantiated in our dataset by our token frequency analysis, where it can be seen that some tokens are preferred by either humans or LLMs. While this may explain one possible factor for the variation in detection performance between different programming languages, further work investigating aspects such as the influence of training data or programming ability of an LLM in a specific language is required.

6   Conclusions

In this paper, we investigated the efficacy of existing zero-shot methods for detection of AI-generated code by assembling a large corpus of 113 776 samples generated by 17 LLMs. Using the generated code samples, we then produced 12 294 388 detection samples, by applying five detection methods and 16 of the LLMs.

We achieved this by developing a framework called "Codetector" that facilitates the modular integration of new data sources and LLMs for generating and detecting AI-generated content. The goal of our unified framework is to benefit the task of AIGC detection by allowing for an easier and more reliable comparison between detection methods. Using our framework, we gathered a mixture of newly labelled and existing human-generated code samples from various online sources, ensuring similar programming languages and length distributions. These samples were then used to generate new ones with the integrated LLMs. This careful scrutiny of collected and generated samples allowed us to not only show the feasibility of detecting AI-generated code but also to closely compare the detection results between subsets of our large set of samples.

We found that both old and new zero-shot detection methods often perform better when using new, unseen samples during generation and detection, with some detector-generator combinations showing an absolute AUROC increase of over 40%. This was achieved by using samples created after the model cut-off date or by relabelling existing samples. By doing so, we achieved AUROC scores of up to 97% with an average of around 83% using newer detection methods, despite using high temperature and top-p values of 0.97 and 0.95, respectively, during generation. In addition to the impact of diverse datasets and possible data leakage, we showed the significant impact of generation parameters on detection performance.

We hope that our findings on how various factors bearing on dataset creation impact detectability will further motivate the development of unified benchmarks for AIGC detection. Additionally, we aim for our unified framework to facilitate a more reliable comparison between detection methods, ultimately accelerating the development of novel detection techniques.

Acknowledgments. All simulations were performed with computing resources granted by RWTH Aachen University under project thes1707. This work was supported by the Alexander-von-Humboldt Foundation. Finally, we would like to thank Marie Anastacio for helpful feedback on the paper.

References

1. Alemohammad, S., et al.: Self-consuming generative models Go MAD. In: The Twelfth International Conference on Learning Representations (ICLR), pp. 1–13 (2024)
2. Bao, G., et al.: Fast-DetectGPT: efficient zero-shot detection of machine-generated text via conditional probability curvature. In: The Twelfth International Conference on Learning Representations (ICLR), pp. 1–13 (2024)
3. Carlini, N., et al.: Extracting training data from large language models. In: 30th USENIX Security Symposium, pp. 2633–2650 (2021)
4. Chakraborty, S., et al.: Position: on the possibilities of AI-generated text detection. In: International Conference on Machine Learning (ICML), pp. 1–15 (2024)
5. Finlayson, M., et al.: Closing the curious case of neural text degeneration. In: The Twelfth International Conference on Learning Representations (ICLR), pp. 1–12 (2024)

6. Gehrmann, S., et al.: GLTR: statistical detection and visualization of generated text. In: 57th Conference of the Association for Computational Linguistics (ACL), pp. 111–116 (2019)
7. Hans, A., et al.: Spotting LLMs with binoculars: zero-shot detection of machine-generated text. In: International Conference on Machine Learning (ICML), pp. 1–20 (2024)
8. Hendrycks, D., et al.: Measuring coding challenge competence with APPS. In: Neural Information Processing Systems Track on Datasets and Benchmarks, pp. 1–12 (2021)
9. Hoq, M., et al.: Detecting ChatGPT-generated code submissions in a CS1 course using machine learning models. In: 55th ACM Technical Symposium on Computer Science Education (SIGCSE), pp. 526–532 (2024)
10. Hu, X., et al.: RADAR: robust AI-text detection via adversarial learning. In: Advances in Neural Information Processing Systems (NeurIPS), pp. 1–12 (2023)
11. Husain, H., et al.: CodeSearchNet challenge: evaluating the state of semantic code search. arXiv preprint arXiv:1909.09436, pp. 1–6 (2019)
12. Khoury, R., et al.: How secure is code generated by ChatGPT? In: IEEE International Conference on Systems, Man, and Cybernetics (SMC), pp. 2445–2451 (2023)
13. Li, B., et al.: Resilient watermarking for LLM-generated codes. arXiv preprint arXiv:2402.07518, pp. 1–18 (2024)
14. Mitchell, E., et al.: DetectGPT: zero-shot machine-generated text detection using probability curvature. In: International Conference on Machine Learning (ICML), pp. 24950–24962 (2023)
15. OpenAI, et al.: GPT-4 technical report. arXiv preprint arXiv:2303.08774, pp. 1–78 (2023)
16. Ouyang, L., et al.: Training language models to follow instructions with human feedback. In: Advances in Neural Information Processing Systems (NeurIPS), pp. 1–15 (2022)
17. Pan, W.H., et al.: Assessing AI detectors in identifying AI-generated code: implications for education. In: 46th International Conference on Software Engineering, pp. 1–11 (2024)
18. Qiang, Y., et al.: Learning to poison large language models during instruction tuning. arXiv preprint arXiv:2402.13459, pp. 1–14 (2024)
19. Sadasivan, V.S., et al.: Can AI-generated text be reliably detected? arXiv preprint arXiv:2303.11156, pp. 1–16 (2023)
20. Shi, Y., et al.: Between lines of code: unraveling the distinct patterns of machine and human programmers. arXiv preprint arXiv:2401.06461, pp. 1–12 (2024)
21. Solaiman, I., et al.: Release strategies and the social impacts of language models. arXiv preprint arXiv:1908.09203, pp. 1–71 (2019)
22. Tian, H., et al.: Is ChatGPT the ultimate programming assistant - how far is it? arXiv preprint arXiv:2304.11938, pp. 1–22 (2023)

23. Wang, J., et al.: An empirical study to evaluate AIGC detectors on code content. In: 39th IEEE/ACM International Conference on Automated Software Engineering (ASE), pp. 844–856 (2024)
24. Wu, J., et al.: A survey on LLM-generated text detection: necessity, methods, and future directions. Comput. Linguist., 1–66 (2025)
25. Yang, X., et al.: Zero-shot detection of machine-generated codes. arXiv preprint arXiv:2310.05103, pp. 1–11 (2023)

Pushing the Limits of the Reactive Affine Shaker Algorithm to Higher Dimensions

Roberto Battiti[ID] and Mauro Brunato[(✉)][ID]

DISI, Università di Trento, Via Sommarive 9, 38123 Trento, Italy
{roberto.battiti,mauro.brunato}@unitn.it

Abstract. Bayesian Optimization (BO) for the minimization of expensive functions of continuous variables uses all the knowledge acquired from previous samples (x_i and $f(x_i)$ values) to build a surrogate model based on Gaussian processes. The surrogate is then exploited to define the next point to sample, through a careful balance of exploration and exploitation. Initially intended for low-dimensional spaces, BO has recently been modified and used also for very large-dimensional spaces (up to about one thousand dimensions).

In this paper we consider a much simpler algorithm, called "Reactive Affine Shaker" (RAS) [3]. The next sample is always generated with a uniform probability distribution inside a parallelepiped (the "box"). At each iteration, the form of the box is adapted during the search through an affine transformation, based only on the point x position and on the success or failure in improving the function. The function values are therefore not used directly to modify the search area and to generate the next sample. The entire dimensionality is kept (no active subspaces like in [14]).

Despite its extreme simplicity and its use of only stochastic local search, surprisingly the produced results are comparable to and not too far from the state-of-the-art results of high-dimensional versions of BO, although with some more function evaluations.

An ablation study and an analysis of probability distribution of directions (improving steps and prevailing box orientation) in very large-dimensional spaces are conducted to understand more about the behavior of RAS and to assess the relative importance of the algorithmic building blocks for the final results.

1 Introduction

All linear optimization problems are alike; each difficult nonlinear problem is difficult in its own way, more so in high dimensions. Each problem has different structural characteristics, and every specific instance of a problem may present widely different "fitness landscapes". Furthermore, the local landscape of a single instance can depend critically on the position of the current point x in the input space and can therefore vary a lot during the search. The so-called "curse of dimensionality" is well known. The complexity of finding an optimum - especially

Y. Zhang et al. (Eds.): LION 2025, LNCS 15745, pp. 174–190, 2026.
https://doi.org/10.1007/978-3-032-09192-5_12

with simple brute-force methods - tends to grow exponentially with the number of dimensions, apart from rare special problems (like Linear Programming) or special instances of a problem.

Despite the negative theoretical worst-case results, a concrete hope of finding improving solutions of practical interest derives from the fact that many relevant real-world problems possess a high level of internal structure that can be exploited (e.g., the Big Valley hypothesis [9] or variations thereof) and from the fact that the appropriate match between problem (or instance) characteristics and algorithm configuration and parameters can be learned automatically via Machine Learning (ML). ML can be applied in an offline manner to adapt the meta-parameters to a problem like in algorithm configuration or online, by reacting to events occurring during a single search process on a specific instance (e.g., as advocated in Reactive Search Optimization - RSO [1]).

The growing availability of massive amounts of memory, starting from the eighties, opened new windows of opportunity for memory-based Intelligent Optimization techniques. The underlying assumption of a rich internal structure of most relevant optimization tasks makes techniques capable of *gradually learning* that structure potentially more powerful and effective than memory-less techniques.

This paper builds upon a previously proposed Reactive Search Optimization algorithm for global optimization of multivariate functions of continuous variables called RAS [4] (Reactive Affine Shaker). RAS is an adaptive search algorithm based only on point-wise function evaluations in a stochastic local (perturbative) search.

The novel contributions of this paper are:

- A qualitative and quantitative study of the evolution of the search box of RAS during the search, in particular for large-dimensional spaces
- An *ablation study* of RAS to assess the relative contribution of its simple algorithmic building blocks
- An experimental comparison of RAS on some very large dimensional problems previously solved with state-of-the-art Bayesian Optimization techniques.

The remainder of this paper is organized as follows. In Sect. 2 we discuss the state of the art on Bayesian optimization techniques; Sect. 3 summarizes and motivates the proposed RAS heuristic. Experimental evaluations and comparisons are reported and discussed in Sect. 4.

2 Bayesian Optimization for High-Dimensional Problems

In the context of the optimization of functions of continuous variables, we assume that the dominant computational cost is the evaluation of the function f at sample points. This holds in many practical applications, e.g., when the evaluation of f requires running a lengthy simulation, or even running an industrial plant and measuring the output.

Bayesian optimization [8] is a sequential design strategy for global optimization of black-box functions that are expensive to evaluate. It builds a surrogate model for the objective from the sampled points, quantifies the uncertainty by using a Bayesian machine learning technique, (Gaussian process regression), and then uses an *acquisition function* defined from this surrogate to decide where to sample the next point(s). The acquisition function trades off exploration and exploitation to reduce the number of expensive function evaluations.

Although based on Gaussian processes (without guarantees that concrete functions are appropriately modeled by GP) and expensive, BO is considered one of the most promising algorithms for optimization of functions of limited dimensionality (usually not more than 10–20).

A series of breakthroughs have recently pushed the envelope of high-dimensional Bayesian optimization for a wider adoption in science and engineering. For the limited scope of this paper, we concentrate on two recent contributions [7,14] and we refer to the contained bibliography for a review of the state of the art.

The research in [7] argues that the implicit homogeneity of the global probabilistic models in BO tends to overemphasize exploration in high-dimensional problems. The fact that search spaces grow faster than sampling budgets implies the presence of regions with large posterior uncertainty. For common myopic acquisition functions, this results in an overemphasized exploration and a failure to exploit promising areas. To remedy, they propose the TuRBO (Trust-region BO) that considers a population of parallel and independent local models and performs a global allocation of samples across these models via an implicit multi-armed bandit approach (the local models that are more promising get progressively more samples to evaluate). The trust-region (TR) idea [20] is that each local model can be trusted only in a region, typically a ball with a given radius around the current solution, a radius that is adapted during optimization. In TuRBO the TR is a hyperrectangle centered at the best solution found so far $\boldsymbol{x}^*$. The initial base side length L is a portion of the range along each coordinate, the total volume of the TR is kept fixed, while the length along each coordinate is rescaled according to its corresponding lengthscale λ_i in the GP model. The base side length L is then adapted during the run, by keeping it sufficiently large so that the TR contains good solutions but small enough to ensure that the local model is accurate within the TR. In detail, the TR is expanded after many consecutive "successes" and shrunk after many consecutive "failures", in a similar spirit as in [13].

In the assumption of the existence of an *active subspace*, with a projection matrix T so that the value of the function to optimize depends only on the projected $\boldsymbol{x}$ value ($F(\boldsymbol{x}) = g(T\boldsymbol{x})$) REMBO (Random embedding BO) [19] and HeSBO (Hashing-enhanced subspace BO) [12] try to capture this active subspace by a randomly chosen linear subspace. SaasBO [6] uses sparse priors on the GP length scales that is particularly effective if the active subspace is axis-aligned. Alebo [11] uses a Mahalanobis kernel and linear constraints on the acquisition function.

The work in [14] starts from the observation that methods for high-dimensional Bayesian optimization (HDBO) suffer from degrading performance in spaces of growing dimension or risk failure if some assumptions are not met. They propose a new algorithm (BAxUS -Bayesian optimization with adaptively expanding subspaces) that uses the idea of nested low-dimensional random subspaces of growing dimensionality to adapt the space it optimizes over to the problem.

3   The RAS Heuristic

We summarize and motivate in this section the RAS local search algorithm which is the method considered in this work.

---

f Function to minimize
x Initial point
$\mathcal{R}$ Search region
Δ Current displacement

---

1. **function RAS** $(f,\, x)$
2. $\mathcal{R} \leftarrow$ small isotropic set around x
3. **while** (local termination condition is not met)
4. Pick $\Delta \in \mathbb{R}^d$ such that $x + \Delta, x - \Delta \in \mathcal{R}$
5. **if** $f(x + \Delta) < f(x)$
6. $x \leftarrow x + \Delta$;
7. Extend $\mathcal{R}$ along Δ
8. Center $\mathcal{R}$ on x
9. **else if** $f(x - \Delta) < f(x)$
10. $x \leftarrow x$ - Δ;
11. Extend $\mathcal{R}$ along Δ
12. Center $\mathcal{R}$ on x
13. **else**
14. Reduce $\mathcal{R}$ along Δ
15. **return** x;

Fig. 1. The RAS algorithm.

The Reactive Affine Shaker Heuristic [4], RAS for short, is a self-tuning local search algorithm based on [18]. No prior knowledge is required on the function f and only evaluations at arbitrary values of the independent variables are allowed. The RAS heuristic tries to rapidly move towards better objective values by maintaining and updating a "search region" $\mathcal{R}$ around the current point x.

The use of memory in RAS is limited: the entire previous history of the search (the trajectory of the generated sample points and the outcome of the evaluations) is summarized through the *dynamic search region*, intended to zoom in on the promising areas where to find points better than the current record point.

RAS adapts the search region $\mathcal{R}$ depending on the occurrence or lack of success during the last step. If a step in a certain direction is improving, then $\mathcal{R}$ is expanded along that direction; it is reduced otherwise. Once a promising direction is found, the probability that subsequent steps will follow the same direction is increased, and the search will proceed more and more aggressively in that direction until bad results reduce its prevalence. The algorithm is outlined in Fig. 1.

RAS starts with an isotropic search region centered around the initial point (line 1). Next, new sample points are generated (line 1). If the sample point $x + \Delta$ yields a lower objective value (line 1 and following), then the current position is updated and $\mathcal{R}$ is expanded along the direction of Δ. To increase the probability of finding a better point, if $x + \Delta$ does not lead to an improvement, also $x - \Delta$ is tried (line 1 and following). If both points fail at improving f, then the search region is reduced along the direction of Δ (line 1) and the current position is kept. The above steps are repeated until a local termination condition is verified. Common termination criteria are the number of iterations, the size of the search region, or a large number of iterations without further improvement.

3.1   Implementation of the Anisotropic Search Box

The main rationale for the introduction of a non-isotropic search box lies in the increasing difficulty of finding improving directions, even for very smooth and regular functions, when displacements are chosen along uniformly random directions. These difficulties arise from two main facts, discussed below: in high dimensions, random directions tend to be mutually orthogonal and improving moves tend to be scarce.

Random Directions Tend to Be Mutually Orthogonal—Following [3], the average angle $\bar{\theta}_d$ in radians between two random directions in $d \geq 2$ dimensions is

$$\bar{\theta}_d = \frac{J_{d-2}}{I_{d-2}},$$

where the numerator and denominator are recursively defined as

$$I_d = \begin{cases} \frac{\pi}{2} & \text{if } d = 0 \\ 1 & \text{if } d = 1 \\ \frac{(d-1)I_{d-2}}{d} & \text{if } d > 1, \end{cases} \qquad J_d = \begin{cases} \frac{\pi^2}{8} & \text{if } d = 0 \\ 1 & \text{if } d = 1 \\ \frac{(d-1)J_{d-2}}{d} + \frac{1}{d^2} & \text{if } d > 1. \end{cases}$$

The first row of Table 1 reports the values of $\bar{\theta}$ computed for an increasing number of dimensions.

For $d = 2$ the average angle is $45°$, but for higher and higher dimensionalities the average tends to $90°$. In the simplifying hypothesis that the function's gradient is unknown and that we generate a move in a random direction, this means

Table 1. Average angle between random vectors in d dimensions (top row) and expected number of double-shot successes for a random displacement depending on the search box radius $r(B')$ relative to the average curvature of the function's isosurface.

	Dimension d								
	1	2	3	5	10	50	100	500	1000
Avg angle $\bar{\theta}_d$ (degrees)	0.00	45.00	57.30	66.85	74.64	83.46	85.40	87.95	88.55
$r_{B'}/r_B = 1.000$	1.00	0.78	0.62	0.42	0.16	0.00	0.00	0.00	0.00
$r_{B'}/r_B = 0.500$	1.00	0.89	0.81	0.69	0.50	0.08	0.01	0.00	0.00
$r_{B'}/r_B = 0.100$	1.00	0.98	0.96	0.94	0.89	0.73	0.62	0.26	0.11
$r_{B'}/r_B = 0.050$	1.00	0.99	0.98	0.97	0.95	0.86	0.81	0.58	0.43
$r_{B'}/r_B = 0.010$	1.00	1.00	1.00	0.99	0.99	0.97	0.96	0.91	0.87
$r_{B'}/r_B = 0.005$	1.00	1.00	1.00	1.00	0.99	0.99	0.98	0.95	0.94
$r_{B'}/r_B = 0.001$	1.00	1.00	1.00	1.00	1.00	1.00	1.00	0.99	0.99

that in high dimensions there is a progressively smaller probability of moving along the gradient.

Improving Moves Tend to Be Scarce—The effectivenes of the double-shot strategy in the limit for small search boxes was proved in [3]. However, given a desired success probability, as the number of search-space dimensions grows the required search box size might become impractically small. Consider the situation described in Fig. 2. Let $\hat{x} \in \mathbb{R}^d$ be the current point during a local search for the minimum of function $f : D \to \mathbb{R}$ on domain $D \subseteq \mathbb{R}^d$. We define the isosurface of function f at point $\hat{x}$ to be the locus of all points x in its domain with the same function value:

$$\text{iso}_f(\hat{x}) = f^{-1}(f(\hat{x})) = \{x : f(x) = f(\hat{x})\}.$$

The isosurface has dimension $d - 1$ and is embedded in $\mathbb{R}^d$. It locally divides the domain into two sides, the "good" one where $f(x) < f(\hat{x})$ (left side of the isoline in Fig. 2) and the "bad" one where $f(x) > f(\hat{x})$. If f is locally convex and smooth enough, we can approximate the curvature radius of the iso-surface at point $\hat{x}$ by considering the radius of the largest sphere B tangent to the iso-surface in $\hat{x}$ and lying on the same side of it[1]. Suppose now that the local move consists of generating a random displacement Δ in a ball B' centered on $\hat{x}$. The move succeeds if $\hat{x} + \Delta$ falls inside B (as a local approximation of the "interior" of the isosurface) or, due to the double-shot strategy, in the symmetrically opposite direction. The probability of success is therefore given by the ratio between the area of the greyed-out portion of B' in Fig. 2 and the full area of B', and it depends on the ratio between the radius of B' and the

---

[1] For $d > 2$, we actually have different curvature radii on different directions; let us consider the smallest one.

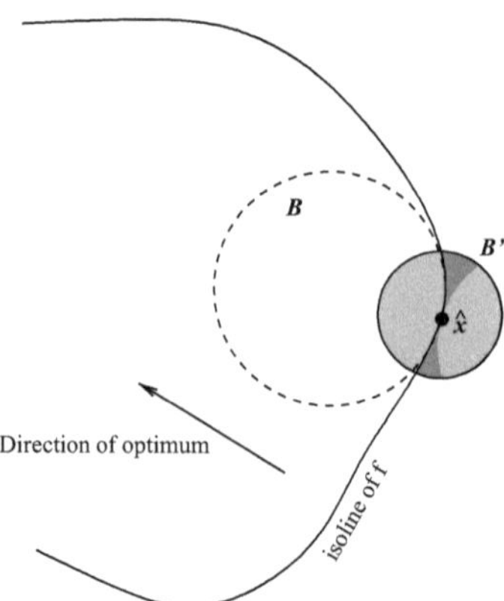

Fig. 2. Estimation of the probability of success for the double-shot strategy with simplifying hypotheses (locally smooth function with isolines locally approximable with a spherical surface B, round search box B'). The light grey area defines a successful displacement (considering the double shot strategy); the ratio between the light grey area and the total area of B' gives the success probability of the double-shot strategy. (Color figure online)

radius of B. We can consider r_B as a normalization factor for this discussion. Table 1 shows the experimental success probability of the double-shot strategy at different dimensions d and for different ratios between the search area radius $r_{B'}$ and the isosurface's curvature radius r_B. In particular, we can see that for high dimensionalities the success probability of the double-shot strategy is negligible and only becomes significant for small search box radii, forcing a heuristic based on an isotropic search box to adopt small displacements that slow the progress down. The RAS heuristic tries to compensate this problem by introducing an anisotropic search region that elongates in an orthogonal direction with respect to the isosurface.

Implementation of the Search Region—The anisotropic search region $\mathcal{R}$ is implemented as a box defined by d independent base vectors $(\boldsymbol{b}_1 \ldots \boldsymbol{b}_d)$, where d is the number of dimensions of the search domain. The search region $\boldsymbol{\Delta}$ is determined as a random linear combination of the base vector with independent coefficients uniformly distributed in $[-1, 1]$:

$$\boldsymbol{\Delta} = \sum_{i=1}^{d} r_i \boldsymbol{b}_i, \qquad r_i \in [-1, 1].$$

Shape modifications are implemented as affine transformations of these vectors by contracting (if the step fails) or dilating (if the step succeeds) their components along the direction of $\boldsymbol{\Delta}$ as shown in Fig. 3. Analytically, we want to add to every base vector $\boldsymbol{b}_j$ a contribution in the direction of $\boldsymbol{\Delta}$ proportional to the

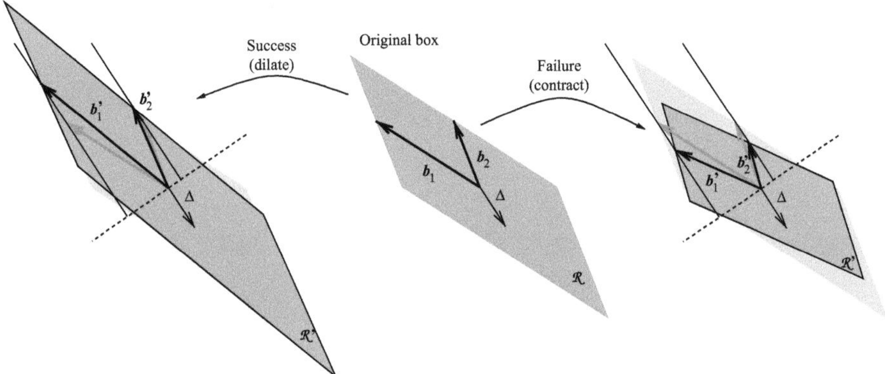

Fig. 3. Evolution of the search box $\mathcal{R}$ under transformation (1): from the current shape (center), the base vectors $\boldsymbol{b}_i$ are dilated in the direction of $\boldsymbol{\Delta}$ if the step succeeds (left), or contracted (right) if the step fails.

base vector's size:

$$(\rho - 1)\frac{\boldsymbol{\Delta} \cdot \boldsymbol{b}_j}{\|\boldsymbol{\Delta}\|^2}\boldsymbol{\Delta},$$

where ρ controls the amount of contraction ($0 < \rho < 1$, making the contribution opposite wrt to the projection of $\boldsymbol{b}_j$ along $\boldsymbol{\Delta}$) or dilation ($\rho > 1$, making the contribution positive along the projection). This can be rewritten as the following linear transformation:

$$\forall j \quad \boldsymbol{b}_j \leftarrow A\boldsymbol{b}_j, \qquad \text{where} \qquad A = \boldsymbol{I} + (\rho - 1)\frac{\boldsymbol{\Delta}\boldsymbol{\Delta}^T}{\|\boldsymbol{\Delta}\|^2}. \tag{1}$$

The RAS heuristic depends therefore on three parameters,

$$0 < \rho_{\mathrm{con}} < 1, \qquad \rho_{\mathrm{dil}} > 1, \qquad 0 < \eta < 1$$

with the following meaning, assuming a search hyperinterval $[\min_i, \max_i]$, $i = 1, \ldots, d$:

- the initial search box vectors $\boldsymbol{b}_i$, $i = 1, \ldots, d$, are aligned along the domain axes, with length $\|\boldsymbol{b}_i\| = \eta \cdot (\max_i - \min_i)$;
- the affine transformation factor in (1) is $\rho = \rho_{\mathrm{dil}}$ for dilations (upon improving step) and $\rho = \rho_{\mathrm{con}}$ for contractions (upon double-shot failure).

4   Experimental Results

The following parameter values are used in all experiments unless otherwise noted:

$$\eta = \frac{1}{5} = 0.2, \qquad \rho_{\mathrm{dil}} = 5, \qquad \rho_{\mathrm{con}} = \frac{1}{\rho_{\mathrm{dil}}} = 0.2.$$

Let's note that the dilation and contraction parameters are far from 1. In a high-dimensionality setting, the affected direction Δ is, with high probability, almost perpendicular to all base vectors $\boldsymbol{b}_i$, therefore the effect of milder parameters (i.e., close to 1) would be very limited.

4.1   Behavior of the Search Box

In Fig. 4 we show a sample RAS run on the 2D Rosenbrock function [16], an unimodal function with a narrow, curved valley and with a selectable dimensionality. The global minimum is in $(1, 1)$, and the search proceeds from right to left. The right-hand side plot shows the trajectory superimposed on the evolving search boxes. In the initial phase, the search proceeds downhill (we see the path crossing many function isolines); the search box is wide and slightly elongated in the search direction. Once the valley is reached around point $\boldsymbol{x} = (2.00, 4.00)$, the trajectory proceeds with a narrower search box aligned with the valley direction.

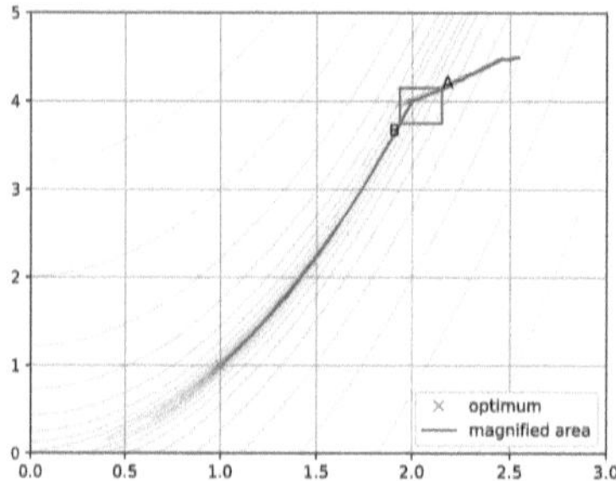
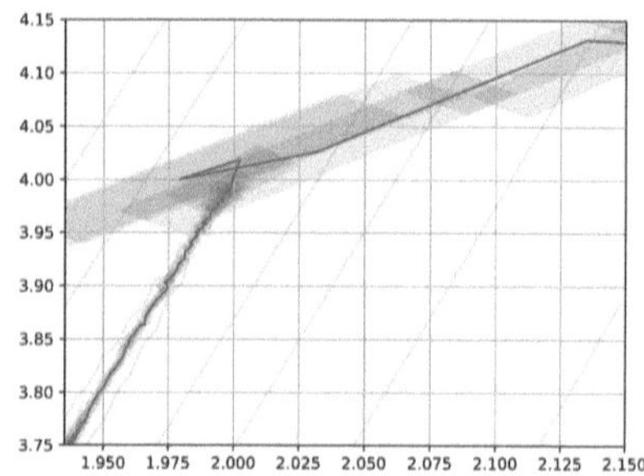

Fig. 4. Evolution of the search box under transformation (1), 2D Rosenbrock function; on the right, a detailed view of the AB rectangle with the direction change.

The analysis of the 2D Rosenbrock function search is expanded in the left column of Fig. 5, where the evolution of the search is shown vs. the number of function evaluations. The top chart represents the best value found so far; the band delimited by the A and B markers corresponds to the portion of trajectory visible in the AB rectangle. The second chart from the top shows the size of the box, represented as a colored band spanning the size in the different directions; a wider band represents a more elongated search box. In the third chart, the evolution of the ratio between the minimum and maximum search box sizes is shown. Finally, the bottom chart displays the angle between the longest vector $\boldsymbol{b}_i$ in the search box (i.e., the dominating search direction) and the direction from the current search position to the global optimum. The angle is given in radians, from 0 (search box perfectly aiming at the optimal point) to $\pi/2 \approx 1.57$ (box aimed at orthogonal direction).

For the 2D Rosenbrock search, in the interval AB, we can see an initial fast improvement of the best value (the downhill phase). Once the valley is reached

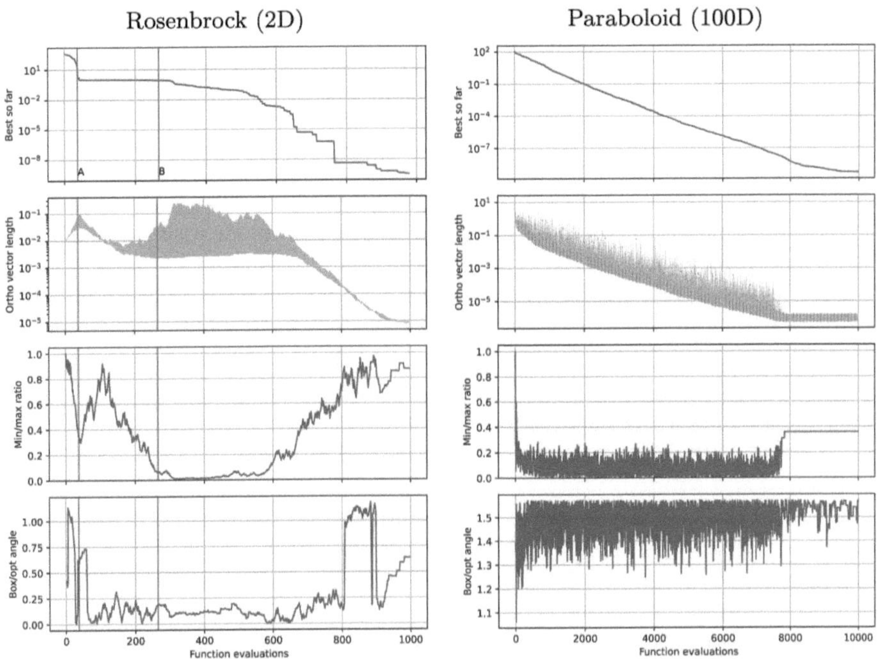

Fig. 5. Evolution of the search box under affine transformation wrt number of function evaluations for two test benchmarks, see the text for details.

(very close to point A), the search box starts evolving in two ways: it becomes smaller (second row), because the shape of the valley causes many steps to fail, and more "square" (the ratio in the third plot approaches 1.0). At some point, close to 100 evaluations, the search box starts elongating (ratio in third plot decreasing), and it points in the correct direction (fourth plot: oriented towards the global minimum). The dominant vector in the search box grows, encouraging steps in the right direction.

The right column of Fig. 5 shows the evolution of the search on a 100-dimensional paraboloid $f(x) = \|x\|^2$ on domain $x \in [-1.5, 1.5]^{100}$. Given the larger number of dimensions, more evaluations are required to achieve similar results. As the optimum point is approached, failures due to overshooting cause the box to shrink consistently (second row), while because of the regular shape, the box tends to stay quite elongated in a good direction (the third plot shows consistently small ratios between the shortest and the largest box vector). Although the bottom plot shows the dominant direction not to be aligned with the local optimum (and hence the gradient), this is expected due to the high dimensionality of the search space and doesn't prevent the generation of improving moves.

4.2   Comparison with State-of-the-Art Heuristics

We compare the performance of RAS with a selection of heuristics covered in Sect. 2 (TuRBO, SaasBO, Alebo, and HeSBO), the popular CMA-ES [10] and random search [2] on the following six high-dimensionality benchmark problems with continuous variables on hyper-rectangular domains.

Mopta08 [6]—a 124-parameter vehicle design problem where the objective is the minimization of the vehicle's mass; the original problem is subject to several constraints transformed into soft penalties.

SVM [6]—a Support Vector Machine training problem with 388 parameters.

Branin2 [19]—the classical 2D test function with 3 global minimizers, embedded in a 500D space.

Lasso-Hard, **Lasso-High** [17]—two synthetic hard problems with 1000 and 300 parameters respectively from the LASSOBENCH benchmark suite.

Hartmann6 [19]—the classical 6D multimodal test function, embedded in a 500D space and rotated.

All functions are implemented in the publicly available BAxUS test suite[2], and all tests have been performed on the noiseless versions.

The results are shown in Fig. 6. The top and third rows contain results from [14], while the second and bottom rows show the RAS results on the corresponding benchmark (names on top of each column) on comparable horizontal and vertical scales. For every function, all 30 RAS runs are shown as lightly colored lines, while the average is superimposed as a black dotted line.

RAS outperforms the Alebo, HeSBO, and SaasBO heuristics in all benchmarks with the exclusion of Branin2, where the identification of the 2D active subspace aligned with the domain axes kicks in very soon, while RAS keeps elongating and reducing the search box in irrelevant directions, and with the further exclusion of Hartmann6 for SaasBO. CMA-ES is outperformed in all cases except the Lasso-High benchmark.

On the other hand, the more sophisticated BAxUS and TuRBO heuristics generally behave better than RSA, with a significant advantage in the SVM, Branin2, and Lasso benchmarks. In the SVM and Lasso-High case, we note that RAS is still significantly decreasing at the end of the allotted number of evaluations, and further examination shows that $1.5\ldots2\times$ evaluations are needed to achieve comparable results; in the three bottom benchmarks, RSA is further penalized by a few runs stuck in a slowly improving path (Lasso benchmarks) or in the attraction basin of a local minimum (Hartmann6, the dashed line represent the local minima).

Since RAS is a local optimization heuristic, the average line in the Hartmann6 plot only considers the trajectories in the global minimum attraction basin, corresponding to those that, at the 1000 evaluations mark, are below the dashed line. However, should the excluded trajectories be considered, the result would still outperform all heuristics except BAxUS and SaasBO.

---

[2] https://baxus.papenmeier.io/.

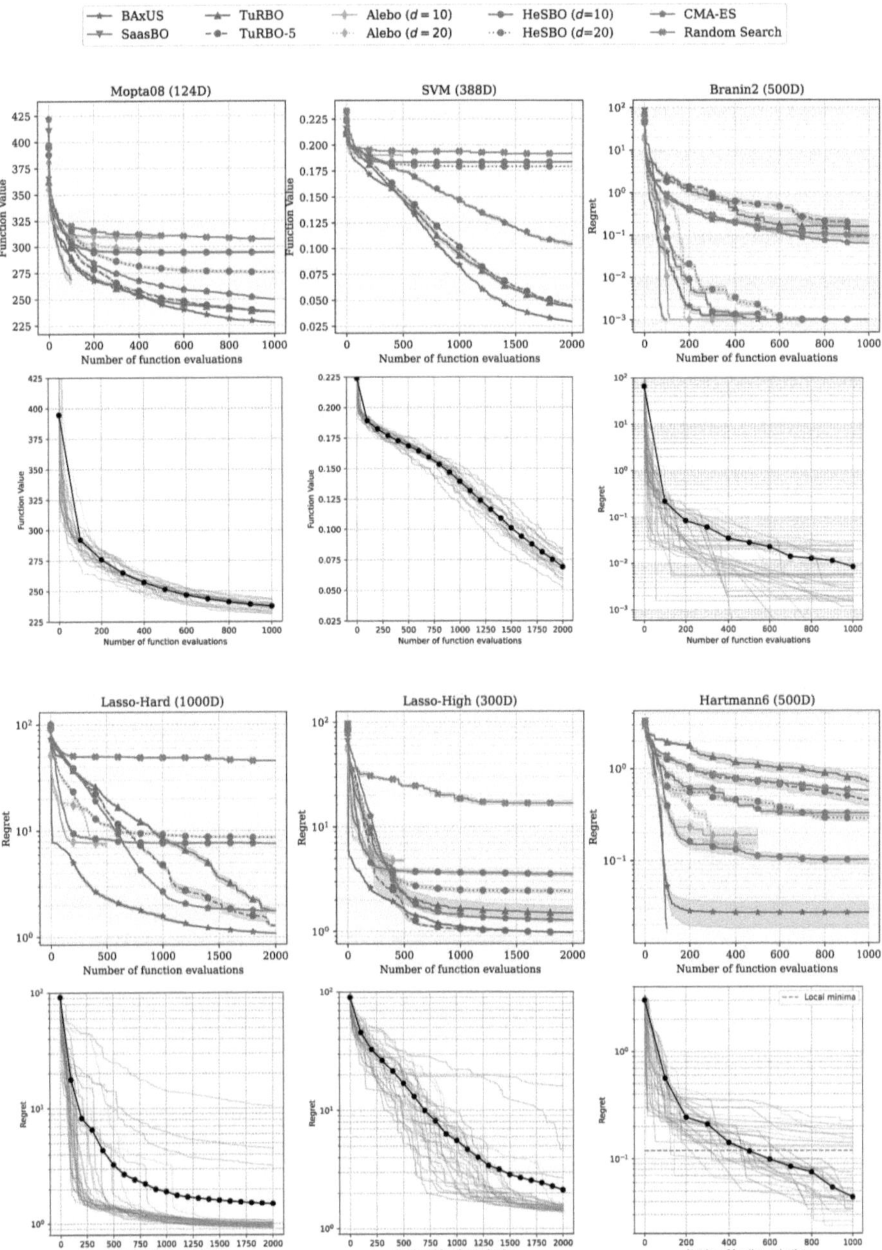

Fig. 6. Test results on high-dimensionality functions. First and third row: TuRBO, SaasBO, Alebo, HeSBO, CMA-ES, random search (reprinted from [14] with permission). Second and fourth row: results on the same functions from 30 runs of RAS, plotted on comparable scales.

4.3   Ablation Study

To motivate the main algorithmic choices in the design of RAS, we performed a series of ablation studies in which specific aspects of the algorithm were silenced in order to assess their impact on test results.

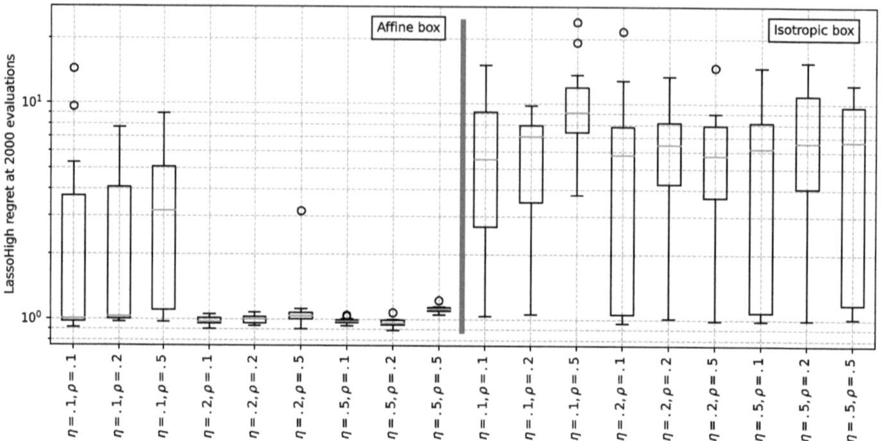

Fig. 7. Effect of the search box affine transformation (left) compared with similarly-sized isotropic transformations (right).

Isotropic Search Region. To assess the impact of the affine transformation (1) on search efficiency, we performed a series of tests in which the affine transformation (1) has been replaced with a uniform ("isotropic") resizing by the same factor: $b_i \leftarrow \rho b_i$. Figure 7 shows the results of a series of comparisons on the 300D Lasso-High function for different values of the search parameters η (initial width of the search box wrt domain size) and contraction factor ρ (the corresponding dilation factor is its reciprocal). Every boxplot collects the results of 30 tests. The results on the left-hand side of Fig. 7 have been obtained with the unmodified RAS algorithm; the corresponding tests on the right-hand side of the same plot result from the isotropic version.

The effect of the affine transformation is very significant: with a high number of dimensions, the chance of finding an improving direction tends to reduce, and the anisotropic nature of transformation (1) becomes the fundamental factor to guide the search without wasting function evaluations.

Single-Shot. Another important component of the RAS search algorithm is the "double-shot" strategy: whenever a step in direction Δ fails, the opposite direction $-\Delta$ is tested. The rationale of this strategy is that a reasonably smooth function is locally approximable with a linear function, for which the double shot

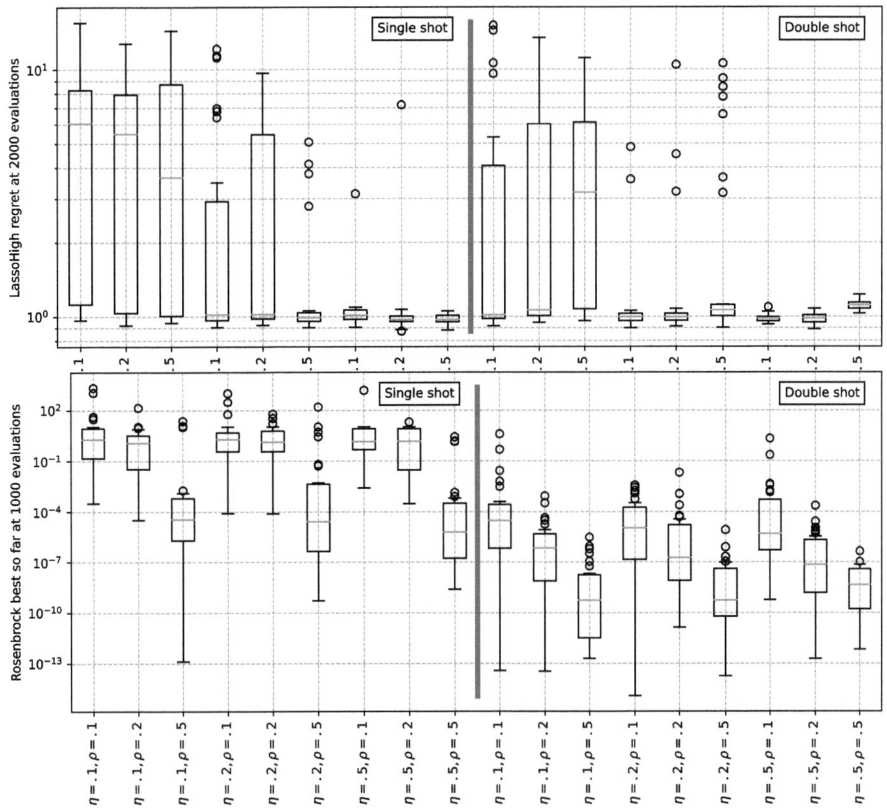

Fig. 8. Effect of the double-shot strategy (top right) compared with a single-shot step (top left) on a high-dimensionality function. For reference, the bottom chart reports the same effect on the low-dimensionality Rosenbrock 2D function.

always works. However, it is expected that in high dimensionality functions, this advantage is reduced: because most randomly-generated search directions tend to be almost normal to the gradient, linear approximations only work in very small neighborhoods, and a second function evaluation before recalibrating the search box might be a waste of time.

In Fig. 8, we compare the full RAS algorithm (left-hand side of the plots) with a version where lines 1–1 of Fig. 1 are removed (right-hand sides) for various parameter values and two different settings: the 300-dimensional Lasso-High function (top) and the 2D Rosenbrock function (bottom). While the advantage of the double-shot strategy is clear in the 2D experiment, its impact in many dimensions is quite interesting: although the results slightly worsen in some cases, its adoption seems to increase the algorithm's robustness concerning parameter changes; see in particular the results for $\eta = .2$, $\rho \in \{.1, .2\}$, where both the 3rd quartile and the inter-quartile range are greatly reduced in the double-shot case. These results call for further analysis on a wider parameter range and more diverse test functions.

5   Conclusion

The main conclusions of this study are:

- RAS is surprisingly effective on the benchmark of very high dimensional problems despite its extreme simplicity, and its use of only success-failure information during the run (no consideration of the detailed function values $f(\boldsymbol{x}_i)$ to build surrogate models), its stochastic local search nature (while BO is intended for global optimization).
- The results on the benchmark are comparable to those of the best BO algorithms (some more function evaluations are necessary but the difference is within a factor of two in most cases)
- RAS is superior to many alternative techniques including popular Genetic and Evolutionary Algorithms like CMA-ES [10], although based on a very simple scheme (affine transformations of a search box and uniform probability for generating the next point)
- The ablation study confirms the extreme relevance of the affine transformation. Instead of subspaces, a search box that is elongated and compressed with an affine transformation along successful or failure directions seems sufficient to reach interesting performance levels.
- Most of the problems in the considered benchmark do not seem to require global search methods (with the exception of the Hartmann6 function).

This preliminary investigation opens interesting questions about whether and when complex surrogate models are critical for high-dimensional black-box function optimization. We plan to extend this research through the integration of BO models on the box produced by RAS and to understand which (learnable) characteristics of the functions are appropriate for the selection, configuration and tuning of different methods in the Intelligent Optimization paradigm.

At the same time, as already explored in low dimensions for the basic RAS scheme (like for the Repeated RAS of [5]), we plan to investigate parallel search streams to deal with multi-modal functions with more local optima, with a possible multi-armed bandit allocation strategy with Bayesian modeling of the potential of different starting points for initializing the local search. A modification of RAS to deal with global optimization called M-RAS via multiple parallel runs and bandit-like allocation has been presented in [5]. M-RAS is an extension of RAS in which promising starting points for local search trails are suggested online by using Bayesian Locally Weighted Regression.

The extension to combinatorial and mixed spaces (like the Bounce algorithm of [15]) is also on the stack.

References

1. Battiti, R., Brunato, M., Mascia, F.: Reactive Search and Intelligent Optimization, Operations research/Computer Science Interfaces, vol. 45. Springer, New York (2008). https://doi.org/10.1007/978-0-387-09624-7
2. Bergstra, J., Bengio, Y.: Random search for hyper-parameter optimization. J. Mach. Learn. Res. **13**(2) (2012)
3. Brunato, M., Battiti, R.: RASH: a self-adaptive random search method. In: Cotta, C., Sevaux, M., Sörensen, K. (eds.) Adaptive and Multilevel Metaheuristics. Studies in Computational Intelligence, vol. 136. Springer (2008)
4. Brunato, M., Battiti, R.: The reactive affine shaker: a building block for minimizing functions of continuous variables. Technical report. DIT-06-012, Università di Trento (2006)
5. Brunato, M., Battiti, R., Pasupuleti, S.: A memory-based RASH optimizer. In: Felner, A., Holte, R., Geffner, H. (eds.) Proceedings of AAAI-06 workshop on Heuristic Search, Memory Based Heuristics and Their Applications, Boston, Mass., pp. 45–51 (2006). iSBN 978-1-57735-290-7
6. Eriksson, D., Jankowiak, M.: High-dimensional bayesian optimization with sparse axis-aligned subspaces. In: Uncertainty in Artificial Intelligence, pp. 493–503. PMLR (2021)
7. Eriksson, D., Pearce, M., Gardner, J., Turner, R.D., Poloczek, M.: Scalable global optimization via local bayesian optimization. Adv. Neural Inf. Process. Syst. **32** (2019)
8. Frazier, P.I.: A tutorial on Bayesian optimization. arXiv preprint arXiv:1807.02811 (2018)
9. Hains, D.R., Whitley, L.D., Howe, A.E.: Revisiting the big valley search space structure in the TSP. J. Oper. Res. Soc. **62**(2), 305–312 (2011)
10. Hansen, N., Ostermeier, A.: Adapting arbitrary normal mutation distributions in evolution strategies: the covariance matrix adaptation. In: Proceedings of IEEE International Conference on Evolutionary Computation, pp. 312–317. IEEE (1996)
11. Letham, B., Calandra, R., Rai, A., Bakshy, E.: Re-examining linear embeddings for high-dimensional Bayesian optimization. Adv. Neural. Inf. Process. Syst. **33**, 1546–1558 (2020)
12. Nayebi, A., Munteanu, A., Poloczek, M.: A framework for Bayesian optimization in embedded subspaces. In: International Conference on Machine Learning, pp. 4752–4761. PMLR (2019)
13. Nelder, J.A., Mead, R.: A simplex method for function minimization. Comput. J. **7**(4), 308–313 (1965)
14. Papenmeier, L., Nardi, L., Poloczek, M.: Increasing the scope as you learn: adaptive Bayesian optimization in nested subspaces. Adv. Neural. Inf. Process. Syst. **35**, 11586–11601 (2022)
15. Papenmeier, L., Nardi, L., Poloczek, M.: Bounce: reliable high-dimensional Bayesian optimization for combinatorial and mixed spaces. Adv. Neural. Inf. Process. Syst. **36**, 1764–1793 (2023)
16. Rosenbrock, H.: An automatic method for finding the greatest or least value of a function. Comput. J. **3**(3), 175–184 (1960)
17. Šehić, K., Gramfort, A., Salmon, J., Nardi, L.: LassoBench: a high-dimensional hyperparameter optimization benchmark suite for Lasso. In: First International Conference on Automated Machine Learning, vol. 188, pp. 2/1–24. PMLR (2022)

18. Solis, F.J., Wets, R.J.B.: Minimization by random search techniques. Math. Oper. Res. **6**(1), 19–30 (1981)
19. Wang, Z., Hutter, F., Zoghi, M., Matheson, D., De Feitas, N.: Bayesian optimization in a billion dimensions via random embeddings. J. Artif. Intell. Res. **55**, 361–387 (2016)
20. Yuan, Y.X.: A review of trust region algorithms for optimization. In: International Council for Industrial and Applied Mathematics, vol. 99-1, pp. 271–282 (2000)

Convex Quadratic Programming-Based Predictors: An Algorithmic Framework and a Study of Possibilities and Computational Challenges

Linnéa Gyllberg, Shudian Zhao, and Jan Kronqvist

Department of Mathematics, KTH Royal Institute of Technology, Stockholm, Sweden
jankr@kth.se

Abstract. We present a class of predictive models for forecasting time-series data, referred to as convex quadratic programming-based (CQPB) predictors. The predictions are computed from the minimizer of a convex quadratic problem, where previous observations are integrated as parameters. The remaining parameters, including constraints and objective coefficients, are trainable parameters. This work investigates the predictive capabilities of CQPB predictors and the computational challenges in their training. We analyze their properties and prove that this class of predictors includes classical autoregressive (AR) models, thus forming a generalization of AR models. The training problem is formulated as a bilevel optimization problem. To solve these training problems efficiently, we propose a two-stage heuristic algorithm based on the block coordinate descent approach. The results highlight the potential of CQPB predictors. Although training is challenging, our approach efficiently computes good solutions for moderate-size datasets.

Keywords: Time-series prediction · Inverse optimization · Bilevel optimization · Optimization-based predictive models

1 Introduction

Prediction, or forecasting, is one of the fundamental challenges in data analysis, machine learning (ML), and artificial intelligence (AI). We focus on the prediction of time-series data, where the goal is to compute a prediction, or forecast, of the next point using a set of previous observations from the time series. However, the proposed framework is not limited to time-series data. Prediction of time-series data has been an active research area for more than fifty years [25]. A wide range of methods have been proposed over the years, ranging from classical methods such as autoregressive (AR) models [3] and autoregressive integrated moving average (ARIMA) models [7] to more modern approaches based on methods, such as support vector machines (SVM) [22], random forests (RF) [8], neural networks (NN) [12], and recurrent neural networks (RNN) [16].

Y. Zhang et al. (Eds.): LION 2025, LNCS 15745, pp. 191–202, 2026.
https://doi.org/10.1007/978-3-032-09192-5_13

In this paper, we present a class of predictors that are based on a convex quadratic program (QP). The prediction is given by the minimizer of a convex QP, and we refer to these as convex quadratic programming-based models. Some of the parameters in the QP are given by previous observations in the time series, and some are trainable parameters. We show that this forms a versatile class of predictive models and that it forms a generalization of AR models for certain parameter configurations. For certain applications, the time series can originate from a sequence of optimization problems, *e.g.*, prices in energy markets [5]. Even though the time series originates from the solutions of more complex optimization problems, it seems natural to compute the prediction from an optimization problem. Compared to, *e.g.*, neural network models, the QP-based prediction model can also offer a somewhat higher degree of interpretability.

Recently, there has been research on integrating convex optimization in neural networks [1,17], *e.g.*, forming layers where the output is given by the minimizer of a convex problem. This line of work is related, but they focus on more general structures and different training procedures. Training the QP-based models can also be viewed as an inverse optimization problem [9], and there are similarities with studies (*e.g.*, [2,10,13,14]) where they learn optimization problems.

The main contribution and goals of this paper can be summarized as:

- We propose a class of optimization-based predictors where the prediction is computed by a quadratic programming problem.
- We investigate the predicting capabilities of the proposed predictors and show that this class of predictors forms a generalization of AR models.
- We investigate the computational challenges of training convex optimization-based predictors through a so-called KKT reformulation. We propose some simple symmetry-breaking constraints to improve the computational performance and a simple heuristic decomposition technique to compute good solutions with a reasonable computational burden.

The goal is not to claim that the proposed class of predictors is superior but to investigate their potential. Training the convex optimization-based predictor through a KKT reformulation is obviously going to be challenging. However, due to the remarkable progress in state-of-the-art MIP solvers, it is not clear how much data or how large the predictive model can be for the training problem to remain solvable. To the authors' best knowledge, there is not much existing in the literature about predictors of this type nor about their predictive capabilities. One of the main goals has, therefore, been to analyze the predictive capabilities of this class of predictive models to determine if further research on solving the training problems more efficiently is motivated. The short answer is yes, and in the numerical experiment section, we show that the models have good predictive capabilities, but training them is challenging.

2   Convex Optimization-Based Predictors

Here we focus on the prediction of time-series data, where the goal is to estimate $\hat{\mathbf{x}}_{t+1} \in \mathbb{R}^d$ given a set of observations $\mathbf{x_t}, \mathbf{x_{t-1}}, \ldots, \mathbf{x_{t-k}} \in \mathbb{R}^d$. We propose a somewhat unusual class of predictors where the prediction is computed from the minimizer of a convex optimization problem, whose parameters include the previous observations $\mathbf{x_t}, \mathbf{x_{t-1}}, \ldots, \mathbf{x_{t-k}}$. We are not the first to propose integrating convex optimization for prediction/estimation types of task, *e.g.*, see [2]. However, as it is a fairly recent methodology, we start by formally defining what we mean by a convex optimization-based prediction.

Definition 1. *A prediction* $\hat{\mathbf{x}}_{t+1}$ *computed by* $\hat{\mathbf{x}}_{t+1} = h(\mathbf{y}^*)$ *and*

$$
\begin{aligned}
\mathbf{y}^* = \underset{\mathbf{y} \geq \mathbf{0}}{\arg\min} \quad & f(\mathbf{y}) \\
s.t. \quad & g(\mathbf{y}, \mathbf{x_t}, \ldots, \mathbf{x_{t-k}}) = \mathbf{0},
\end{aligned}
\tag{1}
$$

where (1) *is a convex optimization problem, is regarded as a convex optimization-based prediction.*

Inference of the convex optimization-based predictor is computationally cheap if problem (1) is "well-behaved" and relatively small (convexity alone does not guarantee easy to solve). Compared to, *e.g.*, neural network models, inference can be more expensive as (1) does not have a closed-form solution. However, the main challenge is clearly in training the predictive model (1), *i.e.*, optimizing the functions f, g, and their coefficients to get an optimal prediction. Training is far from trivial as the training task results in a nonconvex optimization problem, more specifically, a bilevel optimization problem. For more details on bilevel optimization, we refer to [4,18]. We refer to problem (1) as the inner optimization problem and the task of optimizing f, g as the training problem.

For simplicity and ease of presentation, we only consider single-variate time-series prediction, *i.e.*, $d = 1$, and $\hat{x}_{t+1} \in \mathbb{R}_+$. However, the techniques can also be applied to multi-variate prediction and can consider multiple input variables by simply restructuring the inner optimization problem (1) to incorporate these features. Also, note that the data can always be shifted to make all points positive. To keep the training problem tractable, we only consider *convex quadratic programming-based (CQPB)* predictors given by

$$
CQPB(\{x_i\}_{i=t-k+1}^{t}; \mathbf{A}, \mathbf{b}, \mathbf{c}) := \hat{x}_{t+1} = h(\mathbf{y}^*),
$$

$$
\begin{aligned}
\mathbf{y}^* = \underset{\mathbf{y} \in \mathbb{R}^n}{\arg\min} \quad & \mathbf{c}^\top \mathbf{y} + \frac{1}{2} \sum_{i=2}^{n} y_i^2 + \frac{1}{2}(y_1 - x_t)^2 \\
s.t. \quad & \mathbf{A}\mathbf{y} = \begin{bmatrix} x_{t-k+1} \\ \vdots \\ x_t \\ \mathbf{b} \end{bmatrix}, \; \mathbf{y} \geq \mathbf{0},
\end{aligned}
\tag{2}
$$

where $\mathbf{c}, \mathbf{b}$, and $\mathbf{A}$ are trainable parameters with $k > 1$. Furthermore, we restrict the predictions to be given by

$$
\hat{x}_{t+1} = h(\mathbf{y}^*) = [1 \;\; 0 \ldots \;\; 0]\mathbf{y}^*.
\tag{3}
$$

These restrictions are mainly to keep the resulting training problem computationally easier, but we will show that this is still a fairly general class of predictors capable of good predictive performance.

Before we continue, we want to briefly motivate this class of predictive models, *i.e.*, the structure of inner problem (2). The quadratic terms in the objective are included to ensure a unique optimizer of the problem, which avoids additional difficulties in training the parameters. The quadratic terms also act as a regularization and favors predictions close to the last observation.

Next, we briefly discuss the versatility of the predictive model given by (2). A simple but interesting result is that we can show that it forms a generalization of AR models [3]. We formalize this property in the following proposition.

Proposition 1. *The class of predictors given by (2), where* $\mathbf{c}, \mathbf{b}$, *and* $\mathbf{A}$ *are model parameters, can be viewed as a generalization of autoregressive (AR) models.*

Proof. Consider the case with $\mathbf{b} = 0$, an arbitrary vector $\mathbf{c} \in \mathbb{R}^n$, and

$$
\mathbf{A} = \begin{bmatrix} 1 & -1 & -1 \ldots -1 \\ 0 & \omega_{n-1} & 0 & \ldots & 0 \\ \vdots & & \ddots & & \vdots \\ 0 & \ldots & & & \omega_0 \end{bmatrix}. \tag{4}
$$

The prediction given by (2) with this choice of $\mathbf{A}$ matrix, is simply $\hat{x}_{t+1} = \sum_{i=0}^{n-1} \omega_i x_{t-i}$. This shows that by choosing a certain structure of the inner problem, *i.e.*, the size of the $\mathbf{A}$ matrix, the predictor can be equivalent to an AR model with the same number of previous observations. But, when specifying the size of the $\mathbf{A}$ matrix, the training problem is not restricted to the specific $\mathbf{A}$ matrix given in (4), and can, therefore, be considered a generalization. $\qquad\square$

Understanding the model complexity is important to prevent over-fitting. The number of parameters in (2) gives some insight into the model complexity, but to get a better understanding, we need to analyze the structure of problem (2). For example, if the RHS of the equation system in (2) contains as many old data points as the dimension of $\mathbf{y}$ and $\mathbf{A}$ has full rank, then the prediction is always given by a specific linear combination of old data points. We mainly consider the case where there is at least one more variable than the number of rows in $\mathbf{A}$, where the minimization operator plays a clear role in the prediction. With more variables in problem (2), we get a higher dimensional feasible set and more degrees of freedom. For a certain number of rows in $\mathbf{A}$ and a given number of old data points, more degrees of freedom in (2) generally result in a more complex predictive model.

3   Training CQPB Predictors

Given the data $\mathbf{x} = \begin{bmatrix} x_1 \cdots x_N \end{bmatrix} \in \mathbb{R}_+^N$, the task of training the convex quadratic programming-based (CQPB) predictor, *i.e.*, determining the coefficients in $\mathbf{A}, \mathbf{b}$, and $\mathbf{c}$, can be formulated as

$$\min_{(\mathbf{A},\mathbf{b},\mathbf{c})\in\Omega} \quad \sum_{t=k}^{N-1} (x_{t+1} - \hat{x}_{t+1})^2 \tag{5}$$

$$\text{s.t. } \hat{x}_{t+1} = CQPB(x_{t-k+1},\dots,x_t;\mathbf{A},\mathbf{b},\mathbf{c}), \ \forall t = k,..,N-1,$$

where k is the number of observations used for each prediction. The set Ω is a simple box defined by upper and lower bounds of each entry of $\mathbf{A}, \mathbf{b}$ and $\mathbf{c}$. For simplicity and to keep the training easier, we set the bounds as ± 1. The training problem (5) is a so-called QP-QP bilevel problem. As the lower-level problem has a strictly convex quadratic objective function and only linear constraints, it satisfies the linear constraint qualification (LCQ) and has a unique minimizer. We can, therefore, use the standard approach of replacing the inner problems with their KKT conditions, $e.g.$, see [11, 14, 18]. The training problem can then be written as the single-level optimization problem

$$P_{train}(\mathbf{x}, k, n, n_b) := \underset{\mathbf{A},\mathbf{c},\mathbf{y}_k,\dots,\mathbf{y}_{N-1}}{\arg\min} \sum_{t=k}^{N-1} (x_{t+1} - \begin{bmatrix} 1 & 0 & \cdots & 0 \end{bmatrix} \mathbf{y}_t)^2 \tag{6a}$$

$$\begin{aligned}
\text{s.t. } -1 &\leq A_{j\ell} \leq 1, & \forall j \in [k+n_b],\ \forall \ell \in [n], & \tag{6b}\\
-1 &\leq b_j \leq 1, & \forall j \in [n_b], & \tag{6c}\\
-1 &\leq c_\ell \leq 1, & \forall \ell \in [n], & \tag{6d}\\
\mathbf{A}\mathbf{y}_t &= \bar{\mathbf{b}}_t, & \forall t \in \{k, k+1, \dots, N-1\}, & \tag{6e}\\
\mathbf{s}_t^\top \mathbf{y}_t &= 0, & \forall t \in \{k, k+1, \dots, N-1\}, & \tag{6f}\\
\mathbf{c} + \bar{\mathbf{y}}_t &= \mathbf{A}^\top \boldsymbol{\lambda}_t + \mathbf{s}_t, & \forall t \in \{k, k+1, \dots, N-1\}, & \tag{6g}\\
\mathbf{y}_t, \mathbf{s}_t &\in \mathbb{R}_+^n, \boldsymbol{\lambda}_t \in \mathbb{R}^{k+n_b}, & \forall t \in \{k, k+1, \dots, N-1\}, & \tag{6h}\\
A_{1,2} &\geq \cdots \geq A_{1,n}, & & \tag{6i}
\end{aligned}$$

where $\bar{\mathbf{b}}_t^\top = \begin{bmatrix} x_{t-k+1} & \cdots & x_t & \mathbf{b}^\top \end{bmatrix}$ and $\bar{\mathbf{y}}_t = \mathbf{y}_t - \begin{bmatrix} x_t \\ \mathbf{0}_{n-1} \end{bmatrix}$. To clarify the notation, $[n] = \{1, 2, \dots, n\}$ and $\mathbf{0}_{n-1}$ is a vector with $n-1$ zeroes. Moreover, since we only care for trainable parameters and the prediction for the training dataset, we denote $(\mathbf{A}^*, \mathbf{c}^*, \hat{\mathbf{x}}) = P_{train}(\mathbf{x}, k, n, n_b)$.

The constraints (6b)–(6d) enforces our bound restrictions for the model parameters. Constraints (6e)–(6h) originate from the KKT conditions of each inner optimization problem. The last group of constraints (6i) are so-called symmetry-breaking constraints and are included to prevent multiple equivalent solutions. Without these constraints, we could reorder the variables $y_{t,2}, y_{t,3}, \dots, y_{t,n}$ and the corresponding rows of $\mathbf{A}$ to get technically different solutions, but resulting in the same predictions. Eliminating such symmetries typically results in an easier problem.

Problem (6) is challenging and contains both complementarity constraints (6f) and nonconvex constraints with bilinear terms (6e) and (6g). The number of nonconvex terms increases with the size of the inner problem (2). A copy of all these constraints is also generated by each data point used for training. The number of data points N, thus, strongly affects the tractability of problem (6). We do not detail how such problems can be solved, but they can be handed directly by MIP solvers such as Gurobi [15] and SCIP [6]. However, directly solving problem (6) is intractable even with relatively little data and small CQPB models. To handle problems of meaningful size and get a good, but potentially suboptimal solution, we use a two-stage strategy to solve the problem. First, we determine an initial solution by solving (6) with a small subset of the data. Then we fix all but one row of the $\mathbf{A}$ matrix and solve problem (6) with most elements of $\mathbf{A}$ fixed. Such a variables-fixing approach eliminates most of the bilinear terms. We then iterate by unlocking another row of $\mathbf{A}$ and fix all other rows in a block coordinate descent-type fashion. Algorithm 1 presents the complete procedure.

Algorithm 1. The two-stage alternating optimization-based training algorithm

---

Input: A time-series $\mathbf{x}$, a consecutive subset $\mathbf{x}_{init}$, the window size k, model complexity parameters n and n_b, the tolerance $\varepsilon \geq 0$, and time limit T_{max}.

Output: $\hat{\mathbf{A}}$, $\hat{\mathbf{c}}$, and $\hat{\mathbf{x}}$.

1: **procedure** INITIALIZATION STAGE
2:     $(\mathbf{A}^0, \mathbf{c}^0, \hat{\mathbf{x}}_{init}) \leftarrow P_{train}(\mathbf{x}_{init}, k, n, n_b);$ $\triangleright$ Solve within T_{max}
3: **end procedure**
4: **procedure** ALTERNATING STAGE
5:     $j \leftarrow 1, \Delta \leftarrow +\infty, \text{OBJ}^0 \leftarrow +\infty;$
6:     **while** $\Delta \geq \epsilon$ **do**
7:         **for** $i = 1, \ldots, k + n_b$ **do** $\triangleright$ Fix all but the value of $\mathbf{A}_i.$ with $\mathbf{A}^{j-1}$
8:             $(\mathbf{A}^j, \mathbf{c}^j, \hat{\mathbf{x}}) \leftarrow P_{train, \mathbf{A}_{s.} = \mathbf{A}_{s.}^{j-1}, \forall s \neq i}(\mathbf{x}, k, n, n_b) ;$ $\triangleright$ Solve within T_{max}
9:             $\text{OBJ}^j \leftarrow \sum_{t=k}^{N}(x_{t+1} - \hat{x}_{t+1})^2 ;$
10:            $\Delta \leftarrow \text{OBJ}^{j-1} - \text{OBJ}^j ;$
11:            $j \leftarrow j + 1;$ $\triangleright$ Count iterations
12:        **end for**
13:    **end while**
14:    $(\hat{\mathbf{A}}, \hat{\mathbf{c}}, \hat{\mathbf{x}}) \leftarrow (\mathbf{A}^j, \mathbf{c}^j, \hat{\mathbf{x}}^j) .$
15: **end procedure**

One can also initiate Algorithm 1 by forming an initial $\mathbf{A}$ matrix and $\mathbf{b}$ vector based on a trained AR model. The $\mathbf{A}$ matrix and $\mathbf{b}$ vector are then formed from the AR predictor similarly as in the proof of Proposition 1, and potentially filled out with zeroes if the QCPB predictor is of higher dimensionality. This provides a starting point with the same RMSE as the AR predictor on the training set.

4 Numerical Experiments

The implementation of data preprocessing, proposed methods, and other base-line approaches are carried out in Python. The optimization problems solved in Algorithm 1 are solved by Gurobi 11.0.3 [15], and ARMA and RNNs, as baseline approaches, are implemented and trained by `Statsmodels` [23] and `PyTorch` [21], respectively. All the experiments were run on a laptop with an Intel(R) Core(TM) i7-7500U CPU @ 2.70 GHz processor and 16 GB of RAM. To keep the training problems more manageable, we have forced bounds of ± 5 on dual variables (*i.e.*, λ and s) for unbounded directions in problem (6). It is common to assume bounds on the dual variables when using the KKT reformulation, but one should be careful, as too restrictive bounds may lead to infeasibility. A more rigorous approach should be used to determine the bounds. However, here the goal has only been to investigate the potential of CQPB predictors.

In the initialization stage of Algorithm 1, we have used as many data points as possible, while keeping problem (6) solvable within a minute. The number of data points in the initialization varied between 5 and 13. We initially set a time limit T_{max} of 30 s for the subproblems, and we set T_{max} to 60 s when Gurobi could not find an initial solution in 30 s. To keep the problems simpler, we set $n = k + 1$ and $n_b = 0$, *i.e.*, one more variable in problem (2) than the number of data points used for prediction and the number of constraints. Preliminary results with $n = k + 2$ did not show a clear difference in the models' predicting performance. For all experiments, we used $\varepsilon = 1e - 8$.

4.1   Data Sets

We have considered the following data sets that vary in prediction difficulty:

1. **Nordpool** [20]: This dataset contains system prices for the Nordic energy market collected and published by the Nord Pool. We choose a consecutive sequence of daily prices measured at midnight from 2022-01-01 and ending with a given length. All the prices are scaled with a factor of 0.01.
2. **Global Temperature** [19]: This data set contains changes in the annual average global surface temperature, from which we use the data between 1909 and 2023. We specifically consider the yearly mean's locally weighted smoothing (lowess). The data points are shifted upward by the absolute value of the minimum value to make all values nonnegative.
3. **Sunspots** [26]: This dataset contains the monthly mean sunspot number from 1749-01-01 to 2017-08-31. The data are scaled by a factor of 0.01.

4.2   Reference Models

To evaluate the predicting performance, we compared the following commonly used predictors as baselines:

– Autoregressive models (AR): Autoregressive models are created with the idea that the current value x_t can be explained as a function of the k past values $x_{t-1}, x_{t-2}, \ldots, x_{t-k}$. An autoregressive model of order k, abbreviated as AR(k), is defined as $x_t = \phi_1 x_{t-1} + \phi_2 x_{t-2} + \ldots + \phi_k x_{t-k} + w_t$, where $\phi_1, \phi_2, \ldots, \phi_k$ are constants and w_t is white noise with mean zero and variance σ_w^2. As AR models require processes to be (weakly) stationary, the non-stationary time-series **Global Temperature** and **Sunspots** have been made stationary through first-order differencing [25].

- Recurrent neural networks (RNN): A simple RNN predicts the output at each time step depending on the current input and the hidden state from the previous time steps. They excel in simple tasks with short-term dependencies due to the ability to exploit the influence of previous inputs, such as time-series data [16,24]. We chose small RNNs with one hidden layer of size 32 on each recurrent cell with a varying window size $m = \{2,4,6\}$. All the instances are trained with a batch size of 2 and 100 epochs, and the datasets are normalized in the preprocessing.

5   Results

First, we investigate what size of problem (6) is directly solvable with Gurobi. We tested two sizes for the CQPB model on the data set **Global Temperature**. We found that problem (6) could not be solved by Gurobi for either model under 10 min when using more than 6 and 8 data points for $k = 2$ and $k = 4$, respectively. This clearly shows the difficulty of directly optimizing (6).

To illustrate the simple training algorithm, we plot how the solution improves over iterations with a CQPB model with parameters $N = 50$, $N_{init} = 12$, and $k = 6$ for the data set **Nordpool** in Fig. 1. Here Algorithm 1 ran 12 iterations taking 367 s (with $\varepsilon = 1e - 8$). Algorithm 1 is a simple heuristic, but based on our experiments it is effective at producing good solutions relatively fast.

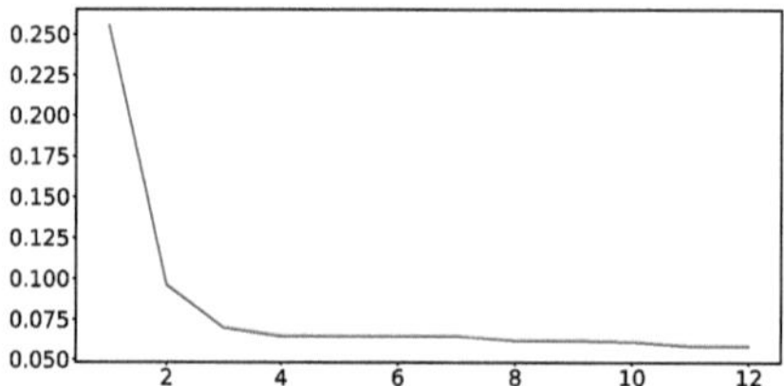

Fig. 1. The iterations of Algorithm 1 plotted against the MSE for the **Nordpool** data with $N = 50$ and $k = 6$.

5.1   Comparison with Other Predictive Models

Table 1 presents the mean square error (MSE) for CQPB, AR(k), and RNN trained with the datasets and various parameters as well as the prediction mean square error (MSE$_p$) for the P one-step ahead predictions outside of the training data by the models. We have also included results with Algorithm 1 initiated with the AR solution, and this is refered to as QCPB-AR. Table 1 shows that CQPB overall performs similarly to both AR(k) and RNN. When it performs worse, it is not by a large margin and sometimes it is the best-performing model.

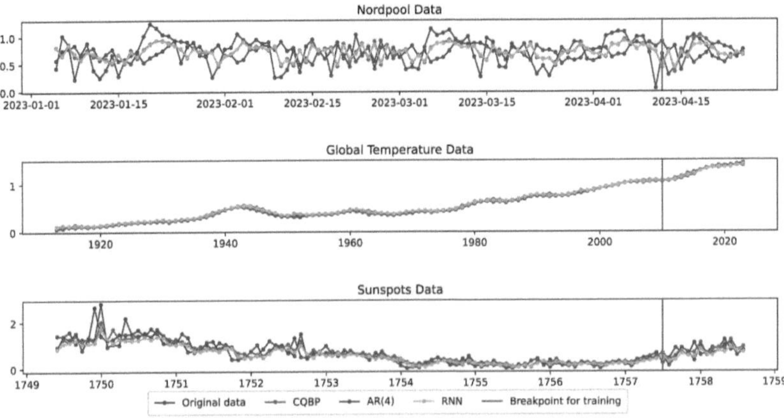

Fig. 2. Comparison of CQPB, AR(k), and RNN with $N = 100$, $P = 15$, and $k = 4$. The last 15 points show one-step predictions beyond the training data.

Figure 2 presents three plots comparing results from different models for each dataset separately, with $N = 100$, $N_{init} = 8$ (**Nordpool** and **Global Temperature**) or $N_{init} = 9$ (**Sunspots**), and $k = 4$ All three models show similar performance.

Whilst the performance of the CQPB predictor shows promise, the difficult training is clearly a drawback. For both AR and RNN, the training of the models is practically instantaneous, but the same cannot be said for the CQPB. The training times varied greatly and were highly dependent on the number of iterations to converge, with the fastest model taking just over a minute (two iterations with a 30 s limit each) and the slowest a bit over 15 min (15 iterations with a 60 s time limit). Most models took between 2 and 6 min to train.

From the experiments, it was clear that using the AR solution as an initialization was favorable. Otherwise the initialization in Algorithm 1 was sensitive with regards to the parameters, e.g., the number of data points used in the initialization. A more rigorous approach for determining bounds on the dual variables should also be investigated. Finally, we also want to mention that the performance of all models can most likely be improved by parameter tuning.

Table 1. The performance of predictors trained with various parameters for the different data sets, given as mean squared error (MSE). MSE_p is the mean squared error for P one-step ahead predictions beyond the training data. For CQPB-AR, Algorithm 1 is initiated with the AR solution.

NORDPOOL DATA											
$N = 50, P = 10$						$N = 100, P = 15$					
$k = 2$		$k = 4$		$k = 6$		$k = 2$		$k = 4$		$k = 6$	
MSE	MSE_p	MSE	MSE_p	MSE	MSE_p	MSE	MSE_p	MSE	MSE_p	MSE	MSE_p
CQPB	0.0649	0.0709	0.0608	0.0733	0.0582	0.0581	0.0571	0.0728	0.0548	0.0642	0.0549	0.0529
CQPB-AR	0.0630	0.0644	0.0592	0.0772	0.0533	0.0736	0.0560	0.0573	0.0533	0.0642	0.0508	0.0649
AR(k)	0.0649	0.0534	0.0608	0.0572	0.0521	0.0419	0.0570	0.0581	0.0544	0.0614	0.0503	0.0529
RNN	0.0640	0.0631	0.0604	0.0604	0.0540	0.0507	0.0608	0.0572	0.0592	0.0553	0.0543	0.0507

GLOBAL TEMPERATURE DATA ($\times 10^{-2}$)											
$N = 50, P = 10$						$N = 100, P = 15$					
$k = 2$		$k = 4$		$k = 6$		$k = 2$		$k = 4$		$k = 6$	
MSE	MSE_p	MSE	MSE_p	MSE	MSE_p	MSE	MSE_p	MSE	MSE_p	MSE	MSE_p
CQPB	0.0326	0.0316	0.0113	0.0236	0.0274	0.0387	0.0140	0.0216	0.0264	0.0350	0.0262	0.0227
CQPB-AR	0.0126	0.0833	0.0112	0.1525	0.0094	0.0318	0.0319	0.0285	0.0115	0.0752	0.0324	0.0268
AR(k)	0.0142	0.0192	0.0115	0.0164	0.0097	0.0173	0.0135	0.0138	0.0116	0.0146	0.0107	0.0161
RNN	0.0642	0.0617	0.0273	0.0263	0.0233	0.0224	0.0763	0.0757	0.0419	0.0396	0.0308	0.0355

SUNSPOTS DATA											
$N = 50, P = 10$						$N = 100, P = 15$					
$k = 2$		$k = 4$		$k = 6$		$k = 2$		$k = 4$		$k = 6$	
MSE	MSE_p	MSE	MSE_p	MSE	MSE_p	MSE	MSE_p	MSE	MSE_p	MSE	MSE_p
CQPB	0.1109	0.0616	0.1033	0.0431	0.1021	0.0405	0.0679	0.0953	0.0681	0.0796	0.1168	0.1043
CQPB-AR	0.1128	0.0614	0.1012	0.0425	0.1013	0.0507	0.0665	0.0944	0.0595	0.0899	0.0584	0.0895
AR(k)	0.1155	0.0212	0.1061	0.0251	0.1033	0.0261	0.0653	0.0641	0.0610	0.0718	0.0591	0.0766
RNN	0.1071	0.0999	0.0990	0.0931	0.0947	0.0911	0.0806	0.0818	0.0703	0.0711	0.0657	0.0682

6 Conclusions

Overall, the CQPB predictor shows good capabilities for forecasting time-series data. Our simple heuristic training approach manages to compute good solutions, with predicting performance comparable to well-established methods for moderate-size datasets. However, the training process is still computationally challenging. Further research on CQPB predictors is, in our opinion, clearly motivated and needed to make them practically applicable.

Acknowledgments. The authors gratefully acknowledge financial support from a grant by Göran Gustafsson Stiftelser to Jan Kronqvist, and support from Digital Futures at KTH.

Disclosure of Interests. The authors have no conflicts of interest.

References

1. Agrawal, A., Amos, B., Barratt, S., Boyd, S., Diamond, S., Kolter, J.Z.: Differentiable convex optimization layers. Adv. Neural Inf. Process. Syst. **32** (2019)
2. Agrawal, A., Barratt, S., Boyd, S.: Learning convex optimization models. IEEE/CAA J. Automatica Sinica **8**(8), 1355–1364 (2021)
3. Akaike, H.: Fitting autoregreesive models for prediction. In: Selected Papers of Hirotugu Akaike, pp. 131–135. Springer (1969)
4. Bard, J.F.: Practical Bilevel Optimization: Algorithms and Applications, vol. 30. Springer, New York (2013). https://doi.org/10.1007/978-1-4757-2836-1
5. Biggar, D.R., Hesamzadeh, M.R.: The Economics of Electricity Markets. Wiley, New York (2014)
6. Bolusani, S., et al.: The SCIP optimization suite 9.0. arXiv preprint arXiv:2402.17702 (2024)
7. Box, G.E., Jenkins, G.M., Reinsel, G.C., Ljung, G.M.: Time Series Analysis: Forecasting and Control. Wiley, Hoboken (2015)
8. Breiman, L.: Random forests. Machine Learn. **45**, 5–32 (2001)
9. Chan, T.C.Y., Mahmood, R., Zhu, I.Y.: Inverse optimization: theory and applications. Oper. Res. (2023)
10. Chan, T.C., Kaw, N.: Inverse optimization for the recovery of constraint parameters. Eur. J. Oper. Res. **282**(2), 415–427 (2020)
11. Dempe, S., Dutta, J.: Is bilevel programming a special case of a mathematical program with complementarity constraints? Math. Program. **131** (2012)
12. Frank, R.J., Davey, N., Hunt, S.P.: Time series prediction and neural networks. J. Intell. Rob. Syst. **31**, 91–103 (2001)
13. Gupta, R., Zhang, Q.: Decomposition and adaptive sampling for data-driven inverse linear optimization. INFORMS J. Comput. **34**(5) (2022)
14. Gupta, R., Zhang, Q.: Efficient learning of decision-making models: a penalty block coordinate descent algorithm for data-driven inverse optimization. Comput. Chem. Eng. **170**, 108123– (2023)
15. Gurobi Optimization, LLC: Gurobi Optimizer Reference Manual (2024). https://www.gurobi.com
16. Hewamalage, H.: Recurrent neural networks for time series forecasting: current status and future directions. Int. J. Forecast. **37** (2020)
17. Katyal, C.: Differentiable convex optimization layers in neural architectures: Foundations and perspectives (2024). arXiv preprint
18. Kleinert, T., Labbé, M., Ljubić, I., Schmidt, M.: A survey on mixed-integer programming techniques in bilevel optimization. EURO J. Comput. Optim. **9**, 100007 (2021)
19. NASA: Global surface temperature (2024). https://climate.nasa.gov/vital-signs/global-temperature/?intent=111
20. Nordpool: Day-ahead prices (2025). https://data.nordpoolgroup.com/auction/day-ahead/prices
21. Paszke, A., et al.: Automatic differentiation in PyTorch. In: NIPS-W (2017)
22. Sapankevych, N.I., Sankar, R.: Time series prediction using support vector machines: a survey. IEEE Comput. Intell. Mag. **4**(2), 24–38 (2009)
23. Seabold, S., Perktold, J.: Statsmodels: econometric and statistical modeling with python. In: 9th Python in Science Conference (2010)
24. Sherstinsky, A.: Fundamentals of recurrent neural network (RNN) and long short-term memory (LSTM) network. Phys. D Nonlinear Phenomena **404** (2020)

25. Shumway, R.H., Stoffer, D.S.: Time Series Analysis and Its Applications. Springer Texts in Statistics, 1st edn. Springer, New York (2000)
26. SIDC: Sunspots (2021). https://www.kaggle.com/datasets/robervalt/sunspots/data

Studies on a Bayesian Optimization Based Approach to Tune Hyperparameters of Matheuristics

Sophie Hildebrandt[1,2(✉)] [iD], Sina Nunes[2] [iD], Meik Franke[1] [iD], and Guido Sand[2] [iD]

[1] University of Twente, Drienerlolaan 5, 7522 NB Enschede, Netherlands
`sophie.hildebrandt@hs-pforzheim.de`
[2] University of Applied Science Pforzheim, Tiefenbronner Str. 65, 75175 Pforzheim, Germany

Abstract. Mixed-integer programming can handle optimization problems with complex constraints, but its computational cost often suffers from the combinatorial complexity of the problem. Decomposition-based matheuristics address this issue by splitting large-scale mixed-integer programs (MIPs) into smaller subproblems. Matheuristics typically exhibit hyperparameters that may affect their performance. An analysis of related work reveals that the optimization potential of hyperparameters is often left unexploited, leading to both inferior MIP-solutions and unnecessarily high computational costs.

This paper studies a novel algorithmic approach to tune hyperparameters of matheuristics by Bayesian optimization. Fundamental properties of the algorithmic approach are examined by computational experiments with small- and large-scale instances of the use case. The results exhibit two natural and competing objectives of the tuning problem: optimizing the MIP-objective and the computational cost. While the two objectives can be optimized separately for small-scale instances, they need to be handled jointly for large-scale instances.

In future research the multi-objective aspect of the hyperparameter tuning problem will be examined more deeply, and the single-instance approach will be extended to multiple instances.

Keywords: Mixed-integer programming · matheuristics · hyperparameter tuning · Bayesian optimization

1 Introduction

Mixed-integer programs (MIPs) can be applied to solve real-world optimization problems with complex constraints like scheduling, routing or packing. An MIP comprises four components: Firstly, the degrees of freedom which are reflected by integer- or real-valued variables and whose values are determined during the solution process. Secondly, the algebraic objective function defines how the objective value, that should be minimized or maximized, depends on the variables. Thirdly, the algebraic equality and inequality constraints which restrict the feasible values of the variables. Fourthly, the

Y. Zhang et al. (Eds.): LION 2025, LNCS 15745, pp. 203–217, 2026.
https://doi.org/10.1007/978-3-032-09192-5_14

parameters which define instances of the MIP. Several general-purpose solvers for MIPs exist: CPLEX[1], Gurobi[2] and XPRESS[3] are among the most prominent ones.

Lots of real-world optimization problems are NP-complete such that the solution of large-scale instances is prohibited by the exploding computational cost. Heuristic decomposition schemes aim at reducing the computational cost while losing only a little optimization potential of the MIP. They split a large-scale monolithic MIP into a polylithic MIP comprising several smaller sub-MIPs. However, while solutions of monolithic MIPs provide optimality certificates (proof of optimality or a conservative optimality gap), solutions of polylithic MIPs, that are based on heuristic decomposition schemes, do not. Heuristic decomposition-schemes for MIPs are called matheuristics and typically comprise hyperparameters which steer the decomposition process.

The hyperparameters may affect the performance of matheuristics (see e.g. [1]) such that they should be optimized. However, their optimization potential is often left unexploited, leading to both inferior MIP-solutions and unnecessarily high computational costs. In contrast to the underlying MIP, the hyperparameter tuning is a black-box optimization problem for which no algebraic formulation is known. The authors propose to optimize the hyperparameters of matheuristics using Bayesian optimization (BO). In this paper such an approach is studied based on a particular use case.

The remainder of the paper is structured as follows: In Sect. 2, related work on hyperparameter tuning of matheuristics and the motivation for choosing BO is discussed. In Sect. 3, the BO-based approach for tuning matheuristics detailing the algorithmic framework is presented. In Sect. 4, the MIP-problem and the decomposition-based matheuristic used as case study are described along with the considered hyperparameters. In Sect. 5 computational experiments with small-scale MIP-instances and two separate objectives of the tuning problem are studied. In Sect. 6 computational experiments for large-scale MIP-instances are studied, whereby in contrast to Sect. 5 the two objectives are handled jointly. In Sect. 7, the conclusions from the studies are summarized and future research directions are outlined.

2 Related Work

Hyperparameter tuning and algorithm selection is widely and successfully applied to various machine learning models, including neural networks, support vector machines, decision trees and random forests as well as gradient boosting machines. Numerous studies have demonstrated that proper hyperparameter tuning can enhance machine learning model accuracy, generalization and performance, as well as prevent overfitting (see e.g. [2, 3]).

Common techniques for hyperparameter tuning for machine learning models include grid search and random search, as well as population-based optimization methods such as evolutionary algorithms. More recent approaches such as successive halving, hyperband, and BO have gained traction for their ability to allocate computational resources more efficiently (see e.g. [4–8]).

---

[1] https://www.ibm.com/products/ilog-cplex-optimization-studio/cplex-optimizer.

[2] https://www.gurobi.com/

[3] https://www.fico.com/en/products/fico-xpress-optimization.

Beyond traditional machine learning models, hyperparameter tuning and algorithm selection techniques have also been applied to mathematical programming. At a high level, learning-based approaches can assist in selecting among different algorithmic strategies: For instance, Degroote et al. [9] use machine learning to choose between two tabu search variants and mixed-integer programming for solving the generalized assignment problem. Once an algorithmic strategy is chosen, further decisions arise, for example, whether to apply decomposition schemes or not. For example, Kruber et al. [10] use machine learning to decide if Dantzig-Wolfe decomposition should be applied or not; Basse et al. [11] demonstrate how to tune the pattern used for the Dantzig-Wolfe decomposition. Mitrai and Daoutidis [12] also demonstrated how machine learning can support the decision whether to apply decomposition schemes or not. In a subsequent work, they proposed a hyperparameter tuning framework specifically for Benders decomposition, a general and exact method for solving MIPs [13]. The sub-MIPs resulting from the decomposition can be solved using standard MIP solvers. Extensive research has explored tuning strategies for standard MIP solvers (e.g. [14–17]). Going even deeper into solver internals, machine learning has also been leveraged to guide key components of branch-and-bound algorithms, such as selecting branching variables, nodes, or cutting planes, as surveyed comprehensivcly by Bengio et al. [18].

While hyperparameter tuning is well-established for machine learning models and metaheuristics and is also applied to rigorous mathematical programming in various ways, there is no research on its use in matheuristics known. A key challenge in matheuristic tuning is the high computational cost associated with evaluating each hyperparameter configuration, as it involves solving one or more MIPs. Consequently, exhaustive methods such as grid search, random search, or even population-based approaches like evolutionary algorithms or particle swarm optimization often become impractical due to their high evaluation budgets and low sample efficiency [4]. To address this issue, more efficient techniques such as successive halving, hyperband, and BO have been proposed [4]. Both bandit-based methods (like successive halving and hyperband) and BO-based methods are promising approaches for the hyperparameter tuning of matheuristics. In this work, BO is applied to tune the hyperparameters of matheuristics.

3 Proposed Approach

A novel approach based on BO is proposed for the hyperparameter tuning of matheuristics. Figure 1 visualizes the building blocks of the algorithmic architecture and the information exchanged between them. On the one hand, BO is a well-established method for hyperparameter tuning, particularly for expensive-to-evaluate objective functions. On the other hand, the use of matheuristics that decompose a monolithic MIP into a polylithic MIP, and the tuning of standard MIP solvers have both been widely studied. To the knowledge of the authors, studies on a BO-based approach to tune hyperparameters of matheuristics are not described in literature yet.

In each iteration, the BO proposes new hyperparameter values for both the matheuristic and the MIP standard solver. With the decomposition scheme of the parameterized matheuristic and the model formulation of the monolithic MIP a polylithic MIP is generated. The polylithic MIP consists of several smaller sub-MIPs which are solved by

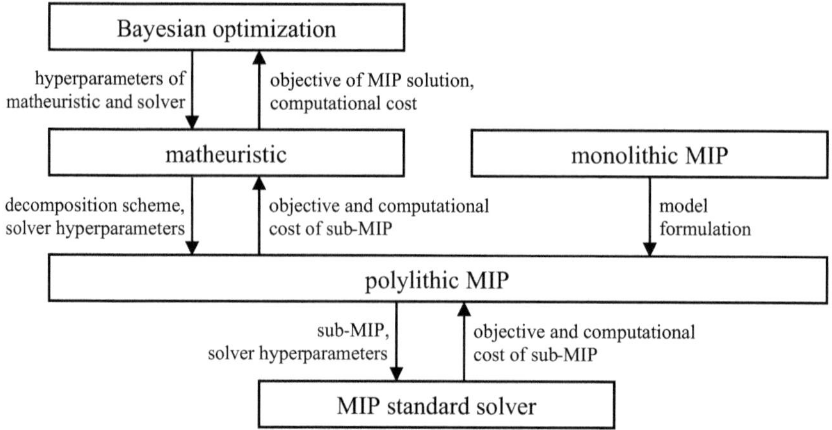

Fig. 1. Algorithmic architecture for the hyperparameter tuning of matheuristics

a MIP standard solver. The MIP standard solver computes a solution for the sub-MIP using the solver hyperparameters given. The objective value of the sub-MIP and the computational cost for solving the sub-MIP are returned to the matheuristic. After all sub-MIPs are solved, the matheuristic returns the objective value of the polylithic MIP solution and the total computational cost used to solve the polylithic MIP to the BO. The BO updates the surrogate model and calculates new promising hyperparameter values using an acquisition function. This is done until a termination criterion for the BO, such as the number of iterations, a time limit or the number of retries to propose new hyperparameter values is met.

The algorithm shown in Fig. 1 is implemented using the python package SMAC3 [20] for the BO (with a gaussian process as surrogate model and the expected improvement as acquisition function), the python package Pyomo[4] is used to implement the MIP and CPLEX[5] is used as MIP standard solver.

4 Case Study

4.1 Hoist Scheduling Problem

The BO-based approach is studied using a single hoist scheduling problem (HSP) described by Aguirre et al. [1] as a use case. HSPs appear in the production scheduling of electroplating plants, see Fig. 2.

These plants consist of different tanks which are arranged in one line and contain different liquids. The products are filled into perforated barrels and immersed in the liquids for a limited time (see Fig. 2 for the lower and upper processing time limits). The products are processed in specific sequences of steps defined by recipes. Several products can be processed simultaneously according to different recipes. The barrels are transported between the tanks by a single hoist with finite speed.

---

[4] https://www.pyomo.org/

[5] https://www.ibm.com/products/ilog-cplex-optimization-studio/cplex-optimizer.

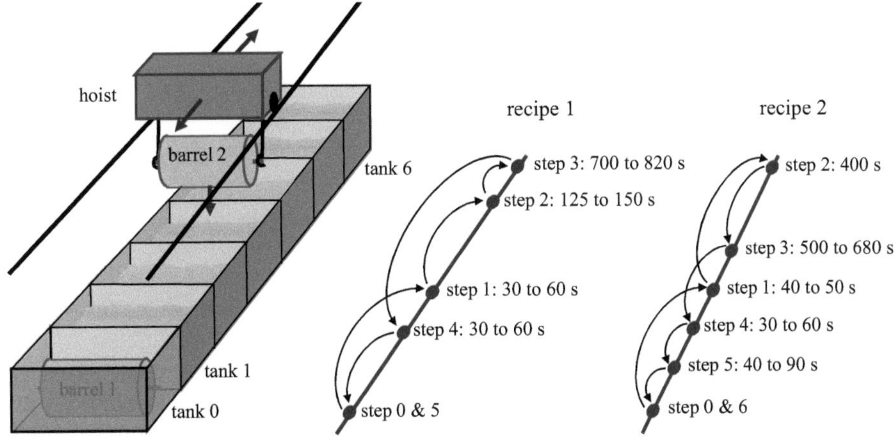

Fig. 2. Schematic visualization of an electroplating plant [19]

The objective of the scheduling problem is to create a hoist schedule with minimum makespan (i.e. time to process all products), specifying which product is processed at which time for how long in which tank [19].

4.2 Mixed-Integer Program

The HSP can be formulated as an MIP. The main variables of the MIP are the starting and the finishing times of each processing step of each product. The values of the variables must meet constraints such as:

- The start time plus the processing time equals the finishing time.
- The processing time must not exceed the minimum and the maximum allowed processing time.
- A processing step can only be started if the previous processing step and the transportation of the barrel to the subsequent tank are finished.
- At most one barrel can be processed within a tank at a time.
- The hoist can only carry one barrel at a time.
- The hoist arrives empty and in time at the tanks to pick up a barrel.

Further details on the formulation of the MIP can be found in the paper by Aguirre et al. [1].

4.3 Matheuristic

The monolithic MIP from above cannot be solved in reasonable times for large-scale instances. Therefore Aguirre et al. [1] developed a matheuristic which decomposes the monolithic MIP into a polylithic MIP with smaller sub-MIPs. The matheuristic comprises two steps: A construction step to create an initial schedule and an improvement step to improve the initial schedule.

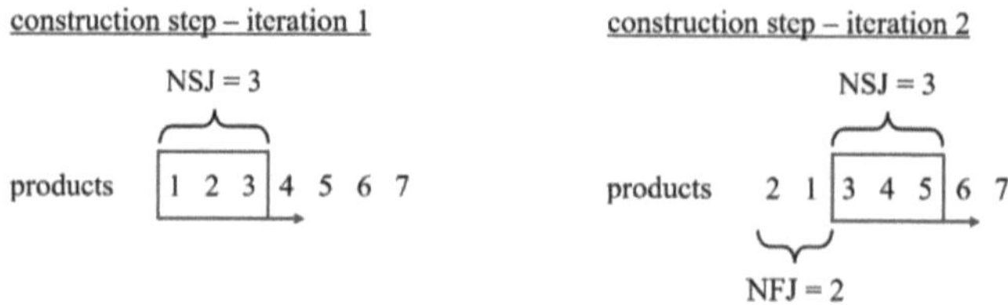

Fig. 3. Schematic visualization of the first two iterations of the construction step

The matheuristic can be explained using a window which is shifted over the *product sequence* (see Fig. 3).

In this example, the construction step starts with the *initial product sequence* 1, 2, 3, … 7. In the first iteration, a window of width *NSJ* (number of scheduled jobs) starts at the first product. All products within the window are scheduled by solving a sub-MIP, while the products to the right of the window are not considered. Afterwards, the window is shifted by *NFJ* (number of fixed jobs) products to the right. In the second iteration the products within and to the left of the window are scheduled by solving a sub-MIP. Again, the products to the right of the window are not considered. While the products within the window can be scheduled freely, the sequence of products to the left of the window are fixed to the result of iteration 1. This procedure is repeated until an initial schedule for all products is constructed.

The improvement step starts with the schedule from the construction step. A window of width *NRJ* (number of rescheduled jobs) is shifted over the product sequence. In each iteration a sub-MIP is solved considering all products. While the products within the window can be freely rescheduled, the sequences of the products outside the window remain fixed. In each iteration the schedule can either be improved or not (but it cannot deteriorate). In the first case the new product sequence is kept, and the window is shifted back to the beginning of the sequence. In the second case the window is shifted by one product to the right. Figure 4 visualizes these two cases.

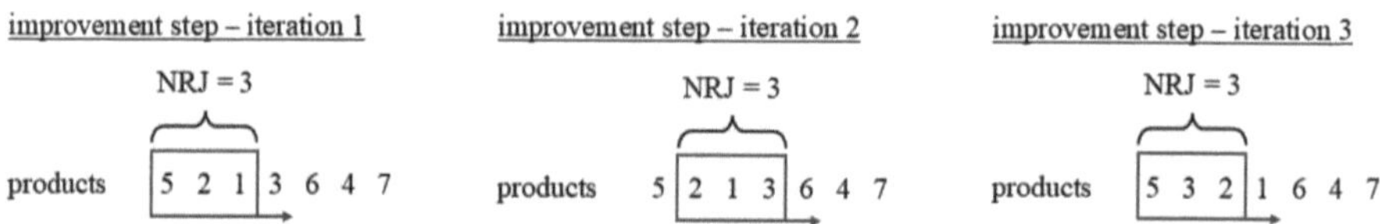

Fig. 4. Schematic visualization of the first three iterations of the improvement step

In iteration 1 the products 5, 2 and 1 can be rearranged during the optimization while the sequence of the other products is fixed, but the schedule is not changed (case 2). Therefor the window is shifted one product to the right. In iteration 2 the products 2, 1 and 3 are rescheduled leading to an improved schedule. Therefore, the window is shifted back to the beginning of the sequence. In iteration 3 the products 5, 3 and 2 can be rearranged. This procedure is repeated until the window is shifted completely over the product sequence with no improvement.

This matheuristic comprises at least four hyperparameters (of which only number 2 and 4 were considered in [1]):

1. The *initial product sequence* for the construction step,
2. the window width *NSJ* in the construction step,
3. the step size of the window *NFJ* in the construction step, and
4. the window width *NRJ* in the improvement step.

Additionally, the accepted optimality gap of the sub-MIP (*"MIPgap"*) is considered as a hyperparameter of the MIP standard solver (see Fig. 1). The optimality gap is the difference between the makespan of the best feasible solution found so far and a conservative lower bound for the makespan. The hyperparameter *MIPgap*[6] is used as a termination criterion for the sub-MIPs. A *MIPgap* of 0% means that the sub-MIPs must be solved to proven optimality. A *MIPgap* of 100% means that the first feasible solution found by the MIP standard solver is used.

The hyperparameters *NSJ*, *NRJ* and *NFJ* are integer valued. The *initial product sequence* is a categorical and the *MIPgap* is a continuous hyperparameter.

5 Computational Experiments with Small-Scale MIP-Instances

The aim of the matheuristic tuning is to find hyperparameter values that lead to a good objective value of the MIP at reasonable computational cost. So, there are two natural objectives: The objective of the MIP, for example minimize the makespan, and the minimization of the computational cost, often measured as the CPU-time used on a given hardware. Intuitively, these two objectives are competing: If the MIP is decomposed into many sub-MIPs, the CPU-time is expected to decrease while the makespan is expected to increase. In this paper, the two objectives are analyzed separately to understand how they vary with the hyperparameters, how the BO behaves and how the MIP-solutions differ for both objectives.

For small-scale MIP-instances the full matheuristic scheme can be applied and the sub-MIPs can be solved to proven optimality with reasonable computational cost for any hyperparameter values. In small discrete or discretized hyperparameter spaces a full grid search can be completed in reasonable computational time and locate all globally optimal hyperparameter values.

To limit the size of the hyperparameter space, a full grid search is conducted with only the two hyperparameters *NSJ* and *NRJ* while the remaining hyperparameters are fixed to specific values. The behavior of the BO is studied in two- and four-dimensional hyperparameter-spaces for the two objectives, makespan and CPU-time, separately. In the four-dimensional space the hyperparameters *NSJ*, *NRJ*, *NFJ* and *MIPgap* are tuned. The BO in the two-dimensional space is evaluated based on the globally optimal solutions known from the grid search. In the four-dimensional space full grid search becomes impractical due to its size and the continuous nature of the hyperparameter *MIPgap*.

All computational experiments were performed on a machine with an Intel Xeon E5-2630 v4 CPU running at 2.20 GHz.

---

[6] https://support.gurobi.com/hc/en-us/articles/8265539575953-What-is-the-MIPGap.

5.1 Grid Search

Four small-scale MIP-instances are studied. They comprise a small number of products (6) and the monolithic MIPs can be solved in less than 3 h CPU-time to proven optimality. Instance s1 is taken from Aguirre et al. [1] while instances s2, s3 and s4 are newly created. In Table 1 the main parameters of the small-scale instances are listed.

Table 1. Main parameters of the small-scale instances

Instance	Number of products	Number of stages	Number of recipes	Number of tanks
s1	6	4–8	3	36
s2	6	4–6	3	7
s3	6	6	2	5
s4	6	6	1	5

For the grid search the hyperparameters NSJ and NRJ are varied while NFJ is fixed to 1, *MIPgap is fixed to 0%* and *initial product sequence* is either ordered or unordered. NSJ and NRJ are integer in nature and bounded between 1 to 6 such that 36 solutions exist. For smaller values of NSJ and NRJ, the MIP is decomposed stronger into more and smaller sub-MIPs. For larger values of NSJ and NRJ, fewer and larger sub-MIPs need to be solved. The largest possible value for NSJ and NRJ is the number of products (here: 6). For this value the monolithic MIPs are actually not decomposed, such that the monolithic MIP is considered as the (one) sub-MIP in the construction or the improvement step, respectively.

Figure 5 shows for the representative instance s2 the results of the grid search for the two objectives, makespan and CPU-time, in the NSJ-NRJ-space. (Note that the integer-values are interpolated for a better visualization.) Two *initial product sequences* are studied: an ordered and an unordered one. In the ordered sequence the products with the same recipe are next to each other, while they are alternated in the unordered sequence.

As expected, the plots show that the two objectives are competing: A strong decomposition leads to a large makespan and to a small CPU-time and vice versa. The makespan exhibits a weak optimum with a lot of optimal solutions. In contrast, the CPU-time is neither monotonically increasing with NSJ nor with NRJ but exhibits local optima: For instance, a decomposition into two sub-MIPs in the improvement step ($NRJ = 5$) leads to a larger CPU-time than solving one monolithic MIP ($NRJ = 6$). The CPU-time is more sensitive to changes in NRJ than to changes in NSJ. This is because the number of iterations in the improvement step (controlled by NRJ) can be higher than the iterations in the construction step (controlled by NSJ), as the window can be reset to the beginning during the improvement step while the window is only shifted forward during the construction step.

A comparison of Fig. 5a and Fig. 5c shows that the *initial product sequence* has a significant impact on the set of optimal hyperparameter values for the makespan objective. With an unordered *initial product sequence* (Fig. 5c) only one of the window sizes (in the construction or the improvement step) needs only to be greater than 1 ($NSJ > 1$

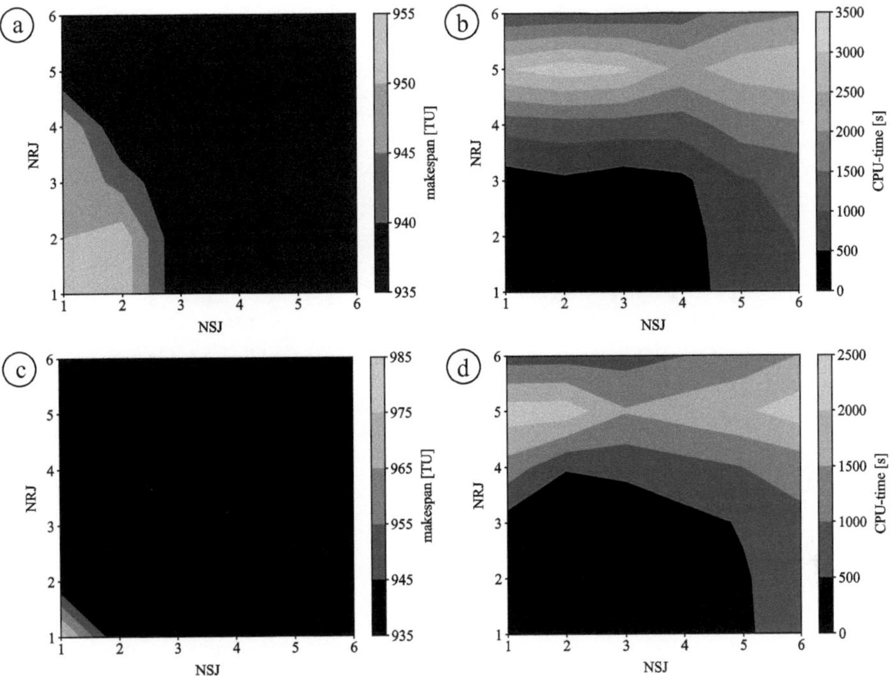

Fig. 5. Results of the grid search for small-scale instance s2 with ordered (top) and unordered (bottom) initial product sequence

or $NRJ > 1$) to achieve the optimal makespan. An ordered product sequence can result in sub-MIPs only comprising products with the same recipe, leading to sub-MIPs with little optimization potential.. If the *initial product sequence* is ordered (Fig. 5a) larger time windows are necessary to achieve the optimal makespan.

5.2 Bayesian Optimization

Figure 6 shows the behavior of the BO in terms of the sequence of its incumbent solutions for instance s2 based on Fig. 5a and b.

The BO is initialized with a single point[7] corresponding to the worst hyperparameter values for both objectives: $NSJ = NRJ = 1$ for the makespan and $NSJ = 2$ and $NRJ = 5$ for the CPU-time. The optimal hyperparameter values are found after the first incumbent solution for the makespan and after the fifth incumbent solution for the CPU-time.

To study the behavior of the BO in a higher-dimensional space, *NFJ* and *MIPgap* are considered as variable hyperparameters in addition to *NSJ* and *NRJ*. Figure 7 and Fig. 8 show how the incumbents of the objective values evolve over the iterations of the BO. The minimum, the maximum and the mean value over all four instances for each iteration are displayed. To make the values of the different instances comparable, they are normalized between 0 and 1.

---

[7] https://automl.github.io/SMAC3/v2.1.0/api/smac.initial_design.html.

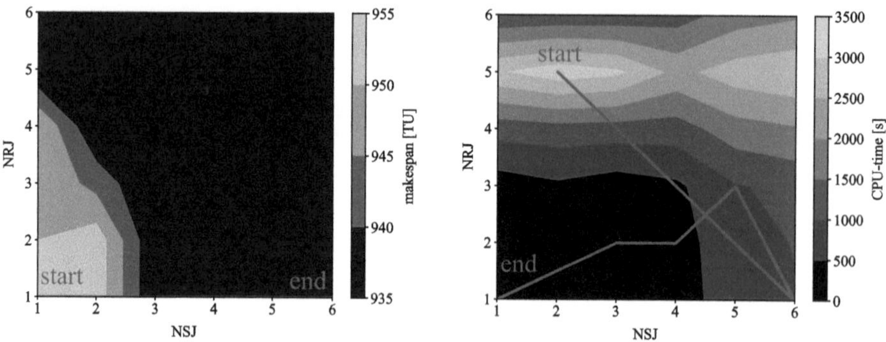

Fig. 6. Behavior of the BO in terms of its incumbent solutions (red lines) for instance s2 (Color figure online)

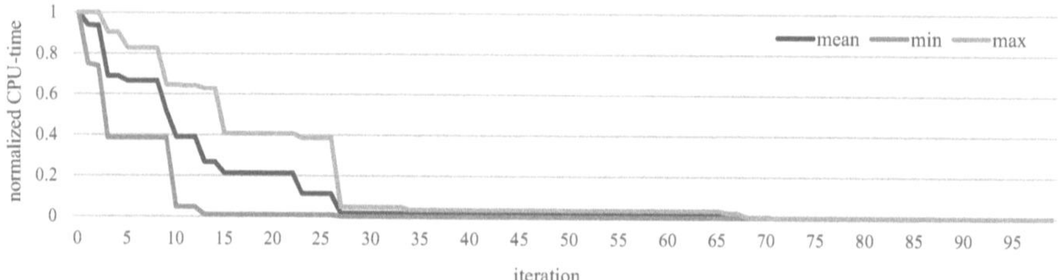

Fig. 7. BO convergence of the CPU-time in the 4D hyperparameter space

As can be seen in Fig. 7, the CPU-time improves mainly in the first 27 iterations. Afterwards it takes more effort – up to 68 iterations – until a better solution is found.

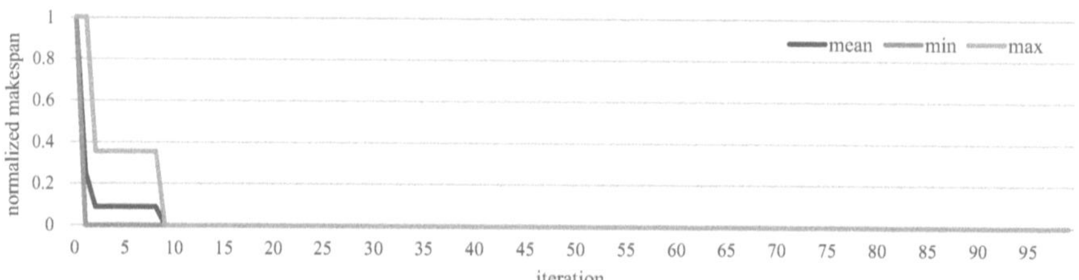

Fig. 8. BO convergence of the makespan in the 4D hyperparameter space

As can be seen in Fig. 8 the makespan improves only in the first 9 iterations. Afterwards no better hyperparameter values can be found.

Table 2 shows the best found hyperparameter values for the two objectives. "Best found" implies that the hyperparameter values are not necessarily optimal and that other hyperparameter values with the same objective values may exist.

The best found hyperparameter values depend on both, the instances and the objective. The dependency on the objectives was intuitively expected, since makespan and

Table 2. Best found hyperparameter values for both objectives

Instance	Minimizing makespan				Minimizing CPU-time			
	NSJ	NRJ	NFJ	MIPgap	NSJ	NRJ	NFJ	MIPgap
s1	6	2	2	0%	1	1	1	78%
s2	3	3	1	2%	4	6	3	86%
s3	3	3	1	2%	1	1	1	100%
s4	5	5	5	9%	6	6	1	91%

CPU-time are competing. The dependency on the instance gives reason to assume that there is no one set of hyperparameters that is optimal for all instances.

Minimizing the CPU-time pushes several hyperparameters of the matheuristic (NSJ, NRJ and NFJ) to their boundaries. This is an intuitive result since the CPU-times are relatively small if either the monolithic MIP is solved ($NSJ = 6, NRJ = 6, NFJ = any$) or the MIP is decomposed as strong as possible ($NSJ = 1, NRJ = 1, NFJ = 1$). In contrast, minimizing the makespan leads to several hyperparameters that are not at their boundaries. This can be explained by the weak optimum which was shown in Fig. 5.

The hyperparameter of the solver ($MIPgap$) is near 0% if the makespan is minimized and close to 100% if the CPU-time is minimized. These results are also intuitive: Larger optimality gaps leave optimization potential unexplored and lead to solutions of the sub-MIPs with longer makespan, while smaller optimality gaps lead to longer CPU-times for the MIP-solver.

6 Computational Experiments with Large-Scale MIP-Instances

For large-scale MIP-instances the full matheuristic scheme cannot be applied or some sub-MIPs cannot be solved to proven optimality with reasonable computational cost for some hyperparameter values. To deal with this property a computational time limit as an additional termination criterion for the matheuristic is introduced. Consequently, three types of terminations are possible:

1. The computational time limit is not reached, and the matheuristic scheme is finished regularly like for the small-scale instances (construction and improvement steps are executed as described in Sect. 4.3).
2. The computational time limit is reached during the improvement step, such that the matheuristic scheme cannot be finished regularly. Nevertheless, a feasible schedule results from the construction step.
3. The computational time limit is reached during the construction step, such that the matheuristic scheme cannot be finished regularly. An infeasible schedule results (unless it terminates in the last iteration after a feasible solution of the sub-MIP was found).

If an infeasible schedule results such that no makespan can be calculated for termination-type 3, the following *infeasible solution value* is returned to the BO instead of the makespan value:

$$\textit{infeasible solution value} = \left(1 + \frac{\#\textit{unscheduled products}}{\#\textit{products}}\right) \cdot \textit{bigM}$$

The degree of infeasibility is reflected by the ratio between the number of unscheduled products and the total number of products. For instance, hyperparameter values leading to only one unscheduled product out of ten are considered better than hyperparameter values that lead to nine unscheduled products out of ten. The value of $\textit{bigM}$ is an upper bound to the makespan, such that the *infeasible solution value* is always greater than the makespan of a feasible solution.

To study this approach, a grid search was performed in analogy to Sect. 5.1. Table 3 shows the main parameters of the four large-scale instances studied. In comparison to the small-scale instances, the number of products increased from six to ten while the other parameters are adapted accordingly. Some rules are applied to ensure that the MIP-instances are feasible.

Table 3. Main parameters of the large-scale instances

Instance	Number of products	Number of stages	Number of recipes	Number of tanks
11	10	6, 9	2	30
12	10	12, 15	2	30
13	10	5–10	5	30
14	10	11–17	5	30

According to some pre-studies the computational time limit was set to 700 CPU-s such that the matheuristic terminates for most of the hyperparameter values during the improvement step (termination-type 2). Figure 9 shows the results of the grid search for the two objectives, makespan (left) and the CPU-time (right), in the $\textit{NSJ-NRJ}$-space for the representative instances 12 (top) and 13 (bottom).

Only solutions with small values of $\textit{NSJ}$ and $\textit{NRJ}$ belong to termination type 1 with CPU-times less than the time limit (700 s). The other solutions are of termination-types 2 or 3 with the CPU-time being equal to the time limit; the additional CPU-time needed to complete the matheuristic scheme is not reflected.

Since for most solutions the improvement phase is not completed (termination-types 2 or 3) the makespan (or the corresponding infeasible solution value) does not depend on $\textit{NRJ}$. In contrast to the makespan (compare the small-scale instances in Fig. 5), the infeasible solution value does not decrease with $\textit{NSJ}$. For instance 12 makespan values can be calculated for $\textit{NSJ}$-values up to 3; for $\textit{NSJ}$-values from 4 to 10 the *infeasible solution value* is significantly larger than the makespan, depending on the number of unscheduled products. For instance 13 an infeasible solution value is calculated for NSJ-values from 6 to 9, while for NSJ-values up to 5 and – in particular – for 10 a makespan can be calculated.

Comparing the results of the grid search for the large-scale instances (Fig. 9) with those of the small-scale instances (Fig. 5) exhibits the following:

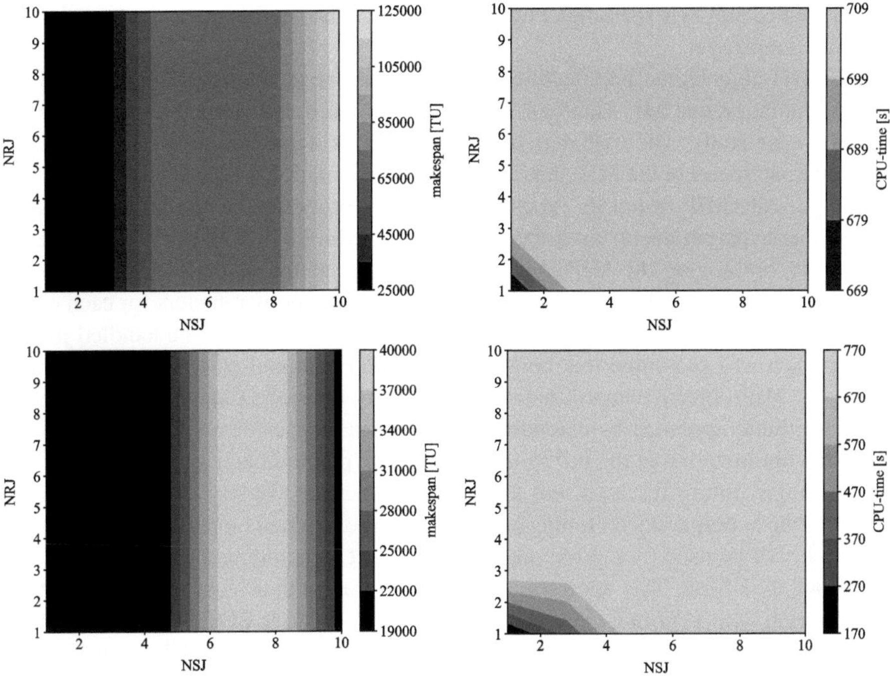

Fig. 9. Results of the grid search for large-scale instances 12 (top) and 13 (bottom)

- The CPU-time increases with the values of the hyperparameters *NSJ* and *NRJ* for small-scale as well as for large-scale instances. The main difference is that the values of the CPU-time are cut off at the given time limit for large-scale instances.
- In contrast to the CPU-time, the makespan behaves differently due to the *infeasible solution value* for the large-scale instances. The *infeasible solution value* reflects that for large-scale instances the makespan cannot be treated independently from the computational cost anymore.

The approach described in this section was motivated by the unreasonable computational cost for some hyperparameter values. However, it can also be considered as a way to deal with multiple objectives of an optimization problem, namely constraining one objective (here: computational cost) while optimizing the other (here: MIP-objective).

7 Conclusions and Future Research Directions

Decomposition of large-scale mixed-integer programs (MIPs) into a series of smaller sub-MIPs using matheuristic schemes is a common approach to deal with high computational cost. Matheuristics comprise hyperparameters which steer the decomposition and affect their performance. The tuning of hyperparameters of matheuristics is a black-box optimization problem with an expensive to evaluate objective function; Bayesian optimization is a well-established method for these types of problems. An analysis of related work revealed that hyperparameter tuning of matheuristics by Bayesian optimization has not been studied yet.

A novel algorithmic architecture combining Bayesian optimization, matheuristic, MIP-formulation and MIP standard solver was proposed and applied to hoist scheduling as a case study. The matheuristic comprises four hyperparameters in addition to one hyperparameter of the MIP standard solver. Computational experiments with small- and large-scale MIP-instances expose fundamental properties of the algorithmic approach: The hyperparameter optimization problem exhibits two natural and competing objectives: optimizing the MIP-objective and the computational cost. For small-scale instances, Bayesian optimization converges to optimal hyperparameters for each objective separately. For large-scale instances the two objectives need to be handled jointly; an approach was presented that constrains the computational-cost-objective and augments the MIP-objective appropriately. Future work aims at validating the findings on the algorithmic approach by extending computational experiments to grid search and Bayesian optimization in the full (5-dimensional) hyperparameter space.

Moreover, future research will aim at two directions: Firstly, the multi-objective aspect of the hyperparameter tuning problem will be examined more deeply. The current approach will be used to approximate the Pareto-set of solutions by varying the computational time limit. This approach will be compared with a "cost-sensitive Bayesian optimization" and a "primal integral" approach. Cost-sensitive BO determines whether a hyperparameter configuration is promising enough to justify a higher CPU-time, leading to a resource-efficient search process. The "primal integral" measures the improvement of solution quality over CPU-time and can be used as an objective function for the BO.

Secondly, the current single-MIP-instance approach will be extended to a multi-MIP-instance approach. Instead of considering one instance at a time multiple instances will be considered in two ways:

1. Hyperparameter values will be optimized on average for multiple instances.
2. The previously solved single instances will be used to predict hyperparameter values for the next instance.

Disclosure of Interests.   The authors have no competing interests to declare that are relevant to the content of this article.

References

1. Aguirre, A.M., Méndez, C.A., Gutierrez, G., De Prada, C.: An improvement-based MILP optimization approach to complex AWS scheduling. Comput. Chem. Eng. **47**, 217–226 (2012). https://doi.org/10.1016/j.compchemeng.2012.06.036
2. Ilemobayo, J.A., et al.: Hyperparameter tuning in machine learning: a comprehensive review. J. Eng. Res. Rep. **26**, 388–395 (2024). https://doi.org/10.9734/jerr/2024/v26i61188
3. Hutter, F., Kotthoff, L., Vanschoren, J. (eds.): Automated Machine Learning: Methods, Systems, Challenges. Springer, Cham (2019). https://doi.org/10.1007/978-3-030-05318-5
4. Bischl, B., et al.: Hyperparameter optimization: foundations, algorithms, best practices, and open challenges. WIREs Data Min. Knowl. Discov. **13**, e1484 (2023). https://doi.org/10.1002/widm.1484

5. Ippolito, P.P.: Hyperparameter tuning. In: Egger, R. (ed.) Applied Data Science in Tourism: Interdisciplinary Approaches, Methodologies, and Applications, pp. 231–251. Springer, Cham (2022). https://doi.org/10.1007/978-3-030-88389-8_12

6. Kim, D., Koo, J., Kim, U.-M.: A survey on automated machine learning: problems, methods and frameworks. In: Kurosu, M. (ed.) Human-Computer Interaction. Theoretical Approaches and Design Methods, pp. 57–70. Springer, Cham (2022). https://doi.org/10.1007/978-3-031-05311-5_4

7. Victoria, A.H., Maragatham, G.: Automatic tuning of hyperparameters using Bayesian optimization. Evol. Syst. **12**, 217–223 (2021). https://doi.org/10.1007/s12530-020-09345-2

8. Baratchi, M., et al.: Automated machine learning: past, present and future. Artif. Intell. Rev. **57**, 122 (2024). https://doi.org/10.1007/s10462-024-10726-1

9. Degroote, H., González-Velarde, J.L., De Causmaecker, P.: Applying algorithm selection – a case study for the generalised assignment problem. Electron. Notes Discrete Math. **69**, 205–212 (2018). https://doi.org/10.1016/j.endm.2018.07.027

10. Kruber, M., Lübbecke, M.E., Parmentier, A.: Learning when to use a decomposition. In: Integration of AI and OR Techniques in Constraint Programming, pp. 202–210. Springer, Cham (2017). https://doi.org/10.1007/978-3-319-59776-8_16

11. Basso, S., Ceselli, A., Tettamanzi, A.: Random sampling and machine learning to understand good decompositions. Ann. Oper. Res. **284**, 501–526 (2020). https://doi.org/10.1007/s10479-018-3067-9

12. Mitrai, I., Daoutidis, P.: Taking the human out of decomposition-based optimization via artificial intelligence, Part I: learning when to decompose. Comput. Chem. Eng. **186**, 108688 (2024). https://doi.org/10.1016/j.compchemeng.2024.108688

13. Mitrai, I., Daoutidis, P.: Taking the human out of decomposition-based optimization via artificial intelligence, Part II: learning to initialize. Comput. Chem. Eng. **186**, 108686 (2024). https://doi.org/10.1016/j.compchemeng.2024.108686

14. Hutter, F., Hoos, H.H., Leyton-Brown, K.: Automated configuration of mixed integer programming solvers. In: Proceedings of the 7th international conference on Integration of AI and OR Techniques in Constraint Programming for Combinatorial Optimization Problems, pp. 186–202. Springer, Heidelberg (2010). https://doi.org/10.1007/978-3-642-13520-0_23

15. Himmich, I., et al.: MPILS: an automatic tuner for MILP solvers. Comput. Oper. Res. **159**, 106344 (2023). https://doi.org/10.1016/j.cor.2023.106344

16. Bonami, P., Lodi, A., Zarpellon, G.: Learning a classification of mixed-integer quadratic programming problems. In: van Hoeve, W.-J. (ed.) Integration of Constraint Programming, Artificial Intelligence, and Operations Research, pp. 595–604. Springer, Cham (2018). https://doi.org/10.1007/978-3-319-93031-2_43

17. Patel, K.K.: Progressively strengthening and tuning MIP solvers for reoptimization. Math. Prog. Comp. **16**, 267–295 (2024). https://doi.org/10.1007/s12532-024-00253-z

18. Bengio, Y., Lodi, A., Prouvost, A.: Machine learning for combinatorial optimization: a methodological tour d'horizon. Eur. J. Oper. Res. **290**, 405–421 (2021). https://doi.org/10.1016/j.ejor.2020.07.063

19. Sand, G., Hildebrandt, S., Nunes, S., Chung-On, Y., Franke, M.: Tune decomposition schemes for large-scale mixed-integer programs by Bayesian optimization. Syst. Control Trans. **4** (2025) (to be published)

20. Lindauer, M., et al.: SMAC3: a versatile Bayesian optimization package for hyperparameter optimization. J. Mach. Learn. Res. **23**, 1–9 (2022)

Local Iterative Algorithms
for Approximate Symmetry Guided
by Network Centralities

David Hartman[✉] , Jaroslav Hlinka , Anna Pidnebesna ,
and František Szczepanik

Institute of Computer Science of the Czech Academy of Sciences,
Pod Vodárenskou věží 271/2, 182 07 Prague, Czech Republic
{hartman,hlinka,pidnebesna,szczepanik}@cs.cas.cz

Abstract. Recently, the influence of potentially present symmetries has begun to be studied in complex networks. A typical way of studying symmetries is via the automorphism group of the corresponding graph. Since complex networks are often subject to uncertainty and automorphisms are very sensitive to small changes, this characterization needs to be modified to an approximate version for successful application. This paper considers a recently introduced approximate symmetry of complex networks computed as an automorphism with acceptance of small edge preservation error, see Liu 2020 [18]. This problem is generally very hard with respect to the large space of candidate permutations, and hence the corresponding computation methods typically lead to the utilization of local algorithms such as the simulated annealing used in the original work. This paper proposes a new heuristic algorithm extending such iterative search algorithm method by using network centralities as a heuristic. Centralities are shown to be a good tool to navigate the local search towards more appropriate permutations and lead to better search results.

Keywords: Approximate symmetry · automorphism · graph matching · simulated annealing · complex network · centrality

1 Introduction

Complex networks are graphs that represent various natural and man-made systems such as, for example, the Internet [1], public transportation [16], or the human brain [27]. A typical way of analyzing such systems is to use several characteristics, both local (bound to vertices or edges) and global (bound to the whole graph). One notable global property is network symmetry, which received attention since MacArthur et al. [19] showed that many real-world networks contain high symmetry. Network symmetry has proven useful in revealing patterns of synchronized node clusters, helping understand forms of collective behavior [6].

© The Author(s), under exclusive license to Springer Nature Switzerland AG 2026
Y. Zhang et al. (Eds.): LION 2025, LNCS 15745, pp. 218–232, 2026.
https://doi.org/10.1007/978-3-032-09192-5_15

All types of network characteristics can be burdened with uncertainty problems. This uncertainty can be due to the construction of the corresponding network, an example is the uncertainty of local characteristics caused by non-linearity in time series, see [10,11,14]. The second type of uncertainty is the sensitivity of the corresponding characteristics to the overall behavior of the system, exemplified by the sensitivity of the global characteristic called a small-world coefficient, see [13,15,26]. Both types of uncertainty are combined in the graph symmetry represented by the automorphic group mainly because of its sensitivity to small changes.

For this reason, a more robust approximate symmetry based on the distance to the originally introduced automorphic symmetry was recently introduced, see Liu 2020 [18]. The proposed numerical solution was based on simulated annealing omitting the fixed points of the corresponding permutation. In a recent improvement of this method, see [21], this approach has been extended by introducing fixed points and also by using a relaxed problem with continuous optimization.

In practice, simulated annealing struggles to find an optimal permutation in the factorially large solution space. To address these limitations, we introduce a new heuristic method. This heuristic is based on the observation that permutations close to automorphisms tend to display similar vertices to themselves. We measure vertex similarity using graph centralities such as eigenvector centrality, PageRank, or betweenness. We compare the performance of the original and heuristic algorithm on several random network models, building on the experiments carried out by Pidnebesna et al. [21]. Presented results are based on the results of the bachelor thesis of F. Szczepanik, where one can find further details about methods used and some additional observations [25].

2   Preliminaries

2.1   Basic Notation and Definitions

The object of interest is a graph (undirected, unweighted, loop-less), defined by a set of vertices (nodes) V and edges (links) E, denoted as $G = (V, E)$. We use the standard notion of an adjacency matrix denoted as A, where $a_{ij} = 1$ if $\{v_i, v_j\} \in E$ and $a_{ij} = 0$ otherwise. We denote the set of vertices adjacent to a given vertex as $N_i = \{v_j | \{v_i, v_j\} \in E\}$. We call its size as $k_i = |N_i|$ and call it the degree of a node i. For a set of nodes $U \subseteq V$ we denote $E(U)$ as the set of edges induced by U, i.e. set of edges where each edge has at least one node in U. The Cartesian product $G \times H$ of two graphs G and H is a graph such that 1) $V(G \times H) = V(G) \times V(H)$, and 2) $\{(u, u'), (v, v')\} \in E(G \times H)$ if $u = v$ and $\{u', v'\} \in E(H)$ or $u' = v'$ and $\{u, v\} \in E(G)$. The object of interest is a graph (undirected, unweighted, loop-less), defined by a set of vertices (nodes) V and edges (links) E, denoted as $G = (V, E)$. We use the standard notion of an adjacency matrix denoted as A, where $a_{ij} = 1$ if $\{v_i, v_j\} \in E$ and $a_{ij} = 0$ otherwise. We denote the set of vertices adjacent to a given vertex as $N_i = \{v_j | \{v_i, v_j\} \in E\}$. We call its size as $k_i = |N_i|$ and call it the degree of a node i. For a set of nodes $U \subseteq V$ we denote $E(U)$ as the set of edges induced by U, i.e. set of edges where each edge has at least one node in U. The Cartesian product $G \times H$ of two graphs G and H is a graph such that 1) $V(G \times H) = V(G) \times V(H)$, and 2) $\{(u, u'), (v, v')\} \in E(G \times H)$ if $u = v$ and $\{u', v'\} \in E(H)$ or $u' = v'$ and $\{u, v\} \in E(G)$.

An automorphism of a graph $G = (V, E)$ is a permutation π of V preserving adjacencies, i.e.: $\{\pi(v_x), \pi(v_y)\} \in E \iff \{v_x, v_y\} \in E$. We represent such permutations via matrices: the permutation matrix P for a permutation π on n elements is an $n \times n$ matrix where:

$$P_{ij} = \begin{cases} 1 & \text{if } i = \pi(j), \\ 0 & \text{otherwise.} \end{cases}$$

PA permutes A's rows and AP^T permutes its columns according to π. Considering the orthogonality of permutation matrices, we obtain the following [3]: *the permutation π is an automorphism of G with an adjacency matrix A if and only if $A = PAP^T$.* Lastly, the L1 norm of an $n \times m$ real matrix A is: $\|A\|_1 = \sum_i^m \sum_j^n |a_{ij}|$.

2.2 Approximate Symmetry

The approximate symmetry of a graph $G = (V, E)$ with an adjacency matrix A is defined as [18]:

$$E_D(A) = \frac{1}{4} \min_P (\|A - PAP^T\|_1).$$

$E_D(A)$ minimizes the number of mismatches $A - PAP^T$ over permutation matrices P. Due to the constant $\frac{1}{4}$, it directly expresses the number of mismatched edges that differ after the permutation: $A_{ij} \neq A_{\pi(i)\pi(j)}$. So, if $A = PAP^T$, then P corresponds to an automorphism, and $E_D(A)$ is equal to zero.

A parametrized symmetry version $\epsilon(A, P)$ is then defined as follows:

$$\epsilon(A, P) = \frac{1}{4} \|A - PAP^T\|_1,$$

yielding $E_D(A) = \min_P \epsilon(A, P)$. For comparability across graphs, Liu introduced *normalized symmetry* $S(A)$, which normalizes $E(A)$ by the maximal $\epsilon(A, P)$. The maximal value of $\epsilon(A, P)$ is reached when the graph has exactly half of all possible edges, that is $\frac{1}{2}\binom{n}{2}$, and all of them are mismatched by P. The normalized approximate symmetry of a graph $G = (V, E)$ with an adjacency matrix A is thus:

$$S'(A) = \frac{E_D(A)}{\frac{1}{2}\binom{n}{2}} = \frac{\min_P \left(\|A - PAP^T\|_1\right)}{n(n-1)}.$$

Let π be a permutation on the set V. The element $v \in V$ is a fixed point of π if $v = \pi(v)$. Liu originally considered only permutations with no fixed points

(FPs), which reduced the problem space at the risk of excluding some potentially promising solutions. This motivated Pidnebesna et al. [21] to experiment with an annealing variant that allows up to K FPs, where K is a parameter. They set $K = n/2$ in their experiments, i.e., half of all the nodes.

An alternative approach avoids specifying an upper limit on FPs and instead determines the number of FPs dynamically by penalizing them in the objective function, resulting in the following version of approximate symmetry:

$$S(A) = \frac{E_D(A)}{\frac{1}{2}\binom{n}{2}} = \frac{\min_P \left(\|A - PAP^\mathrm{T}\|_1\right)}{n(n-1) - F(P)(F(P) - 1)},$$

where $F(P)$ denotes the number of fixed points in the permutation P. In the aforementioned bachelor's thesis by Szczepanik [25], it is shown that both approaches to FPs in simulated annealing - fixing $K = n/2$ or penalizing them in the objective function - yield values of S' (the normalized symmetry as originally defined by Liu) with no statistically significant differences. Based on this finding, all subsequent experiments use the dynamic FPs penalization method.

Identifying the optimal approximate automorphism P minimizing $\epsilon(A, P)$ is computationally hard as it involves searching through a factorially large solution space of vertex permutations. One of the suitable metaheuristics for this setting is Simulated Annealing, a probabilistic optimization technique [23]. Annealing explores the solution space by transitioning from one state to another while decreasing *energy*, which represents the value of the objective function. The process is initialized at a random state and initially explores worse states to avoid getting stuck in local optima according to the current *temperature*. The temperature is gradually reduced according to a chosen *cooling schedule* until some threshold temperature is reached or all move iterations are carried out.

2.3   Graph Centralities

Graph centralities are characteristics that evaluate the importance of a vertex from a structural perspective. More details about centralities can be found e.g. in [20]. One of the simplest measures is *degree centrality* defined as the number of adjacent edges. This centrality captures mainly local effects around the vertices. The generalization of the node degree is *eigenvector centrality*, which considers the importance of the vertex neighbors in addition to the number of its neighbors. Moreover, we apply this consideration to more distant vertices. This centrality can be defined recursively. If we have defined the initial centrality as $e_i(0)$ defined for a node i, then the total eigenvector centrality for this node can be defined as

$$e_i(t) = \sum_{j=1}^{n} A_{ij} e_j(t - 1).$$

The exact value of the centrality is computed for $t \to \infty$. It is possible to show further that this equation can be rewritten into the following characteristic equation $e_i = \frac{1}{\lambda} \sum_j A_{ij} k_j$, where λ is the greatest eigenvalue of A. Thus eigenvector centrality can be expressed as the eigenvector corresponding to the largest eigenvalue of A [20]. The last generalization in this direction is *PageRank*, which originated as a system for ranking the importance of web pages [5]. The main generalization of this centrality over the eigenvector centrality is the normalization of steps with degrees

$$e_i(t) = \alpha \sum_{j=1}^{n} A_{ij} \frac{e_j(t-1)}{k_j},$$

where k_j denotes the degree of vertex j and α is a damping factor typically set to 0.85, see full conditions in [20].

Instead of studying the number of neighbors, one can also study the density of edges in the neighborhood. This is measured by the *clustering coefficient* c_i of a node i defined as

$$c_i = \frac{2|E(N_i)|}{|N_i||N_i - 1|}.$$

This centrality measures how locally clustered the neighborhood of a vertex is, i.e. how much the vertex is part of the denser graphs [2].

The last centrality is representative of a class of characteristics measuring the role of a vertex in communication through the network. In more detail, *betweenness centrality* assesses a node's importance by counting the number of shortest paths between all pairs of nodes passing through it [22]. The betweenness centrality of a node v_i is defined as:

$$b_i = \sum_{\substack{x,y \in V(G) \\ x \neq y \neq v_i}} \frac{\sigma_{xy}(v_i)}{\sigma_{xy}},$$

where σ_{xy} denotes the number of shortest paths between nodes x and y and $\sigma_{xy}(v_i)$ is the number of shortest paths between x and y passing through v_i [22].

2.4   Random Network Models

Random network models provide abstractions of real-world systems and aim to capture their structural properties. In this section, we introduce several random graphs and random network models that serve as the dataset for comparing and evaluating the original and heuristic versions of simulated annealing. For such testing, we follow the methodologies of Straka [24] and Pidnebesna et al. [21].

The first class of graphs is not random but suitable for symmetry testing due to its high regularity. A *grid graph* $R_{n_1,n_2,\ldots,n_d}$ in d dimensions and lengths n_i in dimension i is defined as a graph product of path graphs $P_{n_1} \times P_{n_2} \times \ldots \times P_{n_d}$ [21].

The basic random graph is the standard *Erdős–Rényi* (ER) model $G(n, p)$ [7]. $G(n, p)$ is generated on n vertices by adding an edge between every pair with probability p. The ER graph is simple to analyze but it represents the real networks poorly, e.g. lacks the small-world as well as scale-free character [20, 26].

A more realistic model is the *Barabási-Albert* model (BA) that is s network growing model [2]. The generation procedure for the selected parameters n and k is as follows. We start with a small ($m \ll n$) graph and then add a vertex at each step, which we connect to k existing vertices, preferring higher degree vertices, namely the probability of connecting to an existing node i is $p_i = k_i / \sum_j k_j$. This model captures some characteristics such as power-law degree distribution (scale-free) and the emergence of important high-degree nodes (hubs) [2].

The *duplication-divergence* (DD) model simulates the evolution of protein-protein interaction (PPI) networks. PPI networks evolve by the mechanism of gene duplication, which generates new proteins that are initially identical to pre-existing ones but gradually diverge, with most of the new proteins not surviving. The duplication–divergence is described by the following growth mechanism [17]:

- Duplication: A node is randomly selected and duplicated. The newly created node inherits all links of the duplicated node.
- Divergence: Each link of the new node is retained with *divergence probability* σ. If no links survive, the duplication fails, and the node is discarded.

This model represents a real network but has some form of symmetry directly embedded in the generative process.

3   Heuristic Method

Various combinatorial-heuristic methods including simulated annealing, genetic algorithms, artificial neural networks, and others can be used to search the space of permutations to minimize $E_D(A)$ or $S(A)$. In this study, however, we concentrate on improving previously proposed and tested methods (see [18,21]). The reason for this is a basic investigation of the effect of centralities on search efficiency for well-constructed approaches. It can be assumed that the results will be translatable to more robust methods.

In particular, our primary objective is to investigate how guiding annealing with graph centralities improves the detection of approximate network symmetry. Thus, in this study, we introduce a modified annealing algorithm, *guided annealing*, which uses centrality metrics as a heuristic for selecting transpositions when moving across states.

Roughly speaking, a basic approach to exploring the solution space of approximate automorphisms involves applying random transpositions to the current permutation. Enhancing the transition strategy by incorporating some form of guidance could improve the efficiency of the annealing process. An optimal approximate automorphism should align vertices with similar structural roles in the graph. For instance, mapping a hub node onto a peripheral node will likely result in a large number of mismatched edges and therefore suboptimal approximate symmetry. Centralities allow us to quantify vertex "similarity" and thus identify vertices that should be aligned. An important property of graph centralities is that *automorphisms preserve centralities* [9,22]. Therefore, we expect nodes with similar centrality values to be aligned in an optimal permutation.

3.1 Move Function in Guided Annealing

We reimplement the *move* function, which generates a new permutation π' from the current permutation π and thus determines the next state. In unmodified simulated annealing, the transposition is constructed by randomly swapping the images of two vertices under the current permutation (moving from $a, \pi(a)$ and $b, \pi(b)$ to $a, \pi(b)$ and $b, \pi(a)$, where a and b are vertices and π the current permutation). In our guided version, the transpositions aligning similar vertices are chosen with higher probability.

Similarity Matrix and Implementation Description. We express the similarity of vertex pairs for a given centrality Γ in an $n \times n$ matrix M, indexed by vertices. To construct M, we start by computing a *difference matrix* D, where each element is defined as $d_{ij} = |\Gamma(i) - \Gamma(j)|$, i.e. the absolute difference of the centrality of v_i and v_j. To transform this into a similarity measure where higher values imply higher similarity, we compute inverses over the elements of D. Specifically, the entries of M are defined as $m_{ij} = (d_{ij} + \beta)^{-1}$, where $\beta > 0$ is a division constant preventing the denominator from being zero and moderating the variance of the matrix values; as β increases, the entries of M become more uniform, guiding the annealing less aggressively.

Re-implementing the Move Function. We now describe the re-implemented move function. We start by randomly choosing the first vertex a^1, where $\pi(a)$ is its image under the current permutation and $m_{a\pi(a)}$ is their similarity. We evaluate a potential similarity increase by considering swaps with each vertex b and its image $\pi(b)$. The difference in the similarity of the original and proposed setting can be calculated as $\Delta M_b = m_{a\pi(b)} + m_{b\pi(a)} - m_{a\pi(a)} - m_{b\pi(b)}$.

Before normalizing ΔM into a probability distribution over vertices, we again introduce a smoothing parameter $\phi > 0$, which we call the probability constant, and set $\Delta E_b = \max(\Delta M_b, \phi)$ for each b. Similarly to the division constant β, higher values of ϕ imply all vertices (even dissimilar pairs) have higher chances of being chosen. We then randomly draw b according to the created probability distribution. We present the pseudocode of the re-implemented move function, where we assume π is the current permutation represented by the matrix P:

Time Complexity of the New Version. The similarity matrix is computed only once at the beginning of a run for each graph, and most of the complexity in this process lies in computing the centralities themselves.

---

[1] We also experimented with a "one-step" approach, which does not select the first vertex randomly but instead considers all possible transpositions. However, this requires $O(n^2)$ evaluations in each step or a one-time pre-computation for all pairs of pairs, which has a complexity of $O(n^4)$, both practically unusable for large graphs.

Algorithm 1. Move function in guided annealing.

1: $a \leftarrow$ random vertex
2: **for** every vertex b **do**
3:    $\Delta E_b = max(m_{a,\pi(b)} + m_{b,\pi(a)} - m_{a,\pi(a)} - m_{b,\pi(b)}, \phi)$
4: **end for**
5: $\Pi \leftarrow$ probability distribution obtained by normalizing ΔE
6: $b \leftarrow$ vertex randomly drawn from Π
7: $P' \leftarrow$ permutation generated from P by transposing images of a, b
8: return P'

Optimization of Parameters by Grid Search. The values of the two introduced parameters, β and ϕ, i.e., the division and probability constants, were determined via grid search on about instances 400 of various random network models with varying sizes (50 to 150 vertices) and parameters. ER, BA, and Stochastic Block Models were included in the dataset used in grid search (we later excluded Stochastic Block Model graphs from further analysis due to their structural similarity to ER graphs). The examined values of β and ϕ ranged from 0.001 to 10. For every combination of the mentioned parameters, 30 simulations were conducted. We then selected the parameters with the best average results. When working with PageRank, we also conducted a grid search to optimize the damping factor α, but without significant differences in results; therefore, we keep using the default value of $\alpha = 0.85$. Generally, the few top combinations of parameters performed similarly, whereas the other combinations performed significantly worse. We do not claim the global optimality of the parameters we selected and recognize that different datasets could produce different parameters. However, the parameters we selected were sufficient to answer our hypothesis about guided annealing's improvement in symmetry calculation.

3.2   Evaluation Methodology

We now describe statistical methods used to evaluate the performance of different annealing versions. Violin plots are utilized to visually represent the distribution of measured symmetry (i.e. values of S, where lower values imply higher symmetry) across different instances of the models. Paired t-tests are used to assess whether the mean values of symmetry values obtained from two different annealing versions differ. The null hypothesis assumes that there is no significant difference, and the standard significance level of 0.05 is used to reject the null hypothesis. Finally, Cohen's d quantifies the effect size.

4   Results

Grid. We evaluate improved annealing on grid graphs with $50, 100, 150$ vertices in 2 and 3 dimensions. In 2D, the lengths of sides are always set to $5 \times x$ and in 3D to $2 \times 5 \times x$. For all grid parameters, guided versions significantly improved computed symmetry, with betweenness yielding the highest improvements, followed by eigenvector centrality (see Fig. 1, the first subplot).

Erdős–Rényi Model. For ER graphs with $n = 20, 50, 100$ and edge densities $p = 0.1, 0.3, 0.5$, we observe that there were no statistically significant differences between results produced by the different annealing versions, as presented in the second subplot of Fig. 1. A possible explanation would be that the ER model is entirely random; it lacks any internal structure stemming from a generation process and has no predictable substructures that can be reflected in each other.

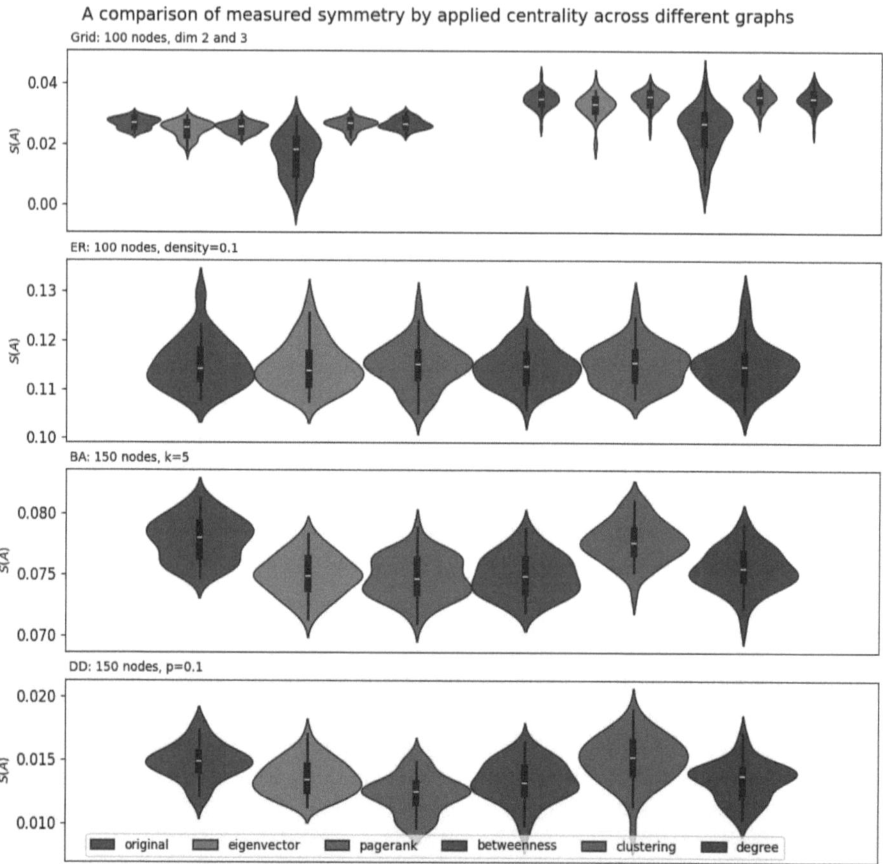

Fig. 1. Comparison of the performance of simulated annealing guided by different centralities across various random networks. The results are visualized in violin plots, where each color represents a version of annealing guided by a different centrality. For each configuration of parameters, we conduct 50 simulations. The first subplot presents results for grid graphs with 100 nodes in 2D and 3D. The second subplot presents results for Erdős–Rényi graphs with 100 nodes and edge density of 0.1. The third subplot presents results for Barabási–Albert graphs with 150 nodes and $k = 5$. The final subplot presents results for the Duplication–Divergence model with 150 nodes and a divergence probability 0.1. (Color figure online)

Barabási–Albert Model. We conduct evaluations on instances with 50, 100, and 150 nodes and k values of 3, 5, and 7. The results for BA graphs on 150 nodes and $k = 5$ are presented in the third subplot of Fig. 1. While not significant for 50 nodes, as graph size increases to $n = 100$ and 150, the improvements grow to become more significant. Versions of annealing guided by eigenvector centrality, betweenness, and PageRank yield some level of improvement over original annealing.

We select eigenvector centrality, which by eye seems to perform similarly or better than other centralities, and evaluate its improvements in more detail. We compute paired t-tests to determine whether the two algorithms (eigenvector-centrality-guided and original annealing) compute symmetry values with the same mean, along with Cohen's d to measure the effect size (see Fig. 2). After applying the Bonferroni correction for multiple hypothesis testing, annealing guided by eigenvector centrality produces significantly better outcomes compared to original annealing, apart from two combinations of parameters. We also see that the larger and sparser the graph is, the greater the improvement, as suggested by the increasing absolute value of Cohen's d.

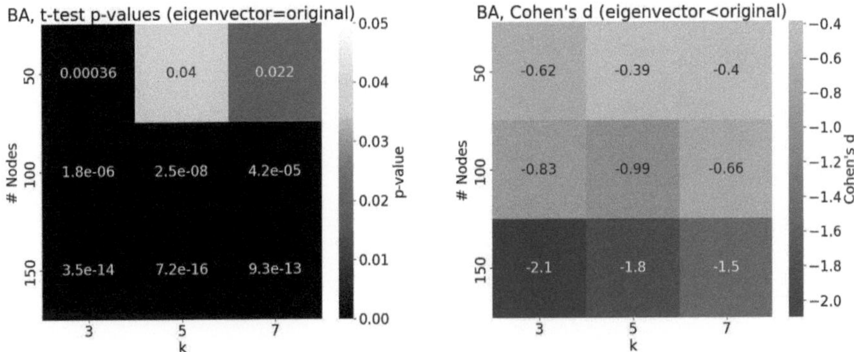

Fig. 2. Comparison of the performance of original simulated annealing and simulated annealing guided by eigenvector centrality, BA model, where k is the number of links of a new node.

Duplication–Divergence Model. Finally, we conduct evaluations on the DD model. The dataset consists of DD graphs with node counts $n = 50, 100, 150$, and divergence probabilities 0.1 and 0.3. The results for DD graphs on 150 nodes and divergence probability 0.1 are presented in the fourth subplot of Fig. 1.

Our findings indicate that guided annealing significantly outperforms original annealing. It also holds that instances with divergence probability 0.3 are less symmetric than those with divergence probability 0.1, which we attribute to the fact that as σ decreases, the graph becomes sparser and starts to resemble trees, which are intuitively somewhat regular (at least in the sense of having many leaf nodes likely sharing similar properties).

For a more rigorous evaluation, we select PageRank, which seems to perform at least as well as other centralities. Calculating paired t-tests and Cohen's d reveals that for smaller graph sizes, we cannot reject the null hypothesis. If we

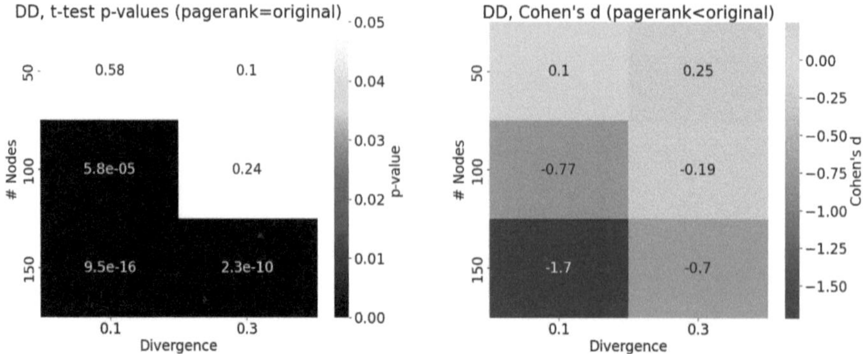

Fig. 3. Comparison of the performance of original simulated annealing and simulated annealing guided by eigenvector centrality, DD model.

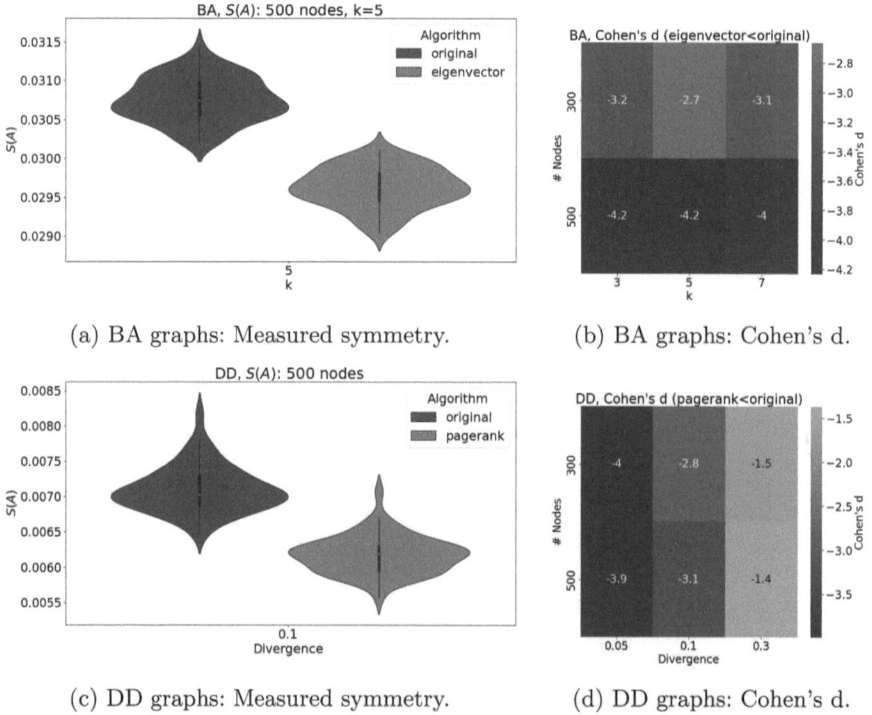

(a) BA graphs: Measured symmetry.

(b) BA graphs: Cohen's d.

(c) DD graphs: Measured symmetry.

(d) DD graphs: Cohen's d.

Fig. 4. Comparison of measured symmetry and Cohen's d values for larger BA and DD graphs.

move to larger graph sizes, however, we see PageRank-guided annealing outperforms the original, unguided version, as evident in Fig. 3, which also holds after applying the Bonferroni correction on the multiple (8 in this case) hypothesis testing. The improvements are more significant in sparser DD instances with lower σ and also become more pronounced with growing graph size.

Larger Graphs. In the previous section, we observed a trend in both the BA and DD models; improvements achieved by guided annealing became more pronounced with growing graph size. To further explore this trend, we extend our analysis to graphs of sizes 300 and 500. Specifically, we focus on the performance of PageRank-guided annealing on larger DD graphs, and eigenvector-centrality-guided annealing on larger BA graphs. Figure 4 shows the distribution of measured symmetry for $n = 500$ and Cohen's d effect size for other combinations of the parameters. In both cases, the guided versions yield even more significant improvements over original annealing than in smaller graph instances.

5   Discussions

This study investigated the impact of guiding simulated annealing with graph centralities as heuristics to improve the computation of approximate network symmetry. Given that graph centralities are preserved by automorphisms, we hypothesized that aligning vertices with similar centrality values would lead to improved values of approximate symmetry.

We introduced a new heuristic algorithm for finding approximate symmetry extending the existing simulated annealing approach. As a heuristic, we used network centralities, namely degree centrality, eigenvector centrality, betweenness centrality, and clustering coefficient. We also defined an efficient way of selecting similar vertices to be transposed. The effectiveness of this new method was evaluated across several graph classes or random models, including the grid, the Erdős–Rényi random graph, the Barabási–Albert model, and the duplication-divergence model.

Our results demonstrated that centrality-guided annealing improved approximate symmetry calculation concerning the model used. In grid graphs, betweenness centrality significantly improved computed approximate symmetry. This can be explained by the fact that this centrality considers communication using the shortest paths and in a grid graph this centrality characterizes vertices relatively uniquely. This graph, however, is not a proper test instance of the principle.

In contrast, no significant improvements were observed for Erdős–Rényi graphs, which we attribute to some of their known properties, e.g. small clustering of ER model implies no improvement using clustering coefficient, whereas degree or even PageRank are potentially confused by many similar nodes. Similarly, given the short distances and generally good connectivity of the ER model, neither betweenness centrality is a suitable discriminating characteristic. This may seem like a disappointing result, however, given the small reproducibility of real networks using the ER model, this result is not crucial.

In the Barabási–Albert and duplication-divergence networks, eigenvector centrality, and PageRank led to measurable improvements, particularly in instances that were inherently more symmetric. With graph size increasing from 100–150 vertices to 300–500 vertices, the improvements in Barabási–Albert and duplication–divergence networks became even more pronounced. These results correspond to intuition in the BA model due to the specific connection of vertices to each other and the creation of a certain hierarchical structure characteristic also typical for real networks [2]. For the duplication-divergence model, the results obtained are more surprising but can be attributed to some structure in the repeated generation of the duplicated vertex. Considering the model parameters, the goal is to test possible further variations in the future. It should also be mentioned that for both networks betweenness centrality works almost similarly to the above-mentioned spectral characteristics. This is probably due to the better resolution of this centrality in terms of its more comprehensive use of more distant areas from the corresponding node. On the other hand, the clustering coefficient performed poorly, likely due to the high locality of this characteristic and the fact that the networks were highly clustered.

There are several ways to extend the results mentioned here. One weakness seems to be the use of simulated annealing itself. Therefore, it would be benefitial to explore other metaheuristic approaches in the context of approximate symmetry computation, such as evolutionary algorithms, and examine whether centrality-based guidance can improve their performance.

In the BA model, for example, simple degree centrality performs comparably to the more complex eigenvector centrality. Since this centrality is a generalization of the vertex degree, it would be interesting to distinguish for each model the degree of locality needed for the centrality used. Similarly, betweenness centrality could be considered in its more local versions, such as k-betweenness [4].

Instances of graphs that have uniform centralities across vertices, such as regular graphs having identical vertex degrees, can cause a certain problem for the whole algorithm. These graphs can often be quite specific. However, for possible use in algorithms, it would be beneficial to identify a well-described class of such uniform graphs, especially for the characteristics yielding good results as a heuristic for approximate symmetry computation. In the case of graphs with uniform betweenness centralities, this is ongoing research [8,12]. The contribution of such a potential classification can be seen in the fact that such betweenness uniform graphs need not to be regular.

6   Conclusion

We introduced centrality-based heuristics into the computation of approximate graph symmetries, enhancing the guidance of local search algorithms such as simulated annealing. By leveraging structural cues from centralities, the method more effectively navigates the vast space of vertex permutations, especially in complex or larger networks. This approach lays the groundwork for making approximate symmetry computation more practical, and future exploration with more advanced algorithms—such as evolutionary methods—could further expand its applicability.

Acknowledgement. This work was supported by the Czech Science Foundation Grant No. 23-07074S.

Disclosure of Interests. The authors have no competing interests to declare that are relevant to the content of this article.

References

1. Abani, N., Braun, T., Gerla, M.: Betweenness centrality and cache privacy in information-centric networks. In: ACM SIGCOMM, pp. 106–116 (2018). https://doi.org/10.1145/3267955.3267964
2. Barabási, A.-L.: Network Science. The Royal Society Publishing. http://networksciencebook.com/
3. Biggs, N.: Algebraic Graph Theory. Cambridge University Press, Cambridge Mathematical Library (1974)
4. Borgatti, S.P., Everett, M.G.: A graph-theoretic perspective on centrality. Soc. Netw. **28**(4), 466–484 (2006). https://doi.org/10.1016/j.socnet.2005.11.005
5. Brin, S., Page, L.: The anatomy of a large-scale hypertextual web search engine. Comput. Netw. ISDN Syst. **30**(1–7), 107–117 (1998). https://doi.org/10.1016/S0169-7552(98)00110-X
6. Cho, Y., Nishikawa, T., Motter, A.: Stable Chimeras and independently synchronizable clusters. Phys. Rev. Lett. **119**, 084101 (2017). https://doi.org/10.1103/PhysRevLett.119.084101
7. Erdös, P., Rényi, A.: On random graphs I. Publ. Math. Debrecen **6**, 290–297 (1959). https://static.renyi.hu/~p_erdos/1959-11.pdf
8. Ghanbari, B., Hartman, D., Jelínek, V., Pokorná, A., Šámal, R., Valtr, P.: Structure of betweenness uniform graphs with low values of betweenness centrality. arXiv preprint arXiv:2401.00347 (2023)
9. Ghorbani, M., Dehmer, M., Lotfi, A., Amraei, N., Mowshowitz, A., Emmert-Streib, F.: On the relationship between PageRank and automorphisms of a graph. Inf. Sci. **582**, 874–885 (2021). https://doi.org/10.1016/j.ins.2021.08.013
10. Hartman, D., Hlinka, J., Paluš, M., Mantini, D., Corbetta, M.: The role of nonlinearity in computing graph-theoretical properties of resting-state functional magnetic resonance imaging brain networks. Chaos Interdiscipl. J. Nonlinear Sci. **21**(1), 013119 (2011). https://doi.org/10.1103/PhysRevLett.119.084101
11. Hartman, D., Hlinka, J.: Nonlinearity in stock networks. Chaos Interdiscipl. J. Nonlinear Sci. **28**(8), 083127 (2018). https://doi.org/10.1063/1.5023309
12. Hartman, D., Pokorná, A., Valtr, P.: On the connectivity and the diameter of betweenness-uniform graphs. Discret. Appl. Math. **342**, 27–37 (2024). https://doi.org/10.1016/j.dam.2023.08.017
13. Hlinka, J., Hartman, D., Paluš, M.: Small-world topology of functional connectivity in randomly connected dynamical systems. Chaos Interdiscipl. J. Nonlinear Sci. **22**(3), 033107 (2012). https://doi.org/10.1063/1.4732541
14. Hlinka, J., Hartman, D., Vejmelka, M., Novotná, D., Paluš, M.: Non-linear dependence and teleconnections in climate data: sources, relevance, non-stationarity. Clim. Dyn. **42**, 1873–1886 (2014). https://doi.org/10.1007/s00382-013-1780-2

15. Hlinka, J., Hartman, D., Jajcay, N., Tomeček, D., Tintěra, J., Paluš, M.: Small-world bias of correlation networks: from brain to climate. Chaos Interdiscipl. J. Nonlinear Sci. **27**(3), 035812 (2017). https://doi.org/10.1063/1.4977951
16. Hong, J., Tamakloe, R., Lee, S., Park, D.: Exploring the topological characteristics of complex public transportation networks: focus on variations in both single and integrated systems in the Seoul metropolitan area. Sustainability **11**(19), 5404 (2019). https://doi.org/10.3390/su11195404
17. Ispolatov, Y., Krapivsky, P., Yuryev, A.: Duplication-divergence model of protein interaction network. Phys. Rev. E **71**(6), 061911 (2005). https://doi.org/10.1103/physreve.71.061911
18. Liu, Y.: Approximate network symmetry. arXiv preprint arXiv:2012.05129 (2020)
19. MacArthur, B., Sanchez-Garcia, R., Anderson, J.: Symmetry in complex networks. Discret. Appl. Math. **156**(18), 3525–3531 (2008). https://doi.org/10.1016/j.dam.2008.04.008
20. Newman, M.: Networks: An Introduction. Oxford University Press (2010). https://doi.org/10.1093/acprof:oso/9780199206650.001.0001
21. Pidnebesna, A., Hartman, D., Pokorná, A., Straka, M., Hlinka, J.: Computing approximate global symmetry of complex networks with application to brain lateral symmetry. Inf. Syst. Front. (2025). https://doi.org/10.1007/s10796-025-10585-3
22. Pokorná, A.: Characteristics of network centralities. Prague: Charles University. Faculty of Mathematics and Physics, Computer Science Institute (2020). https://dspace.cuni.cz/handle/20.500.11956/118612
23. Kirkpatrick, S., Gelatt, D.C., Vecchi, M.P.: Optimization by simulated annealing. Science **220**(4598), 671–680 (1983). https://doi.org/10.1126/science.220.4598.671
24. Straka, M.: Approximative symmetries of complex networks, Bachelor thesis, Prague: Charles University, Faculty of Mathematics and Physics, Computer Science Institute (2022). https://dspace.cuni.cz/handle/20.500.11956/174639
25. Szczepanik, F.: Centralities in computation of approximate symmetries, Prague: Charles University, Faculty of Mathematics and Physics, Computer Science Institute (2024). http://hdl.handle.net/20.500.11956/192809
26. Watts, D., Strogatz, S.: Collective dynamics of 'small-world' networks. Nature **393**, 440–442 (1998). https://doi.org/10.1038/30918
27. Zhao, X., et al.: Disrupted small-world brain networks in moderate Alzheimer's disease: a resting-state FMRI study. PLoS ONE **7**(3), e33540 (2012). https://doi.org/10.1371/journal.pone.0033540

Addressing Over-Fitting in Passive Constraint Acquisition Through Active Learning

Vasileios Balafas[1](✉), Dimos Tsouros[2], Nikolaos Ploskas[1], and Kostas Stergiou[1]

[1] University of Western Macedonia, Campus ZEP, 50100 Kozani, Greece
`{v.balafas,nploskas,kstergiou}@uowm.gr`
[2] KU Leuven, Celestijnenlaan 200a, 3001 Leuven, Belgium
`dimos.tsouros@kuleuven.be`

Abstract. Constraint Programming (CP) is a powerful approach to solving complex combinatorial problems. However, formulating combinatorial problems as CP models typically demands substantial expertise. *Constraint acquisition* (CA) seeks to assist in model building by deriving constraints from data. In *passive learning*, the system relies on a pre-labeled set of examples (solutions or non-solutions) to infer constraints, whereas *active learning* engages a domain expert or software system through targeted queries that classify newly proposed assignments to the variables of the problem. Hybrid CA frameworks that combine both strategies have also emerged to leverage the strengths of both approaches. However, when training data are scarce or noisy, passive methods may overfit the observed examples—appearing valid on the training set but failing to generalize to other, unseen solutions—and thereby introduce invalid constraints into the model. To address this issue that has been overlooked, we propose a new *query-driven refinement* approach that systematically challenges suspicious acquired constraints, using "violating assignments" designed to refute them while preserving all other constraints. Focusing on the `AllDifferent` constraint, we integrate this refinement into an existing hybrid CA system and experimentally demonstrate that our approach facilitates convergence to a correct final model.

Keywords: Constraint Programming · Active Learning · Constraint Acquisition · Global Constraints · Over-fitting

1 Introduction

Constraint Programming (CP) is a powerful paradigm for solving a variety of combinatorial optimization problems in domains such as scheduling, timetabling, and configuration [20]. As in other approaches to combinatorial optimization (e.g. MIP, SAT), modeling is of primary importance when solving a new problem [10]. It is well known that an efficient model can result in exponentially

Y. Zhang et al. (Eds.): LION 2025, LNCS 15745, pp. 233–249, 2026.
https://doi.org/10.1007/978-3-032-09192-5_16

shorter run times compared to a bad one. A CP model consists of variables with finite domains and constraints that collectively define feasible assignments to these variables. *Global constraints*, such as the `AllDifferent` [5], which states that the variables in a given set must take different values, are ubiquitous, because they compactly encode patterns across multiple variables and can significantly speed up solving by exploiting specialized propagators.

Despite the effectiveness of CP, constructing reliable and efficient models often requires specialized domain expertise, and this is perhaps the most significant obstacle to the wider adoption of CP technology. To address this shortcoming, (semi)-automated modeling is attracting ever-increasing interest, with *Constraint Acquisition* (CA) being one of the most promising approaches [3,13,26]. CA is an area where combinatorial optimization, and CP in particular, meets Machine Learning (ML), as CA methods make use of inductive learning as well as statistical learning techniques [8,17,24].

Broadly, CA methods can be categorized into *passive* and *active*. In *passive learning* (PL), the system acquires constraints from a provided dataset of example solutions and non-solutions, without further interaction [8]. Often, passive CA aims at learning constraints that fit specific patterns [13] or global constraints [4]. Conversely, *active learning* (AL) involves an interactive process: the system proposes specific variable assignments (queries) to an oracle (e.g., a human expert or a software system encoding the ground-truth model) to confirm whether those assignments are valid solutions. The feedback received allows the system to refine its constraints more accurately. However, active CA systems cannot deal with global constraints, and are thus limited to learning constraints over small sets of variables (usually binary ones). Recently, *hybrid* CA frameworks have emerged, aiming to harness the complementary strengths of both approaches [2]. In such a framework, an initial set of global constraints is derived passively from available data, and an active learning phase then completes the model by learning fixed-arity constraints, which are often instance-specific.

However, there is a crucial shortcoming that has not received much attention and therefore remains unresolved: when the training data are limited or noisy, *passive learning can introduce invalid constraints*, i.e., constraints that do not appear in the true model, over-fitting on the specific examples provided to the system. For instance, consider a Sudoku puzzle where, based on the few provided solution examples, a passive learner may incorrectly infer an `AllDifferent` constraint on the main diagonal, if by chance the variables in the diagonal take different values in all given training examples. However, this does not hold for all valid Sudoku solutions, meaning that the model will be over-fitted to the observed data and an invalid constraint will be introduced. The problem of over-fitting becomes more severe when only a few solutions are available or when they are not diverse [1], as is usually the case with real-world problems.

In this paper, we propose a *query-driven refinement* methodology, specifically designed to address the problem of over-fitting by identifying and discarding from the model global constraints that were incorrectly acquired by passive learning.

For each constraint acquired through a passive learning system, we first compute a probabilistic confidence score using a Random Forest classifier trained on existing CP models. Then, our method generates *violating assignments*, i.e. assignments that violate a specific constraint while satisfying all others, and queries the oracle. If a violating assignment is labeled valid, it contradicts the constraint, which is therefore removed. Otherwise, the confidence score of this constraint is strengthened through Bayesian reasoning.

Hence, by maintaining and updating confidence scores, the system can make informed decisions about which constraints to retain or discard.

For the purposes of this study, we demonstrate the applicability and effectiveness of our violation-based approach focusing on the `AllDifferent` constraint, which is the most common global constraint. However, the same methodology can be applied to other global constraints (e.g., `Sum`, `Count`), given suitable strategies to generate violating assignments. Although our method does not theoretically guarantee to remove all over-fitted constraints, an experimental evaluation on benchmarks such as Sudoku variants and Exam Timetabling demonstrates that our refinement step manages to do so, reducing the overall number of queries compared to purely active learning methods, while improving model accuracy compared to purely passive or hybrid learning methods.

2   Related Work

Research in CA can be broadly divided into methods that focus on learning *fixed-arity constraints* and those that learn *global constraints or structural patterns*.

Fixed-arity approaches typically rely on active learning to iteratively query an oracle and validate candidate assignments. A recent family of interactive CA methods is based on the QUACQ [6] algorithms, using (partial) membership queries to prune the search space and acquire constraints. Extensions of this paradigm, such as *MQuAcq-2* [25], and *GROWACQ* [23,24] leverage structural properties or probabilistic cues to optimize query generation. Although effective, these active learning techniques often require a large number of queries.

Passive CA approaches targeting fixed-arity constraints have also been given some attention in the literature. One of the first passive CA approaches is CONACQ.1 [7], which is a SAT-based version space algorithm for acquiring CP models from given training examples. Recently, approaches that are robust to noise have been introduced, integrating ML techniques [17,18] (e.g., classifiers, unsupervised learning) or statistical methods [16].

Methods targeting global constraints or structural patterns (e.g., *ModelSeeker* [4] and *COUNT-CP* [13]) are typically passive learners that rely on pattern matching or frequent pattern mining to capture recurring global properties from data; however, they too may over-fit in scenarios with sparse or unrepresentative examples. Hybrid CA methods combine the strengths of both paradigms by first using passive learning to capture global patterns and then applying active learning to refine and complete the model. For instance, [2] exploits passive learning for initially learning global constraints using structural patterns and then uses active learning to complete the model with missing fixed-arity constraints. Our work builds on this hybrid methodology by introducing a query-driven refinement phase that actively identifies and removes invalid constraints from the over-fitted CP model before the active learning phase finalizes the model.

3   Background

A *Constraint Satisfaction Problem (CSP)* is a triple (X, D, C), where:

- X is a set of variables,
- D is a set of domains corresponding to the variables, and
- C is a set of constraints restricting the allowable assignments.

A *constraint* c is a pair $(rel(c),\ var(c))$, where $var(c) \subseteq X$ is the *scope* of the constraint, and $rel(c)$ is a relation over the domains of the variables in $var(c)$, restricting their allowed value assignments. The *arity* of the constraint, denoted as $|var(c)|$, indicates the number of variables involved. An assignment $A : X \to \bigcup_{x \in X} D_x$ is a *solution* iff it satisfies all constraints in C. Global constraints capture common substructures across multiple variables. For instance, the `AllDifferent` constraint requires that all variables in a set S take distinct values:

$$\texttt{AllDifferent}(S) = \{A : S \to \bigcup_{x \in S} D_x \mid \forall\, x, y \in S,\ x \neq y :\ A(x) \neq A(y)\},$$

and can be *decomposed* into a semantically equivalent set of binary inequalities:

$$\texttt{AllDifferent}(S) \equiv \{x_i \neq x_j \mid x_i, x_j \in S,\ i < j\}.$$

Many CSPs exhibit inherent structural patterns that can be effectively exploited through *matrix modelling* [9]. In matrix models, decision variables are organized into one or more matrices—much like the grid structure of a Sudoku puzzle—that naturally reflect the problem's structure.

In *Constraint Acquisition (CA)*, the goal is to learn a target CSP (X, D, C^*) where the vocabulary X, D is considered known while the target set of constraints C^* is unknown. Besides the vocabulary, typically CA systems are also given a *language* Γ that includes the possible relations that may hold between variables (e.g. `AllDifferent` or $\leq$). CA methods operate over a bias B of (possibly explicitly stored) candidate constraints, which is constructed using the vocabulary (X, D) and the constraint language Γ. During the acquisition process, a set of learned constraints $C_L \subseteq B$ is acquired. The system has correctly converged iff C_L is equivalent to C^*.

In passive CA, the system is also given a set of positive examples a set $\{A_1, A_2, \ldots, A_m\} \subseteq \text{sol}(X, D, C^*)$, which is denoted as E^+. Possibly, a set of negative examples E^- is also given. The goal of passive learning is to acquire a set of constraints C that is consistent with the data. In our study, we focus exclusively on global constraints, denoted by $C_G = \{c_g^1, c_g^2, \ldots, c_g^p\}$, which capture common structural patterns, but may over-fit the training data.

In interactive CA, the candidate constraints in B must have a fixed-arity, as global constraints with unrestricted arity require a *bias* of exponential size. An AL system generates an assignment A^* to all or some of the variables and asks the oracle whether A^* is valid or not. This is known as a *Membership Query*. Based on a Yes (valid) or No (invalid) reply to a query, the system removes constraints from B or adds constraints to C_L.

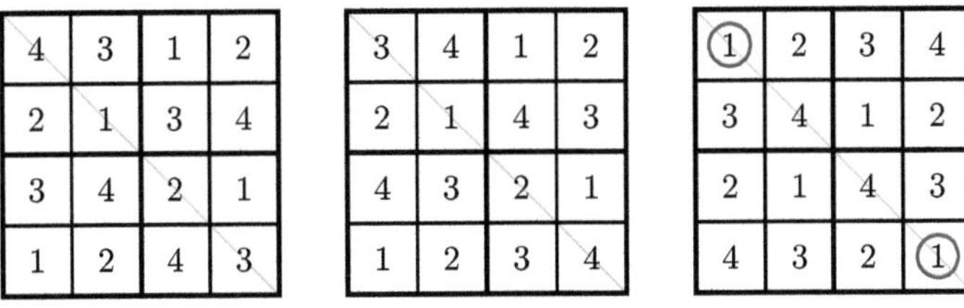

Training Example 1   Training Example 2   Violating Assignment

Fig. 1. Example of over-fitting in passive CA.

Example 1. To motivate our work, Fig. 1 illustrates an example of over-fitting in passive CA using a 4×4 Sudoku puzzle. In the target model there are 16 variables (one for each cell) and 12 `AllDifferent` constraints (one for each row, column, and 2×2 sub-grid).

Assume that a PL system is used to derive the global constraints of this problem given the language $\Gamma = \{\texttt{AllDifferent}\}$. Such a PL system will create the set of candidate `AllDifferent` constraints in B, based on some predefined patterns, and will use the given solutions to extract the constraints that agree with the given examples E^+. Assume that only the two example solutions from Fig. 1 are given. Note that both examples satisfy all 12 `AllDifferent` Sudoku constraints, while their main diagonal (light red shading) also has distinct values ($\{4, 1, 2, 3\}$ in example 1, $\{3, 1, 2, 4\}$ in example 2). Using these two examples, the PL algorithm may over-fit its acquired model to these available examples and erroneously infer an `AllDifferent` constraint on the diagonal if such a pattern is included in the ones it seeks. In the following, we will show how the invalid constraint can be eliminated using a targeted membership query.

4   Methodology

Our approach integrates passive learning with active learning through a novel intermediate phase—Query-Driven Refinement—to eliminate invalid global constraints that were acquired due to over-fitting in the given examples. For the purposes of this study, we focus on the `AllDifferent` constraint.

Figure 2 summarises the data flow among the three phases. From the positive examples E^+ and an initial bias B, the *Passive Learning* step retains only the constraints consistent with the data and extracts a first set of global constraints C_G. These global constraints feed the *Query-Driven Refinement* block, which utilizes the oracle to eliminate any over-fitted constraint, producing C'_G. The refined global constraints—decomposed into binary inequalities—together with the pruned bias bootstrap the final *Active Learning* phase; Finally, the Active Learning phase completes the model by using additional oracle queries to learn any remaining fixed-arity constraints, populating the set C_L. The union $C'_G \cup C_L$ (green box in the figure) is the final model returned by the framework.

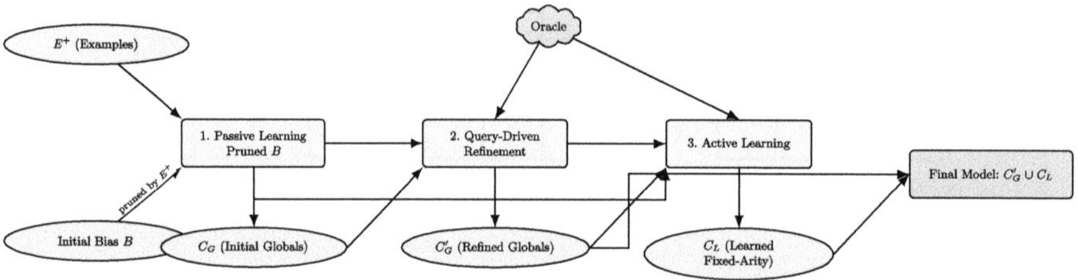

Fig. 2. Overview of the proposed Hybrid Constraint Acquisition framework. E^+: positive examples, C_G: initial global constraints from Passive Learning (PL), B: bias set (pruned during PL), C'_G: refined global constraints after Query-Driven Refinement (QDR), C_L: learned fixed-arity constraints from Active Learning (AL). The Oracle interacts with both QDR and AL phases.

The process proceeds in three sequential stages:

1. **Passive Learning:** Candidate `AllDifferent` constraints are acquired from positive examples E^+ to form the set C_G.
2. **Query-Driven Refinement:** Each constraint in C_G is first assigned a *confidence score*, denoting the probability that the constraint belongs to the target model. This is done using a constraints-level probabilistic classifier, which is trained on existing CP models. Then, each `AllDifferent`(S) in C_G is evaluated via violation queries that assign a common value to a variable pair in S, and it is either refuted or it's confidence probability is updated using Bayesian inference.
3. **Active Learning:** Each `AllDifferent` in the refined set of constraints is decomposed into binary $\neq$ constraints and, together with a bias set B, are passed on to an AL process that finalizes the model.

In the remainder of this section, we first briefly review the Hybrid Passive-Active Learning Framework (stages 1 and 3) and then we detail the novel Query-Driven Refinement phase that we propose in this study.

4.1   Hybrid Passive-Active Learning Framework

In the PL phase, the system is provided with a set E^+ of positive examples. Using E^+, the PL algorithm learns the initial set C_G of `AllDifferent` constraints (see [2] for details). Also, the system constructs a *bias* set B that contains all candidate fixed-arity constraints. For the purposes of this study, B contains all possible binary constraints that can be derived using the basic relations $\{\neq, ==, >, <, \geq, \leq, \}$. This set is refined by removing any candidate constraint that is violated by at least one example in E^+.

In the AL phase, the system refines the constraint network by interactively querying an oracle. First, any `AllDifferent` constraint that remains in the model, after the intermediate refinement step has been applied (as detailed below), is decomposed into a set of binary inequalities. This is necessary for the correct operation of the AL algorithm, which is then initialized with:

- the current learned set C_L, containing the binary inequality constraints derived by the decomposition of the `AllDifferent` constraints, and
- the bias set B of candidate fixed-arity constraints.

The AL component iteratively generates (often partial) assignments and submits them as membership queries to the oracle. If a query is answered *positively* (i.e., the assignment is valid), then any candidate constraints in B that conflict with the query are removed. Conversely, if a query is answered *negatively* (i.e., the assignment is not valid), then the AL algorithm identifies the minimal subset of constraints in B responsible for the violation and adds these to C_L (see [25] for details). Upon termination, the final constraint network is obtained by taking the union of the `AllDifferent` constraints in C_G and the constraints in C_L, excluding the $\neq$ constraints of the `AllDifferent` decompositions.

Although the hybrid framework combines the strengths of both PL and AL, it still carries an important disadvantage: any constraint acquired during the PL phase is by default considered valid, meaning that over-fitted constraints will remain in the final learned model.

4.2   Query-Driven Refinement

To address the over-fitting of passively learned `AllDifferent` constraints, we introduce a Query-Driven Refinement procedure. This method actively tests each learned constraint by generating targeted violation queries and updating our confidence in its validity. Our approach integrates three key components:

1. A probabilistic classification mechanism that assigns prior probabilities to candidate constraints, denoting their likelihood of being over-fitted.
2. A query generation mechanism that constructs violating assignments.
3. A Bayesian update framework that refines our confidence in each constraint.

The overall process, summarized in Algorithm 1, proceeds as follows:

Prior Probability. The procedure takes as input the set C_G of `AllDifferent` constraints learned from the passive learning phase of the Hybrid Passive-Active Learning Framework. To assess whether a constraint c actually belongs to the target model or is a result of over-fitting, we assign to it a prior confidence probability $P_{\text{prior}}(c)$ using a Random Forest classifier.

Machine Learning Estimation from Synthetic Data. To train the classifier, synthetic data are generated in a controlled offline process by simulating known CP models using benchmark problems from the CP benchmark library CSPLib [11]. For each model, candidate constraints are extracted from a large set of solution assignments and labeled as follows:

- *Positive examples:* `AllDifferent` constraints that truly belong to the model.
- *Negative examples:* `AllDifferent` constraints that hold only for the generated solutions (i.e., they are over-fitted), created by randomly selecting subsets of variables or perturbing valid groupings.

For each candidate, we extract a feature vector including: *Scope Size* (number of variables); *Is Full Row* and *Is Full Column* (binary indicators for complete rows or columns); *Sliding Window Pattern* (binary indicator if the candidate forms a sliding window grouping over the grid); *Is Diagonal* (binary indicator); and *Average Row* and *Column Positions* (mean indices). These features were chosen based on domain knowledge of common structural patterns in CSPs and were found to effectively capture the properties relevant for distinguishing valid constraints from over-fitted ones.

The classifier outputs a probability $\hat{p}_{\text{ML}}(c)$ that candidate c belongs to the target model, and we set $P_{\text{prior}}(c) = \hat{p}_{\text{ML}}(c)$.

Violation Using Membership Queries. The goal of the violation phase is to construct an assignment that deliberately violates a candidate constraint $c = \texttt{AllDifferent}(S)$ while satisfying all other constraints, and to post this as a query to the oracle. For each constraint $c \in C_G$, we enter a loop that is terminated if any of the following occurs:

- The oracle deems that a violating assignment, posted as a query, is valid.
- The confidence probability of c reaches a predefined threshold θ_{max}.
- It is not possible to generate an assignment that violates c and satisfies all other constraints.

These cases are explained in detail below. For now, note that in the first case, c is removed from the model, whereas in the other two cases it is preserved.

Pairwise Scoring and Query Generation. We first decompose c into binary inequalities over all variable pairs $(x_i, x_j) \in S$. For grid-based problems where spatial positions are known, we rank these pairs using the scoring function:

$$\text{score}(x_i, x_j) = \alpha \cdot |d(x_i, x_j)| - \beta \cdot \big(I(x_i) + I(x_j)\big),$$

- $d(x_i, x_j) = |r(x_i) - r(x_j)| + |c(x_i) - c(x_j)|$ is the Manhattan distance between x_i and x_j, with $r(x)$ and $c(x)$ denoting row and column positions;
- $I(x)$ is the involvement score (i.e., the number of candidate constraints in which x appears);
- α and β are positive weighting parameters that balance the relative contributions of the Manhattan distance and the involvement score.

We then sort the variable pairs in descending order of this score so that pairs with higher scores, indicating greater spatial separation and lower involvement, are prioritized. For each pair, we randomly select a single value from the intersection of the two variables' domains and we assign it to both variables. Obviously, this forces the `AllDifferent` constraint c to be violated, as now two variables in the scope of the constraint have the same value. We then proceed to extend this assignment to all variables in the problem using a CP solver. To do this we remove c from the model and the modified model M' is given to the solver. Crucially, for efficiency, our current implementation selects and tests only one random value v per pair. This is a heuristic choice trading exhaustiveness for speed; a failure to find a completing assignment A^* with this specific violation (e.g., due to solver timeout or unsatisfiability with v) does not guarantee that the constraint c is valid, only that this specific refutation attempt failed. Testing multiple or all common values is a possible extension but would significantly increase computational cost, potentially impacting the interactivity of the system. The timeout T (line 9) applies *per solver call* when attempting to find a complete assignment A^*. It is necessary to keep the overall runtime practical, as constraint solving is NP-hard. If the solver times out, it is treated as a failure to find a violating assignment for that specific attempt, but it does not prove impossibility.

Generating and Evaluating a Violating Assignment. If the modified model M' is indeed solved and a complete assignment A^* is obtained, this assignment is then submitted as a membership query $(\text{ASK}(A^*))$ to the oracle. Two outcomes are possible:

- If the oracle accepts A^* as a valid solution, this contradicts c; so we immediately remove c from the learned model.
- If the oracle rejects A^*, meaning that it is more likely that c is indeed violated by A^*, we update the probability of c via a Bayesian reasoning, as follows.

Probability Updates and Constraint Removal. Let D denote the event that the oracle rejects the violating assignment. Given the candidate's current probability $P_{\text{prior}}(c)$, and assuming:

$$P(D \mid c \in C^*) = 0.95, \quad P(D \mid c \notin C^*) = 0.05,$$

we update the probability using Bayes' rule:

$$P_{\text{post}}(c \in C^* \mid D) = \frac{P(D \mid c \in C^*)\, P_{\text{prior}}(c)}{P(D \mid c \in C^*)\, P_{\text{prior}}(c) + P(D \mid c \notin C^*)\, (1 - P_{\text{prior}}(c))}.$$

If a violation query results in a"Yes" (i.e., A^* is accepted by the oracle), then the candidate c is immediately refuted (i.e., $P(c)$ is set to 0) and removed from C'_G. Otherwise, additional queries are generated for c until the updated probability exceeds the threshold $\theta_{\max}$ (in which case c is retained). To assess the robustness of our approach to these parameter choices, we performed a sensitivity analysis. The analysis shows that varying $P(D \mid c \in C^*)$ within the range $[0.70, 0.98]$ and $P(D \mid c \notin C^*)$ within $[0.02, 0.30]$ has a negligible impact on the final model accuracy and the number of queries. This validates that our parameter settings yield stable results under moderate variations.

Returning to Example 1, recall that the passive learning phase may erroneously infer an `AllDifferent` constraint on the main diagonal. To refute this invalid constraint, our method will generate a violation query, shown in the third grid of Example 1, where we deliberately force a violation by assigning the same value to two cells along the main diagonal. This assignment, which violates the inferred diagonal `AllDifferent` constraint while satisfying all other constraints, is then submitted to the oracle. The oracle will accept this assignment as valid under the target model, and as a result the invalid constraint will be removed from the model.

Algorithm 1 refines the passively learned set of candidate `AllDifferent` constraints, C_G. First, for each constraint c (line 1), the algorithm collects all variable pairs from the scope of c (line 2) in a set VarPairs, sorts them by a scoring function (line 3), and initializes a boolean flag `violatingFound` to `false` (line 4). Next, it iterates over each pair (x_i, x_j) (line 5) and checks whether the two variables share any common value in their domains (line 6). If so, a random value v is selected (line 7), and a modified model M' is solved (line 8). If a solution (assignment A^*) is found (line 10), `violatingFound` is set to `true` (line 11), and the algorithm breaks out of the inner loop (line 12).

The timeout T is necessary to keep the overall run time manageable. Let us not forget that in order to find an assignment A^* in line 9, we need to solve the CSP that corresponds to M', which is an NP-hard problem. If no solution is found within T, the algorithm proceeds to the next pair of variables in VarPairs. Alternatively, we could try other values that are common to the two variables under examination. If no violating assignment is found after examining all pairs, the algorithm assumes that c is valid and proceeds to the next constraint (line 13). Otherwise, once a violating assignment A^* has been found, it is submitted to the oracle (line 14). If the oracle replies "Yes" (line 15), c is removed from C_G (line 16), since A^* contradicts c. Otherwise, c's prior probability is updated via Bayes' rule (line 19), and if it exceeds the threshold $\theta_{\max}$, the algorithm accepts c without further queries. After all constraints have been processed, the refined set C_G is returned (line 21).

If the inner loop of Algorithm 1 (lines 5–12) completes without finding any violating assignment A^* for constraint c (indicated by 'violatingFound' remaining 'false'), the algorithm conservatively retains c in C_G (line 13's implicit break). This occurs if no pairs had common domain values, or if for all tested pairs and single random values v, the solver timed out or proved unsatisfiability for M'.

Importantly, this failure to refute c via our heuristic search does *not* prove that c is valid or redundant with respect to the other constraints in C_G. It merely signifies that this specific, time-limited search strategy could not find evidence to discard c. Removing c in this situation would risk discarding a potentially valid constraint essential to the target model. Determining true redundancy would require a different, potentially more expensive, analysis.

Algorithm 1. Query-Driven Refinement

Require: C_G: candidate `AllDifferent` constraints with prior probabilities $P_{\mathrm{prior}}(c)$; confidence threshold $\theta_{\max}$; functions $\mathrm{score}(x_i, x_j)$ for pairwise scoring; timeout T for the solver

1: **for all** $c \in C_G$ **do**
2:     Let VarPairs $\leftarrow \{(x_i, x_j) \mid x_i, x_j \in \mathrm{scope}(c),\ i < j\}$.
3:     Sort VarPairs in descending order of $\mathrm{score}(x_i, x_j)$.
4:     violatingFound $\leftarrow$ false.
5:     **for all** $(x_i, x_j) \in$ VarPairs **do**
6:         **if** $\mathrm{dom}(x_i) \cap \mathrm{dom}(x_j) \neq \emptyset$ **then**
7:             Let v be a random value chosen from $\mathrm{dom}(x_i) \cap \mathrm{dom}(x_j)$
8:             $C_G' \leftarrow (C_G \setminus \{c\}) \cup \{x_i = v,\ x_j = v\}$, and let $M' = (X, D, C_G')$.
9:             Solve M' with timeout T to obtain assignment A^*
10:            **if** A^* is found **then**
11:                violatingFound $\leftarrow$ true
12:                **break**
13:     **if** not violatingFound **then continue** $\triangleright$ No violation found; accept c
14:     $feedback \leftarrow \mathrm{ASK}(A^*)$;
15:     **if** $feedback =$ "Yes" **then**
16:         Remove c from C_G.
17:         **continue** $\triangleright$ c is refuted by the oracle.
18:     **else**
19:         Update $P_{\mathrm{prior}}(c)$ via Bayes' rule.
20:         **if** $P_{\mathrm{prior}}(c) > \theta_{\max}$ **then continue** $\triangleright$ c is accepted with high confidence.
21: **return** C_G.

---

5   Experiments

This section presents our experimental evaluation, including the implementation details, the generation of synthetic data for training the Random Forest classifier, the benchmark problems used, the metrics for comparison with baseline methods, and the experimental results obtained.

The PL component is implemented following the hybrid system described in [2], which uses the Choco CP solver [19] to test if certain sets of variables satisfy the `AllDifferent` constraint in the given examples. In the violation-based AL phase, we use the CPMPy modeling library [12] for constructing and manipulating CP models and Google OR-Tools [15] to generate queries (feasible assignments A^*). The final AL phase is implemented using PyConA [22], a Python library for interactive CA, which includes an implementation of MQuAcq-2. To load our benchmark problems for evaluation, we employ the standardized data format specified by the PTHG21 CA Challenge [21]. This format defines the structure of solutions—typically as matrices for grid-based problems—ensuring that the examples are encoded in a consistent and reproducible manner. For the solving of modified models M' in the violation phase, we set the timeout T to 5 s. In our scoring function, we set the parameters to $\alpha = 1.0$ and $\beta = 0.5$. These values were determined through preliminary tuning to balance the contributions of the Manhattan distance and the involvement score.

5.1   Synthetic Data Generation for Classifier Training

To estimate the validity of candidate global constraints, we train a Random Forest classifier on synthetic data generated from several CSPLib benchmarks [11] that are modeled using `AllDifferent` constraints, including Quasigroup Existence (Latin Square), All-Interval Series, Magic Squares, Langford's Number Problem, N-Queens, Quasigroup Completion, Costas Arrays, n-Queens Completion, and Blocked n-Queens Problem.

For each benchmark, we generate a large set of solution assignments using the corresponding CP model and extract candidate `AllDifferent` constraints. Constraints inherent to the target model are labeled as *positive examples*, while those that hold only for the provided solutions (i.e., over-fitted constraints) are labeled as *negative examples*. A feature vector is constructed for each candidate using the attributes described in Subsect. 4.2. From the synthetic CSP instances, we extracted approximately 500 candidate constraints in total, with around 200 positive examples (true constraints from the target model) and 300 negative examples (over-fitted constraints). The dataset was split into an 80/20 training/test partition and subjected to 10-fold cross-validation. Using Scikit-learn [14], our Random Forest classifier achieved an average accuracy of 92%. These results indicate that the classifier reliably discriminates between valid and over-fitted candidate constraints on the synthetic dataset.

5.2   Benchmark Problems for Constraint Acquisition

We evaluate our method on puzzle and realistic benchmark problems. Specifically, 9×9 Sudoku (S9 × 9), Greater-Than Sudoku (GTS), JSudoku (JSud), and two instances of Exam Timetabling (ET1 and ET2). Standard Sudoku features a well-known model with 27 `AllDifferent` constraints—one for each row, column, and sub-grid—while Greater-Than Sudoku and JSudoku incorporate additional binary *greater-than* ($>$) constraints. In the Exam Timetabling problem, `AllDifferent` constraints are critical for ensuring that exams with overlapping student enrollments are scheduled in distinct time slots or rooms. Our method is compared against two baselines:

1. MQuAcq-2, which is a purely active learning method [25].
2. Our earlier hybrid CA framework, which employs PL plus AL, without the query-driven refinement [2].

5.3   Results and Discussion

Table 1 summarizes our experimental results by reporting the following metrics: the number of given solutions (Sols), the number of candidate constraints initially learned by passive learning (StartC), the number of constraints invalidated by our query-driven refinement (InvC), and the number of AllDifferent constraints in the target model (C_T). In addition, we give the size of the generated bias (Bias), the number of violation queries for our method's refinement phase (ViolQ), active learning queries for our method's AL phase (MQuQ), and the total queries for our proposed method (TQ = ViolQ + MQuQ). For comparison with baselines, we report the total queries used by the purely Active Learning baseline (MQuAcq-2), denoted as **ALQ**, and the total queries used by the baseline Passive+Active hybrid approach (which lacks the query-driven refinement phase), denoted as **PAQ**. Timing metrics include durations for our method's violation phase (VT(s)), active learning phase (MQuT(s)), and overall runtime (TT(s)), as well as total runtimes for the purely Active Learning baseline, **ALT(s)**, and the Passive+Active baseline, **PAT(s)**.

Table 1. Experimental results comparing our proposed method against baseline Pure Active Learning and baseline Passive+Active

Prob.	Sols	StartC	InvC	C_T	Bias	ViolQ	MQuQ	TQ	ALQ	PAQ	VT(s)	MQuT(s)	TT(s)	ALT(s)	PAT(s)
S9x9	2	47	20	**27**	4287	141	255	396	6672	36	50.50	47.19	97.69	634.19	67.83
S9x9	5	35	8	**27**	1814	105	199	304		22	32.81	29.19	61.99		39.18
S9x9	10	33	6	**27**	1400	99	181	280		20	27.83	25.07	52.89		31.37
S9x9	50	30	3	**27**	813	90	107	197		17	25.74	19.02	44.77		23.72
GTS	2	44	17	**27**	5498	132	279	411	6612	129	55.97	137.21	193.17	651.3	32.8
GTS	5	35	8	**27**	3036	105	272	377		263	56.42	127.15	183.58		13.59
GTS	20	32	5	**27**	1642	96	243	339		292	48.85	132.66	181.51		117.75
GTS	200	28	1	**27**	666	84	150	234		250	41.43	53.52	94.94		97.55
JSud	2	68	41	**27**	4952	449	3280	3729	6844	746	30.13	236.29	276.68	653.82	308.74
JSud	20	68	41	**27**	4559	449	3054	3503		938	30.04	227.50	258.44		529.89
JSud	200	59	32	**27**	4514	409	2809	3218		951	31.98	249.58	282.28		356.19
JSud	500	52	25	**27**	2373	156	1218	1374		669	27.56	263.66	291.69		257.32
ET1	2	41	25	**16**	1661	89	921	1110	7114	1025	68.14	842.12	910.42	1844.41	1587.62
ET1	5	36	20	**16**	1021	67	883	950		678	62.38	654.89	717.64		984.56
ET1	10	27	11	**16**	874	48	853	901		457	56.42	159.45	215.32		308.72
ET1	50	21	5	**16**	562	35	451	486		312	51.08	112.24	163.09		287.33
ET2	2	64	44	**20**	2864	147	2592	2739	9483	2121	84.32	974.68	1059.24	2286.09	1723.57
ET2	5	48	28	**20**	2156	92	1968	2060		1487	76.42	714.64	791.15		1461.09
ET2	10	29	9	**20**	1687	67	1452	1519		768	69.98	185.22	255.43		897.44
ET2	50	24	4	**20**	952	49	869	918		687	51.43	167.6	219.25		463.61

In our experiments, we observed that each invalid AllDifferent constraint was refuted with a single violation query. For constraints that were retained in the model, the maximum number of violation queries executed was 3. This low query count is not very surprising because: (i) an `AllDifferent` constraint can be refuted by simply forcing any pair of variables within its scope to take the same value, and (ii) the Bayesian update mechanism was designed to aggressively increase the confidence probabilities, thereby avoiding unnecessary queries on valid constraints.

In the standard 9×9 Sudoku experiments, when only 2 training solutions are provided, the passive phase initially infers 47 global constraints. Our query-driven refinement then invalidates 20 of these, yielding the correct final model of 27 `AllDifferent` constraints (C_T). With an increasing number of training solutions (5, 10, and 50), the initial candidate count decreases to 35, 33, and 30 respectively, while the number of invalidated constraints correspondingly decreases to 8, 6, and 3, yet the final model consistently contains the target 27 constraints. Moreover, with more training solutions the bias size is reduced from 4287 to 813 (by eliminating fixed-arity constraints that are inconsistent with the provided examples), and the total number of violation queries declines from 141 to 90. Consequently, the overall query count is reduced from 396 to 197, and the runtime decreases from 97.69 s to 44.77 s.

When compared to the purely active learning approach (i.e. MQuAcq-2), our hybrid CA method demonstrates significant efficiency gains, both in query burden and cpu time, as MQuAcq-2 requires 6672 queries and 634.19 s to learn the model of 9 × 9 Sudoku. The baseline hybrid CA approach yields low query counts (PAQ ranges from 36 down to 17) and shorter passive phase runtimes (PAT decreases from 67.83 s to 23.72 s). However, without the refinement mechanism, this baseline fails to eliminate over-fitted constraints, thereby compromising the correctness of the final model.

In the Greater-Than Sudoku (GTS) experiments, with 2 training solutions the passive phase infers 44 candidates and 17 are then invalidated, producing the target model of 27 constraints. With 5, 20, and 200 training solutions, the candidate counts decrease to 35, 32, and 28, with invalidations of 8, 5, and 1, respectively. Correspondingly, the bias size reduces from 5498 to 666, violation queries decline from 132 to 84, overall queries drop from 411 to 234, and the runtime decreases from 193.17 to 94.94 s. The purely active learning approach requires 6612 queries and 651.3 s, highlighting significant efficiency gains for our method. The baseline hybrid CA uses fewer queries (129 vs. our 411 for 2 training solutions) but fails to eliminate over-fitted constraints.

In JSudoku, with 2 training solutions the passive phase infers 68 candidates and 41 are then invalidated to yield 27 constraints. As training solutions increase to 20, 200, and 500, learned constraint counts become 68, 59, and 52, and invalidations reduce to 41, 32, and 25, respectively. The bias size falls from 4952 to 2373 and overall queries decline from 3729 to 1374, while runtime remains around 277âĂŞ292 s. Although the baseline hybrid CA gives fewer queries and lower runtimes, our query-driven refinement obtains the correct model, with the number of queries ranging from 746 with 2 initial solutions to 669 with 500 solutions. In contrast, the purely active approach requires 6844 queries and 653.82 s.

In the Exam Timetabling instances (ET1 and ET2), with only 2 solutions provided, the passive phase infers 41 and 64 candidate constraints for ET1 and ET2, respectively. Our query-driven refinement then invalidates 25 constraints in ET1 and 44 in ET2, yielding the correct final models of 16 and 20 `AllDifferent` constraints. As more training solutions become available, the number of learned `AllDifferent` constraints decreases, bias sizes shrink (from 1661 to 562 for ET1 and from 2864 to 952 for ET2), and runtimes are substantially reduced (from 910.42 to 163.09 s for ET1 and from 1059.24 to 219.25 s for ET2).

Although the baseline hybrid CA (PAQ) often requires fewer queries, it does not remove over-fitted constraints and thus may produce an incorrect final model. For instance, in ET1 with 2 solutions, PAQ uses 1025 queries—slightly fewer than our 1110 queries—yet fails to eliminate the spurious constraints. A similar pattern arises in ET2, where PAQ again omits the refinement step and consequently runs fewer queries (2121 vs. our 2739 for 2 training solutions, or 687 vs. our 918 for 50 training solutions) but retains invalid constraints. In contrast, a purely active approach requires 7114 (ET1) and 9483 (ET2) queries, with active learning times of 1844.41 and 2286.09 s, respectively, making our query-driven refinement a far more efficient alternative.

Overall, the experimental results demonstrate that through the query-driven violation mechanism, we manage to obtain the correct target `AllDifferent` constraints in the learned model. And importantly, this is achieved while significantly reducing the number of queries and overall runtime compared to the purely active learning method.

6   Conclusion and Future Work

Constraint Acquisition is an area where combinatorial optimization, and in particular CP, meets ML. Despite the numerous recent developments in CA, a shortcoming of existing learning methods that has not been addressed yet, is that of over-fitting the learned constraint model to the available example solutions.

In this paper, we tackle the problem of over-fitting, focusing on learning models with the common `AllDifferent` constraint, by integrating a query-driven refinement mechanism into a hybrid CA framework. Our approach employs a Random Forest classifier to assign prior probabilities to learned constraints and generates targeted violation queries to eliminate constraints that have been wrongly acquired because of over-fitting. Experimental results demonstrate that our method converges to the correct target model while significantly reducing the number of queries and overall runtime compared to purely active learning.

It is important to note a limitation of the current refinement strategy: it focuses on refuting individual candidate constraints $c \in C_G$ by finding an assignment A^* that violates only c while satisfying $C_G \setminus \{c\}$. This approach may not detect more complex over-fitting scenarios, such as when multiple incorrect constraints are symmetrically involved, or where refuting one incorrect constraint requires violating another (potentially also incorrect) constraint simultaneously. Handling such coupled errors, potentially by exploring violations of small subsets of constraints, remains an open challenge and an avenue for future research.

In the future, we plan to extend our query-driven refinement methodology to other global constraint types, such as Sum and Count. This is a non-trivial extension that will require developing specialized violation generation strategies tailored to the semantics of each constraint. For instance, to refute a candidate $\text{Sum}(S, \text{target})$ constraint, one might try to find an assignment where the sum of variables in S slightly deviates from target while satisfying all other model constraints. Similarly, for a Count constraint, the strategy would involve finding an assignment that marginally violates the specified count. Furthermore, a promising direction is to not only attempt to violate the constraint on the full set S, but also to explore violations on its critical subsets. Identifying the smallest subset of variables within S whose values cause the violation could provide more precise feedback for refinement or lead to the discovery of related, more accurate constraints. The pairwise scoring heuristic used for AllDifferent might also need adaptation or replacement with more general or constraint-specific heuristics (perhaps incorporating subset exploration ideas) to guide the search for violating assignments effectively for these and other global constraints. We also intend to explore alternative ML techniques for prior probability estimation, potentially incorporating features specific to different constraint types.

Acknowledgments. The research work was supported by the Hellenic Foundation for Research and Innovation (HFRI) under the 4th Call for HFRI PhD Fellowships (Fellowship Number: 9446).

References

1. Balafas, V., Tsouros, D., Ploskas, N., Stergiou, K.: The impact of solution diversity on passive constraint acquisition. In: Proceedings of the 13th Hellenic Conference on Artificial Intelligence. SETN '24, Association for Computing Machinery (2024)
2. Balafas, V., Tsouros, D.C., Ploskas, N., Stergiou, K.: Enhancing constraint acquisition through hybrid learning: an integration of passive and active learning strategies. Int. J. Artif. Intell. Tools (2024)
3. Barral, H., Gaha, M., Dems, A., Côté, A., Nguewouo, F., Cappart, Q.: Acquiring constraints for a non-linear transmission maintenance scheduling problem. In: CPAIOR 2024. LNCS, vol. 14742, pp. 34–50. Springer (2024)
4. Beldiceanu, N., Simonis, H.: Modelseeker: extracting global constraint models from positive examples. In: Data Mining and Constraint Programming. LNCS, vol. 10101, pp. 77–95. Springer (2016)
5. Beldiceanu, N., Carlsson, M., Demassey, S., Petit, T.: Global constraint catalogue: past, present and future. Constraints **12**, 21–62 (2007)
6. Bessiere, C., et al.: Constraint acquisition via partial queries. In: IJCAI: International Joint Conference on Artificial Intelligence, pp. 475–481 (2013)
7. Bessiere, C., Coletta, R., Koriche, F., O'Sullivan, B.: A SAT-based version space algorithm for acquiring constraint satisfaction problems. In: Gama, J., Camacho, R., Brazdil, P.B., Jorge, A.M., Torgo, L. (eds.) ECML 2005. LNCS (LNAI), vol. 3720, pp. 23–34. Springer, Heidelberg (2005). https://doi.org/10.1007/11564096_8
8. Bessiere, C., Koriche, F., Lazaar, N., O'Sullivan, B.: Constraint acquisition. Artif. Intell. **244**, 315–342 (2017)

9. Flener, P., Frisch, A., Hnich, B., Kiziltan, Z., Miguel, I., Walsh, T.: Matrix modelling. In: Proceedings of the CP-01 Workshop on Modelling and Problem Formulation, p. 223 (2001)

10. Freuder, E.C., O'Sullivan, B.: Grand challenges for constraint programming. Constraints **19**, 150–162 (2014)

11. Gent, I.P., Walsh, T.: CSPLIB: A Benchmark Library for Constraints (1999)

12. Guns, T.: Increasing modeling language convenience with a universal n-dimensional array, cppy as Python-embedded example. In: Proceedings of the 18th Workshop on Constraint Modelling and Reformulation at CP (Modref 2019), vol. 19 (2019)

13. Kumar, M., Kolb, S., Guns, T.: Learning constraint programming models from data using generate-and-aggregate. In: 28th International Conference on Principles and Practice of Constraint Programming (CP 2022). LIPIcs, vol. 235, pp. 29:1–29:16

14. Pedregosa, F., et al.: Scikit-learn: machine learning in Python. J. Mach. Learn. Res. **12**, 2825–2830 (2011)

15. Perron, L., Didier, F.: Cp-SAT. https://developers.google.com/optimization/cp/cp_solver/

16. Prestwich, S.D.: Robust constraint acquisition by sequential analysis. In: 24th European Conference on Artificial Intelligence (ECAI 2020). Frontiers in Artificial Intelligence and Applications, vol. 325, pp. 355–362. IOS Press (2020)

17. Prestwich, S.: Unsupervised constraint acquisition. In: 2021 IEEE 33rd International Conference on Tools with Artificial Intelligence (ICTAI), pp. 256–262 (2021)

18. Prestwich, S.D., Freuder, E.C., O'Sullivan, B., Browne, D.: Classifier-based constraint acquisition. Ann. Math. Artif. Intell. **89**(7), 655–674 (2021). https://doi.org/10.1007/s10472-021-09736-4

19. Prud'homme, C., Jean-Guillaume, F., Xavier, L.: Choco solver documentation (2016). https://choco-solver.org

20. Rossi, F., Van Beek, P., Walsh, T.: Handbook of Constraint Programming. Elsevier (2006)

21. Simonis, H., Freuder, E.: PTHG21 Challenge (2021). https://doi.org/10.5281/ZENODO.5155465

22. Tsouros, D., Guns, T.: A CPMpy-based python library for constraint acquisition - PyCona. In: Proceedings of the AAAI 2025 Bridge on Constraint Programming and Machine Learning (CPML) (2025)

23. Tsouros, D.C., Berden, S., Guns, T.: Guided bottom-up interactive constraint acquisition. In: 29th International Conference on Principles and Practice of Constraint Programming (CP 2023), vol. 280, pp. 36:1–36:20 (2023)

24. Tsouros, D.C., Berden, S., Guns, T.: Learning to learn in interactive constraint acquisition. In: Thirty-Eighth AAAI Conference on Artificial Intelligence, AAAI 2024, February 20–27, 2024, Vancouver, Canada, pp. 8154–8162. AAAI Press (2024)

25. Tsouros, D.C., Stergiou, K., Bessiere, C.: Structure-driven multiple constraint acquisition. In: Principles and Practice of Constraint Programming, pp. 709–725. Springer (2019)

26. Tsouros, D.C., Stergiou, K., Sarigiannidis, P.G.: Efficient methods for constraint acquisition. In: Principles and Practice of Constraint Programming (CP 2018), pp. 373–388. Springer (2018)

Learning to Solve the Skill Vehicle Routing Problem with Deep Reinforcement Learning

Nayeli Gast Zepeda[✉], André Hottung, and Kevin Tierney

Bielefeld University, Bielefeld, Germany
{nayeli.gast,andre.hottung,kevin.tierney}@uni-bielefeld.de

Abstract. Neural combinatorial optimization has proven effective in solving various simple routing problems including the traveling salesperson problem and the vehicle routing problem (VRP). However, real-world routing scenarios are usually significantly more complex, often requiring sophisticated methods to find even a single feasible solution. In this work, we apply neural combinatorial optimization to the more challenging skill VRP, where routes must be constructed for technicians with diverse skill sets while adhering to customer time windows. Due to the limited number of available technicians, finding feasible solutions is usually very challenging. We evaluate several state-of-the-art learning-based approaches on the skill VRP and explore different reward shaping techniques to penalize infeasible solutions during training. Our findings show that while most approaches can effectively solve instances with 20 customers, all approaches struggle to reliably find feasible solutions for instances with 50 customers.

Keywords: Neural Combinatorial Optimization · Deep Reinforcement Learning · Routing Problems

1 Introduction

Deep neural networks (DNNs) can be used to solve a variety of optimization problems, with a particular focus on vehicle routing problems (VRPs) in recent literature. There has been great progress with regard to problem sizes considered, VRP variants studied, and ever-improving architectures and methods proposed [2,10,13,14]. However, the literature has primarily focused on problems for which it is trivial to find feasible solutions. This stands in stark contrast to practical applications, where finding feasible solutions may be difficult due to restrictive real-world constraints.

In this work, we present a neural combinatorial optimization (NCO) approach for solving the skill VRP.[1] The skill VRP involves a set of technicians with varying skill sets that must serve customers with varying skill requirements during

---

[1] Code available at https://github.com/ngastzepeda/lion2025-drl-skillvrp.

Y. Zhang et al. (Eds.): LION 2025, LNCS 15745, pp. 250–266, 2026.
https://doi.org/10.1007/978-3-032-09192-5_17

pre-specified time windows. This combination of constraints is a challenging one for NCO approaches, as it can lead to infeasible solutions, which create challenges for the reinforcement learning (RL) mechanisms used in NCO methods. A straightforward way of handling infeasibility is with a penalty, i.e., infeasible solutions are assigned a value proportional to how much the constraints are violated. This is called *reward shaping* in the RL literature. We investigate penalty functions for NCO methods on the skill VRP in this work to gain insights into whether penalties are sufficient for RL methods to learn effective policies for finding feasible solutions.

The contributions of this work can be summarized as follows:

1. We introduce a Markov decision process (MDP) formulation for the skill VRP, enabling us to tackle this problem with various NCO techniques.
2. We present an instance generator that generates instances guaranteed to have at least one feasible solution. This generator can easily be extended to generate instances for other routing problems.
3. We compare different reward shaping techniques to incentivize feasible solution generation.

The remainder of the paper is organized as follows: Sect. 2 reviews related work on the skill VRP and NCO approaches for routing problems. Section 3 defines the mathematical notation for the skill VRP used in this paper. Section 4 presents our MDP formulation of the skill VRP, introduces the different reward formulations, and describes the generator for feasible instances. Finally, Sect. 5 reports the results of our experiments.

2 Related Work

2.1 The Skill Vehicle Routing Problem

The skill VRP is a well-known resource-constrained routing and scheduling problem. In these types of problems customers have certain demands that can only be met by a subset of the available vehicles or service operators from the start [16]. In the skill VRP, specifically, technicians with certain sets of skills service customers who in turn have specific skill demands, as shown in Fig. 1.

The first flow-based mathematical formulation for the skill VRP was introduced in [5]. The problem formulation assumes different skill levels S, where each customer with skill demand S_j must be serviced by any one technician t that has skill level $S_j \leq S_t$. In [4] the authors extend the model to consider skill types instead of skill levels. A technician t can service a customer node j if $S_j \subseteq S_t$, i.e., if the technician provides all the skills/service types required by the customer. The authors in [6] extend the problem further to include time window constraints for customers (morning vs. afternoon), a special device, precedence and synchronization constraints between services. Travel cost is technician-dependent, i.e., more skilled technicians will be more expensive. In this work we model the skill VRP based on [4], but including time windows and service durations.

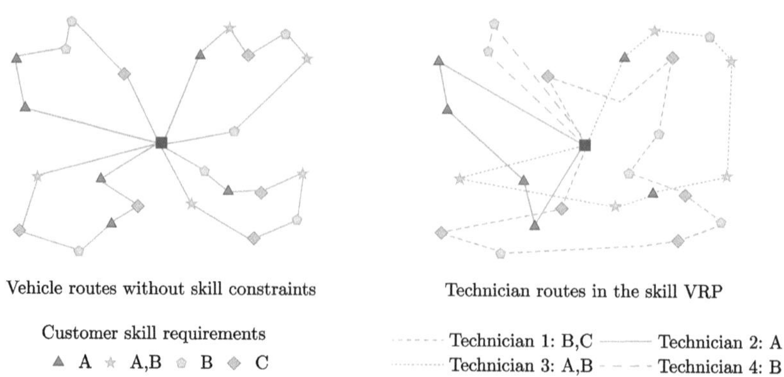

Fig. 1. Example of the skill VRP, based on [16].

2.2 Deep Reinforcement Learning for Routing Problems

Neural Combinatorial Optimization (NCO) approaches apply machine learning methods and other neural techniques to solve combinatorial optimization (CO) problems. Specifically, NCO research has mostly been applied to routing and scheduling problems, which are NP-hard, i.e., no polynomial-time algorithm is known to solve these problems optimally.

The foundation for NCO to routing problems was laid in [18], which introduces Pointer Networks (Ptr-Net). These networks can handle variable-size outputs that depend on the input size and produce a softmax probability distribution to effectively point to positions in the input sequence. The first application to routing problems was in [2], which uses Recurrent Neural Networks (RNNs) to encode city locations and sequentially construct Traveling Salesperson Problem (TSP) tours by predicting the next city to visit. [15] extends this approach to the Capacitated VRP (CVRP) by incorporating Reinforcement Learning (RL) to learn routing policies that respect vehicle capacity constraints while minimizing total route length.

Recent architectural improvements have further advanced the field. The Attention Model (AM) [13], a transformer-based encoder with self-attention, processes all locations in parallel rather than sequentially, allowing the model to better capture global relationships between locations through multi-head attention mechanisms. This architecture also improves training stability and scalability to larger problem instances. Policy Optimization with Multiple Optima (POMO) [14] greatly increases performance while maintaining computational efficiency, by training with multiple starting locations and augmenting solutions through rotations and reflections during inference. The SymNCO method [12] extends this concept of leveraging problem and solution symmetries to enhance performance. Unlike POMO, SymNCO also incorporates augmentations during training to improve the model's generalization capability. In [20], the authors introduce MVMoE, a neural multi-task vehicle routing solver based on the mixture-of-experts (MoE) approach, designed to handle multiple VRP variants simultaneously. They report strong generalization in both zero-shot and few-shot settings. More recently, [11] propose PolyNet, a method that learns multiple diverse solution strategies using a single neural network. PolyNet demonstrates strong performance, achieving near-optimal results, on the TSP with 100 nodes. All of these methods, however, have in common that they have only been evaluated on problems for which it is trivial to find feasible solutions. We go further and apply these methods to the skill VRP which has a challenging combination of constraints that can lead to infeasible solutions.

3   The Skill VRP: Notation

We build upon the skill VRP model introduced in [4]. Our model additionally includes service time windows as in [6] and service durations. Our goal is to provide a set of realistic constraints for investigating NCO methods.

Sets

T	Set of technicians.
N	Set of nodes (customers and depot).
N'	Set of nodes without the depot.
A	Set of feasible arcs between nodes $i, j \in N$ with $i \neq j$.
S_t^T	Set of skills offered by technician $t \in T$.
S_j^C	Set of skills required by customer $j \in N'$.

Parameters

c_t	Travel cost per distance unit for technician $t \in T$.
d_{ij}	Distance between $(i, j) \in A$, which we assume is equal to the travel duration.
s_j	Service time required at node $j \in N'$.
$[e_j, l_j]$	Time window for customer $j \in N'$, where e_j is the earliest start time and l_j is the latest.
H	System end time, which is the maximum route duration per technician.

Variables

$x_{tij} \in \{0, 1\}$	1 iff technician $t \in T$ travels on arc $(i, j) \in A$.
$z_j \in \mathbb{R}_0^+$	Arrival time of a technician at node $j \in N$.

Objective and Constraints

$$\min \sum_{t \in T} \sum_{(i,j) \in A} c_t d_{ij} x_{tij} \tag{1}$$

$$\sum_{(0,j) \in A} x_{t0j} \leq 1 \qquad \forall t \in T \tag{2}$$

$$\sum_{(i,j) \in A} x_{tij} = \sum_{(j,k) \in A} x_{tjk} \qquad \forall t \in T, j \in N \tag{3}$$

$$x_{tij} = 0 \qquad \forall (i,j) \in A, t \in T, S_j^C \nsubseteq S_t^T \tag{4}$$

$$\sum_{t \in T} \sum_{(i,j) \in A} x_{tij} = 1 \qquad \forall j \in N' \tag{5}$$

$$z_0 = 0 \tag{6}$$

$$e_j \leq z_j \leq l_j \qquad \forall j \in N' \tag{7}$$

$$z_i + s_i + d_{ij} \leq z_j + H\left(1 - \sum_{t \in T} x_{tij}\right) \qquad \forall (i,j) \in A, j \neq 0 \tag{8}$$

$$z_j + s_j + d_{j0} x_{tj0} \leq H \qquad \forall t \in T, (j,0) \in A \tag{9}$$

The objective minimizes the total travel cost for all technicians in Term (1). Constraints (2) indicate that every technician can leave the depot at most once. The flow balance Constraints (3) ensure that if a technician visits a customers, they must also leave the customer. Constraints (4) describe the skill constraints, i.e., if a technician t does not offer the skills S_j^C, they cannot visit customer j. Constraints (5) demand that every customer j be visited exactly once. The remaining constraints ensure temporal feasibility of the routes. The depot's start time is set in Constraint (6), Constraints (7) limit the arrival time to fall within the customer's time window. If a technician arrives at customer j before e_j, they have to wait for the time window to start, i.e., $z_j = e_j$. Constraints (8) ensure that when traveling from customer i to j, the arrival time z_j cannot be smaller than the sum of the arrival time at the previous customer plus the customer's service time and the travel distance. Finally, Constraints (9) ensure all technicians arrive back at the depot before the system end time H.

4   An MDP Formulation of the Skill VRP

Given a problem instance x, the solution is generated through a **Markov Decision Process (MDP)** [1] defined as $(\mathcal{S}, \mathcal{A}, \mathcal{T}, \mathcal{R})$. The state space $\mathcal{S}$ describes the problem instance and the current partial solution at each step σ. The action space $\mathcal{A}_s$ includes all available actions at a given state $s \in \mathcal{S}$. Note that the action space can never be empty even if the current state corresponds to an infeasible solution. We describe how we model this in the context of the skill VRP in the following subsection. The state transition function $\mathcal{T}$ updates a state s_σ to the next state $s_{\sigma+1} = \mathcal{T}(s_\sigma, a_\sigma)$. The reward function $\mathcal{R}(s_\sigma, a_\sigma)$ represents the reward after taking action a_σ in state s_σ.

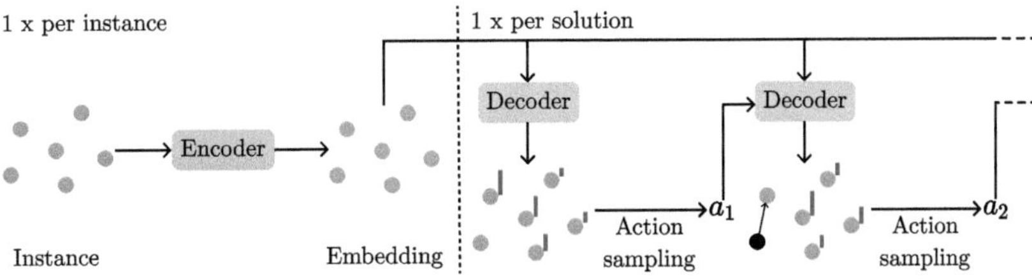

Fig. 2. Visualization of transformer-based RL, based on [7,9]. The encoder maps the problem instance to an initial embedding that includes node locations, time windows, service durations, skill levels, and technicians' travel costs. Based on the current state $s \in \mathcal{S}$, which additionally includes the current time, node, and technician, as well as the partial solution, the decoder iteratively samples from the available actions $\mathcal{A}_s$ until all customers have been visited and the reward for the route can be calculated.

Routes are constructed by taking actions in the environment until all customer nodes have been visited, which is intuitively visualized in Fig. 2. First, a trainable encoder f_θ maps the problem instance x to an embedding $h = f_\theta(x)$. The decoder g_θ iteratively builds a solution by sampling from the available actions, based on the embedding h and the current partial solution, until all customer nodes have been visited. We call a sequence of T actions that defines a feasible route a *rollout* $\rho = (a_1, \ldots, a_T)$. The reward for the rollout is $R(\rho, x) = \sum_{\sigma=1}^{T} \mathcal{R}(s_\sigma, a_\sigma)$. The encoding and decoding process is formalized as:

$$\pi_\theta(\rho|x) \triangleq \prod_{\sigma=1}^{T-1} g_\theta(a_\sigma|a_{\sigma-1}, \ldots, a_0, h), \tag{10}$$

where π_θ is the stochastic policy mapping the problem instance x to a rollout ρ.

We train the policy π_θ to maximize the expected reward $\mathbb{E}_{\rho \sim \pi_\theta(\rho|x)}[R(\rho, x)]$. For the distribution $P(x)$ of problem instances x the training objective becomes:

$$\theta^* = \underset{\theta}{\operatorname{argmax}} \left[\mathbb{E}_{x \sim P(x)} \left[\mathbb{E}_{\rho \sim \pi_\theta(\rho|x)} \left[R(\rho, x) \right] \right] \right]. \tag{11}$$

The REINFORCE algorithm [17] trains the solver π_θ by transforming Eq. 11 into a minimization problem with a loss function, which can be optimized using gradient descent. The gradient for the REINFORCE loss function is given by:

$$\nabla_\theta \mathcal{L}(\theta|x) = \mathbb{E}_{\pi(\rho|x)} \left[(R(\rho, x) - b(x)) \nabla_\theta \log \pi(\rho|x) \right], \tag{12}$$

where $b(\cdot)$ is a baseline function that stabilizes training and reduces variance.

4.1 Specifying the Skill VRP

State Space $\mathcal{S}$. The state space includes both static information about the problem instance x and information about the partial solution, which is needed to determine the available actions at each step. Static problem information includes the node locations in terms of their x and y coordinates, the skills of the technicians S_t^T, skill demands of the customers S_j^C, the customer time windows $[e_j, l_j]$ and service durations s_j, as well as the technician's travel cost c_t. Note that we use the Euclidean distance and consider a fully connected graph.

We encode the technicians each as separate depots, all identical except for the associated skill sets S_t and travel costs c_t. This way, we obtain a total of $|T| + |C|$ nodes, indexed from 0 to $|T| + |C| - 1$, where the first $|T|$ nodes correspond to technicians and the remaining $|C|$ nodes to customers. This allows us to encode the skills for each node j as a binary vector $s_j \in \{0, 1\}^K$, where K is the total number of distinct skills in the system. For technician nodes, $s_{jk} = 1$ indicates technician j offers skill k, and $s_{jk} = 0$ otherwise. For customer nodes, $s_{jk} = 1$ indicates customer j requires skill k, and $s_{jk} = 0$ otherwise. This binary vector representation allows for a straightforward extension to skill levels in the future by generalizing to integer-valued or continuous skill proficiency levels.

Information regarding the partial solution in the state space includes the set of visited nodes V (initialized as $V = \emptyset$), current node ν (initialized as $\nu = 0$), current time τ (determined by the travel durations and time windows of the nodes visited so far, initialized as $\tau = 0$), and the current technician t (chosen in the first step, determines the skills available and travel cost on the route). To facilitate learning, we also encode the remaining routes in the state, defined as $|T|$ minus the current number of routes in the solution. Every time a technician returns to the depot, a route is finished and added to the count. Note that any feasible solution will assign at most one route per technician. Finally, the current partial solution is given as the sequence of nodes visited so far in the rollout.

Action Space $\mathcal{A}$. The action space determines which nodes can be visited at each step. We use an action mask $a \in \{0, 1\}^{|T|+|C|}$, where $a_i = 1$ indicates node i can be selected at the current step and will be part of $\mathcal{A}_s$, and $a_i = 0$ otherwise. In the first step the current technician t is set, i.e., only nodes with index $i < |T|$ are valid actions. To determine the subsequent available actions, we handle the mask separately for customer and depot nodes.

For customer nodes, we check four main conditions. First, we check for customers that have not been visited yet. Second, customers need to be reachable before their time windows end, considering the current time and travel distance to each node. Third, after servicing a customer, technicians need to be able to get back to the depot before the system end time H, i.e., we check for which nodes $\tau + s_j + d_{j0} \leq H$. Finally, only customers j can be visited for which the current technician t offers all required skills, i.e. $S_j^C \subseteq S_t^T$.

We only allow selecting a depot node if it has not been visited before, i.e., every technician should do at most one route, and if ν is a customer node or no more customers are left that can be serviced by t on the current route. This can lead to an empty action space $\mathcal{A}_s = \emptyset$ when all $|T|$ technicians have finished their routes but there are still unserved customers. To still allow training, we relax the condition that every technician can only do one route, i.e., for $\mathcal{A}_s = \emptyset$ all technicians become available again. This will lead to a number of routes $> |T|$, but these infeasible solutions can be penalized in the reward function $\mathcal{R}$.

State Transition Function $\mathcal{T}$. We update the current node ν to the last sampled action. If the last action was to go back to a depot, we update the current technician t and increase the number of routes by 1. The current time is updated to $\tau_\sigma = \max(\tau_{\sigma-1} + d_{ij}, e_j) + s_j$ if ν is a customer node, i.e., s_j cannot start before e_j, and reset to $\tau_\sigma = 0$ at depot nodes.

Reward Function $\mathcal{R}$. We use a penalized objective function consisting of the sum of the route distances and a weighted penalty term, $\mathcal{R}(\rho, \boldsymbol{x}) = -(\lambda c_r + (1 - \lambda)c_p)$. There are multiple options for how to compute c_p and how to set λ, which we describe below.

Penalty Types. We want to discourage the RL model from using more routes than available technicians. We use two simple penalty types for c_p. First, we penalize the number of excess routes E and set $c_p = |E|p_r$, where p_r is a penalty term set by the user. The reward function then is given as $\mathcal{R}^{Routes}(\rho, \boldsymbol{x}) = -(\lambda c_r + (1 - \lambda)|E|p_r)$. Second, we penalize the route costs of all excess routes, setting $c_p = c_E p_r$, where c_E is the total route cost of excess routes. The reward is defined as $\mathcal{R}^{Cost}(\rho, \boldsymbol{x}) = -(\lambda c_r + (1 - \lambda)c_E p_r)$.

Reward Weighing Strategies. The definition of the reward function allows for weighting the route cost and the penalty factor through the parameter λ. We define three strategies for setting the value of λ: (1) *fixed value*, (2) *batch-wise*, and (3) *exp. smoothed*. For *fixed value*, one constant value is used throughout training. For *batch-wise*, the average feasibility ratio is used, i.e., the ratio of feasible solutions in a training batch. In *exp. smoothed* we additionally smoothen the value from *batch-wise* with a smoothing factor of α. Both *batch-wise* and *exp. smoothed* depend on the feasibility ratio during training. The intuition behind these strategies is to lower the prominence of the penalty term once feasible solutions start to be found to avoid the search for feasibility from dominating the search for optimality.

4.2   Generating Feasible Instances

Much of the NCO literature on routing focuses on training models for problems where finding feasible solutions is trivial, but finding feasible solutions for the skill VRP is challenging. In particular, the number of vehicles is usually unlimited, however, in the skill VRP the number of technicians is inherently fixed. We posit that the model ought to be trained on feasible instances, as instances where there is no solution will always result in penalties. Thus, we need an instance generator that guarantees feasible instances.

The core idea is that instead of generating customer locations (x, y) and calculating distances $d_{i,j}$ based on the locations, we do the opposite. Each route in a solution is constructed individually and sequentially such that the total travel duration including service times is always $\leq H$. The procedure is explained in Algorithm 1.

We first partition the $|N'|$ customers into $|T|$ routes in line 2. We simply allocate the customers randomly to routes, ensuring each route has between $[|N'|/|T| - |T|, |N'|/|T| + |T|]$ customers. The technicians are assigned skills on line 3. The number of skills per technician t, n_t^S, is determined by the user. For each technician t, we then build a route separately. For the number of customers $|M_t|$ on the route, we sample service times s in line 5 and subtract their total sum from H to get the time D that is actually available for traveling in line 6. We use the same unit for time and distance, so we split the remaining time D randomly into $|M_t| + 1$ travel legs d in line 7. We then sample the time windows $[e, l]$ in line 8 and customer skills S^C in line 9. Note that the customer time windows are only generated once d and s are available to ensure feasible time windows for each customer.

To generate the coordinates in line 10, we also need travel times d and service durations s. We construct the route sequentially by starting in the depot and iteratively sampling an angle θ out of all available angles such that (1) we do not leave the defined domain for (x, y) and (2) do not get so far away from the depot that the route cannot return to the depot before time H. Based on the angle θ_ι and travel leg d_ι at each step ι we update the current location before sampling the next angle, until we return to the depot and start the next route in the instance.

With this procedure we ensure that instances are random and potentially inefficient, while having at least one feasible solution. Note that the generation process is generic, except for the skill definitions, and can therefore easily be extended to other problems.

Algorithm 1. Feasible Instance Generation

```
1: procedure GenerateInstance(T, |N'|, S, c, H, n^S)
2:     M ← PartitionCustomers(|T|, |N'|)
3:     S_t^T ← a random subset of S of size n^S ∀t ∈ T
4:     for t in T do
5:         s ← SampleServiceTimes(M_t)
6:         D ← H − Σ_{i∈M_t} s_i                    ▷ Available time for travel
7:         d ← SampleTravelLegs(M_t, D)              ▷ Distances between customers
8:         e, l ← SampleTimeWindows(M_t, d, s)
9:         S^C ← SampleCustomerSkills(S, M_t)
10:        x, y ← GenerateCoordinates(d, s)
       return Instance(S^T, S^C, s, d, e, l, x, y)
```

5   Experiments

We compare five NCO approaches with regard to their ability to find feasible, high-quality solutions on the skill VRP and consider the following research questions:

- **RQ1** Can neural solvers effectively find feasible solutions for skill VRP instances as the problem constraints become more restrictive?
- **RQ2** Does training on datasets with only feasible instances significantly improve model performance compared to training on data that includes infeasible instances?
- **RQ3** How should the reward function be designed to effectively incentivize the construction of feasible solutions?

In all experiments, models are trained with the same seed on NVIDIA A100 GPUs, inference is performed on an Nvidia GeForce RTX 4090. We use the open-source RL4CO framework [3], which contains implementations of many of the latest state-of-the-art methods and allows us to build on current research.

5.1   Experimental Setup

Instance Generation. Using the procedure described in Sect. 4.2, we generate two datasets, one with $n = 20$ customers and another with $n = 50$ customers. For both sets we set the system end time $H = 480$ and assume 3 technicians. One technician possesses all six skills and has travel cost $c_t = 2.0$, and two technicians offer four services with travel cost $c_t = 1.0$. The depot is located at the center of the instance, which is 100 by 100 units in size.

For instances with $n = 20$ customers, each customer requires three skills with probability 0.3 or one skill with probability 0.7. The service durations s_j follow a discrete distribution, where $P(s_j = 10) = 0.5$, $P(s_j = 20) = 0.3$, and $P(s_j = 30) = 0.2$. The time windows $[e_j, l_j]$ are assigned from the set $\{[0, 240], [240, 480]\}$. Each customer is assigned a time window with probability 0.5, otherwise they are assigned $[0, H]$.

For $n = 50$ customers, each customer requires three skills with probability 0.7 and one skill with probability 0.3. Service durations s_j follow the discrete distribution, where $P(s_j = 5) = 0.5$, $P(s_j = 10) = 0.35$, and $P(s_j = 20) = 0.15$. Time windows $[e_j, l_j]$ are assigned from the set $\{(0, 120), (120, 240), (240, 360), (360, 480)\}$. Each customer is assigned a time window with probability 0.8, otherwise they are assigned $[0, H]$.

Per dataset we generate 100,000 instances for training, 1,000 for validation, and 1,000 for testing. The models for $n = 20$ are trained for 100 epochs, and for $n = 50$ for 500 epochs. All instances are scaled down such that their locations lie in $[0, 1]$, time windows and service durations are also scaled down accordingly.

Models. The models we use are AM [13], POMO [14], MVMoE-POMO [20], PolyNet [11], and SymNCO [12]. We set the batch size for training, validation and test to 128 and the learning rate to 0.0001. For all models we use the default parameter settings as defined in the RL4CO framework.

During training, the AM approach uses a greedy rollout as baseline in Eq. 12 and SymNCO uses its own baseline (see [12] for details), the other models use a shared baseline, i.e., the mean reward in the training batch is used as the baseline $b(\cdot)$ for the loss calculation.

At test time, all models except AM use instance augmentation. POMO, MVMoE-POMO and PolyNet use dihedral augmentation, i.e., instances are rotated and flipped in the domain, to achieve 8 different representations of the same instance [14, Table 1]. SymNCO uses symmetric augmentation, i.e., it randomly samples 10 rotation angles $\phi \in (0, 4\pi)$ and rotates the instances accordingly. All models except PolyNet perform greedy rollouts during inference. The AM approach generates a single solution per instance, while POMO and MVMoE-POMO generate $8 \times |T|$ solutions, SymNCO generates $10 \times |T|$, and PolyNet produces 8×800 solutions per instance.

Baselines. As baselines we use version Gurobi 11.0.3 [8] and PyVRP 0.11.0a0 [19]. For Gurobi we set a per-instance time limit of 5 h and for PyVRP 10 s. Note that PyVRP uses integer values in its objective function, thus its values are not as exact as for Gurobi. All gaps are computed as the gap to the value found by Gurobi, which is not always the optimal value due to the timeout.

Evaluation Metrics. Our evaluation considers a total of four metrics. The feasibility *count* tells us for how many of the 1,000 instances the neural solvers find feasible solutions. The *cost* is the average route cost over the best feasible solution found for each instance (if any feasible solution was found). For the *gap %* calculation we take the percentage deviation of *cost* from the solutions provided by Gurobi. Lastly, we provide the runtime *time (s)* during inference.

5.2   RQ1: Feasibility Under Tight Constraints

We evaluate five neural solvers on their ability to find feasible solutions for skill VRP instances as the problem constraints become more restrictive, comparing their performance on $n = 20$ and $n = 50$ in Table 4. Note that the $n = 50$ instances also have tighter time windows and higher skill requirements. For an intuition on how difficult it is to find feasible solutions for these instances, we include the training curves reporting the *average* feasibility rates for all five neural solvers while training with $\mathcal{R}^{Routes}$ on $n = 20$ and $n = 50$ in Fig. 3.

Except for AM with $\mathcal{R}^{Cost}$, all models find feasible solutions for $n = 20$, but PolyNet provides considerably lower gaps than the other models. For $n = 50$, the count of feasible solutions decreases for all models, but to different degrees. AM finds feasible solutions for the lowest number of instances, and is the only model that does not use augmentation during inference, while SymNCO finds them for the most instances. The effectiveness of reward functions varies across models. While POMO achieves a higher count and lower gap using $\mathcal{R}^{Cost}$, the

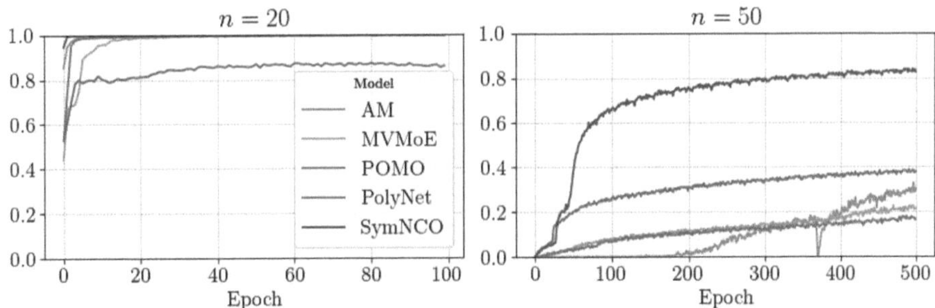

Fig. 3. Average feasibility rates during training using $\mathcal{R}^{Routes}$.

opposite is true for PolyNet, i.e., it achieves a higher count and lower gap with $\mathcal{R}^{Routes}$. Across all experiments in Table 4, however, PolyNet provides the lowest gaps, indicating that solution diversity may be crucial for finding high-quality, feasible solutions to constrained problems as the skill VRP.

5.3   RQ2: Training on Feasible Vs. Purely Random Instances

To investigate the importance of feasible instances for training, we perform two sets of experiments on different training instances, generated such that feasibility is not guaranteed. For the first set of instances we generate locations uniformly in the domain, a standard approach in NCO research [2,3,13,14]. This can be considered *out-of-distribution* for instances generated according to Algorithm 1. We base the second set of instances on Algorithm 1 and consider these *in-distribution*. For validation and testing we use the same instances as in Table 4.

Out-Of-Distribution Location Generation. For the first set of experiments, we generate locations randomly in the domain without enforcing a total travel time $\leq H$, as is commonly done in NCO research. With increasing node count we expect more instances to be infeasible, simply due to increased travel duration with constant H and $|T|$. To get a rough approximation of the feasibility rates in these instances, we solve subsets in PyVRP. For $n = 20$, about 94% of instances are shown to be feasible within 10 s, for $n = 50$ none could be. We note that we have not proven the instances to be infeasible with Gurobi due to the high computational expense of doing so for a large training set.

In Table 5 we find that for $n = 20$ all models find feasible solutions for all instances, with MVMoE-POMO even achieving lower gaps than in Table 4. The other models come close to their previously reported gaps. For $n = 50$, however, the models struggle to find feasible solutions, with POMO finding the most (314) for $\mathcal{R}^{Routes}$ and PolyNet (823) for $\mathcal{R}^{Cost}$. While the POMO-based models perform better with $\mathcal{R}^{Routes}$, the other models clearly perform better with $\mathcal{R}^{Cost}$, with PolyNet achieving the lowest gaps across experiments.

Table 4. Performance of NCO models in finding feasible solutions.

	$\mathcal{R}^{Routes}$							
	$n = 20$				$n = 50$			
	count	cost	gap %	time (s)	count	cost	gap %	time (s)
Gurobi	1000	642.272	-	106.249	1000	986.901	-	12398.907
PyVRP	1000	642.404	0.000	10.001	1000	984.077	−0.003	10.001
AM	**1000**	851.383	35.242	0.047	380	1248.194	27.370	0.090
POMO	**1000**	742.646	15.748	0.140	907	1328.356	35.654	0.282
MVMoE-POMO	**1000**	765.171	19.330	0.175	907	1328.464	35.881	0.370
PolyNet	**1000**	678.310	**5.592**	0.144	974	1043.798	**5.965**	0.300
SymNCO	**1000**	808.252	27.048	0.075	**977**	1318.571	34.253	0.200
	$\mathcal{R}^{Cost}$							
	$n = 20$				$n = 50$			
AM	999	861.726	36.885	0.039	483	1398.502	42.708	0.091
POMO	**1000**	745.470	16.227	0.141	935	1106.688	13.045	0.293
MVMoE-POMO	**1000**	772.037	20.456	0.176	861	1168.011	19.828	0.382
PolyNet	**1000**	677.313	**5.425**	0.150	958	1058.284	**7.533**	0.301
SymNCO	**1000**	808.612	27.082	0.078	**981**	1308.991	33.303	0.200

In-Distribution Location Generation. For the second set of experiments, we generate instances using Algorithm 1 and then shuffle customer attributes until the resulting instances could not be solved by PyVRP within 30 s. We do this for $n = 50$, reporting results in Table 6, as the $n = 20$ instances are too easily solved even after perturbation, not allowing us to generate a set of 100,000 instances within a reasonable time.

In Table 6 we find that for $\mathcal{R}^{Routes}$ SymNCO finds the most feasible solutions and PolyNet has the lowest gap, distinctly lower than the equivalent result in Table 5, while achieving a considerably higher count. For $\mathcal{R}^{Cost}$ PolyNet finds the most feasible solutions, even more than in Table 4, but the gap is larger than with $\mathcal{R}^{Routes}$ and considerably larger than when trained on guaranteed feasible instances (Table 4).

Overall, the models seem to learn more effectively when trained on guaranteed feasible instances. PolyNet is an exception, and also learns effectively from in-distribution instances not guaranteed to be feasible. This holds promise for industry applications where making an instance generator that produces feasible instances may be difficult, even though the gaps are larger than when trained on only feasible instances.

Table 5. Training on randomly generated instances not guaranteed to be feasible.

| | $\mathcal{R}^{Routes}$ | | | | | | |
| | $n = 20$ | | | | $n = 50$ | | |
	count	cost	gap %	time (s)	count	cost	gap %	time (s)
Gurobi	1000	642.272	-	106.249	1000	986.901	-	12398.907
PyVRP	1000	642.404	0.000	10.001	1000	984.077	−0.003	10.001
AM	**1000**	873.002	38.676	0.040	78	1424.730	45.384	0.091
POMO	**1000**	760.313	18.767	0.149	**314**	1383.196	46.314	0.289
MVMoE-POMO	**1000**	763.432	19.054	0.176	258	1399.219	48.727	0.375
PolyNet	**1000**	691.736	**7.899**	0.147	50	1156.012	**25.422**	0.303
SymNCO	**1000**	833.354	31.020	0.070	29	1444.435	53.448	0.206

| | $\mathcal{R}^{Cost}$ | | | | | | |
	$n = 20$				$n = 50$		
AM	**1000**	874.195	38.865	0.039	111	1363.708	39.157	0.093
POMO	**1000**	757.684	18.355	0.142	269	1354.237	42.511	0.291
MVMoE-POMO	**1000**	772.261	20.381	0.176	239	1373.085	43.818	0.378
PolyNet	**1000**	679.317	**5.807**	0.150	**823**	1275.300	**30.397**	0.311
SymNCO	**1000**	833.918	31.121	0.077	664	1339.627	38.003	0.210

5.4   RQ3: Impact of Different Penalty Strategies

All experiments in this section are performed on SymNCO with identical settings, only varying individual parameters to examine their respective impact on solution quality during inference.

Fixed Versus Dynamic λ. In Sect. 4.1 we introduce three simple strategies to determine the penalization factor λ. We compare the three strategies with regard to their performance on $n = 20$ and $n = 50$ in Table 7. For fixed λ we further evaluate different values. For exponentially smoothed λ we set $\alpha = 0.01$. For $n = 20$, both reward functions find feasible solutions more reliably with fixed λ, while the dynamic λ strategies lead to lower gaps. For $n = 50$, $\mathcal{R}^{Routes}$ achieves the highest count and lowest gap with exponentially smoothed λ. $\mathcal{R}^{Cost}$ achieves the highest count with fixed $\lambda \in [0.2, 0.5]$, but the values are lower than the equivalent results for $\mathcal{R}^{Routes}$. Overall, no λ-strategy is clearly dominant.

Effect of Penalty Size. We investigate the effect of different values for the penalty term p_r. We hypothesize that higher values ought to lead to more feasible solutions, but at the expense of finding solutions with low costs. Table 8 provides our results. We find that for $n = 20$ SymNCO indeed achieves the lowest gaps with $p_r = 10$ across both reward functions. For $n = 50$, $p_r = 100$ seems to

Table 6. Training on instances perturbed to be hardly feasible.

	$\mathcal{R}^{Routes}$			$\mathcal{R}^{Cost}$		
	count	cost	gap %	count	cost	gap %
Gurobi	1000	986.901	-	1000	986.901	-
PyVRP	1000	984.077	−0.003	1000	984.077	-0.003
AM	64	1431.158	46.040	170	1365.029	39.292
POMO	744	1367.769	41.200	355	1244.830	31.104
MVMoE-POMO	739	1381.087	42.876	238	1389.658	47.305
PolyNet	673	1102.454	**14.108**	**986**	1210.036	**22.902**
SymNCO	**763**	1364.073	41.040	841	1310.710	34.588

Table 7. Different strategies for setting λ.

n	λ strategy	$\mathcal{R}^{Routes}$			$\mathcal{R}^{Cost}$		
		count	cost	gap %	count	cost	gap %
	fixed 0.1	**1000**	808.252	27.048	**1000**	808.612	27.082
	fixed 0.2	**1000**	805.919	26.650	**1000**	808.728	27.095
20	fixed 0.5	**1000**	797.804	25.371	**1000**	787.483	23.733
	batch-wise	999	723.006	12.864	993	721.490	**12.442**
	exp. smoothed	996	696.706	**8.379**	996	721.505	12.502
	fixed 0.1	977	1318.571	34.253	981	1308.991	33.303
	fixed 0.2	993	1300.029	32.112	**983**	1313.881	33.663
50	fixed 0.5	987	1299.946	32.254	**983**	1307.740	33.009
	batch-wise	979	1289.506	31.235	967	1297.098	32.245
	exp. smoothed	**995**	1277.704	**29.799**	981	1284.873	**30.750**

strike a better balance between finding more feasible solutions and finding low-cost solutions, with results varying slightly between reward functions. Overall, it depends on the difficulty and size of the problem which penalty size is preferred.

Alternative Reward Functions. Neither reward function dominates the other across all problem sizes and models. While $\mathcal{R}^{Routes}$ performs better in many experiments, there are settings where $\mathcal{R}^{Cost}$ clearly outperforms it. In summary, overall performance is highly dependent on both the reward function and the model chosen, and these must thus be selected jointly.

Table 8. The impact of different penalty sizes.

n	p_r	$\mathcal{R}^{Routes}$			$\mathcal{R}^{Cost}$		
		count	cost	gap %	count	cost	gap %
	10	**1000**	781.981	**22.877**	**1000**	790.494	**24.204**
20	100	**1000**	793.697	24.739	**1000**	806.035	26.647
	1000	**1000**	808.252	27.048	**1000**	808.612	27.082
	10	**990**	1303.906	32.614	978	1307.522	**33.120**
50	100	**990**	1301.255	**32.347**	**982**	1312.947	33.660
	1000	977	1318.571	34.253	981	1308.991	33.303

6 Conclusion

In this work, we applied NCO approaches to the challenging skill VRP. We formulated the solution generation process as a MDP, incorporating penalty mechanisms for infeasible solutions, and evaluated various learning-based optimization methods from the literature. To facilitate effective training, we designed an instance generator that ensures at least one feasible solution per problem instance. Furthermore, we explored different penalization and reward strategies to guide the models toward feasible solutions. Our results demonstrate that NCO approaches can effectively tackle complex routing problems, opening avenues for future research on more constrained and real-world VRP scenarios.

Acknowledgements. This work was funded by the Deutsche Forschungsgemeinschaft (DFG, German Research Foundation) - No. 521243122. The authors gratefully acknowledge the computing time provided to them on the high-performance computer Noctua 2 at the NHR Center PC2. These are funded by the Federal Ministry of Education and Research and the state governments participating on the basis of the resolutions of the GWK for the national high-performance computing at universities (www.nhr-verein. de/unsere-partner).

Disclosure of Interests. The authors have no competing interests.

References

1. Bellman, R.: A markovian decision process. Indiana Univ. Math. J. **6**, 679–684 (1957)
2. Bello, I., Pham, H., Le, Q.V., Norouzi, M., Bengio, S.: Neural combinatorial optimization with reinforcement learning. In: 5th International Conference on Learning Representations, ICLR 2017, Toulon, France, 24–26 April 2017. Workshop Track Proceedings. OpenReview.net (2017)
3. Berto, F., et al.: RL4CO: an extensive reinforcement learning for combinatorial optimization benchmark. In: 31st SIGKDD Conference on Knowledge Discovery and Data Mining - Datasets and Benchmarks Track (2025)

4. Cappanera, P., Gouveia, L., Scutellà, M.G.: Models and valid inequalities to asymmetric skill-based routing problems. EURO J. Transp. Logistics **2**(1), 29–55 (2013)
5. Cappanera, P., Gouveia, L., Scutellà, M.G.: The skill vehicle routing problem. In: Pahl, J., Reiners, T., Voß, S. (eds.) Network Optimization, pp. 354–364. Lecture Notes in Computer Science vol. 6701. Springer (2011)
6. Cappanera, P., Requejo, C., Scutellà, M.G.: Temporal constraints and device management for the skill VRP: mathematical model and lower bounding techniques. Comput. Operations Res. **124**, 105054 (2020)
7. Gast Zepeda, N., Hottung, A., Tierney, K.: Deep learning in search heuristics. In: Martí, R., Pardalos, P.M., Resende, M.G. (eds.) Handbook of Heuristics, pp. 1–18. Springer Nature Switzerland, Cham (2025)
8. Gurobi Optimization, LLC: Gurobi Optimizer Reference Manual (2024)
9. Hottung, A., Kwon, Y.D., Tierney, K.: Efficient active search for combinatorial optimization problems. In: International Conference on Learning Representations (2022)
10. Hottung, A., Wong-Chung, P., Tierney, K.: Neural deconstruction search for vehicle routing problems. Trans. Mach. Learn. Res. (2025)
11. Hottung, A., Mahajan, M., Tierney, K.: PolyNet: learning diverse solution strategies for neural combinatorial optimization. In: International Conference on Learning Representations (2025)
12. Kim, M., Park, J., Park, J.: Sym-NCO: leveraging symmetricity for neural combinatorial optimization. In: Koyejo, S., Mohamed, S., Agarwal, A., Belgrave, D., Cho, K., Oh, A. (eds.) Advances in Neural Information Processing Systems. vol. 35, pp. 1936–1949. Curran Associates, Inc. (2022)
13. Kool, W., van Hoof, H., Welling, M.: Attention, learn to solve routing problems! In: International Conference on Learning Representations (2019)
14. Kwon, Y.D., Choo, J., Kim, B., Yoon, I., Gwon, Y., Min, S.: POMO: policy optimization with multiple optima for reinforcement learning. In: Advances in Neural Information Processing Systems. vol. 33, pp. 21188–21198. Curran Associates, Inc. (2020)
15. Nazari, M., Oroojlooy, A., Snyder, L., Takac, M.: Reinforcement learning for solving the vehicle routing problem. In: Bengio, S., Wallach, H., Larochelle, H., Grauman, K., Cesa-Bianchi, N., Garnett, R. (eds.) Advances in Neural Information Processing Systems. vol. 31. Curran Associates, Inc. (2018)
16. Paraskevopoulos, D.C., Laporte, G., Repoussis, P.P., Tarantilis, C.D.: Resource constrained routing and scheduling: Review and research prospects **263**(3), 737–754 (2017)
17. Sutton, R.S., McAllester, D., Singh, S., Mansour, Y.: Policy gradient methods for reinforcement learning with function approximation. In: Advances in Neural Information Processing Systems, vol. 12 (1999)
18. Vinyals, O., Fortunato, M., Jaitly, N.: Pointer networks. In: Cortes, C., Lawrence, N., Lee, D., Sugiyama, M., Garnett, R. (eds.) Advances in Neural Information Processing Systems. vol. 28. Curran Associates, Inc. (2015)
19. Wouda, N.A., Lan, L., Kool, W.: PyVRP: a high-performance VRP solver package. INFORMS J. Comput. **36**(4), 943–955 (2024)
20. Zhou, J., Cao, Z., Wu, Y., Song, W., Ma, Y., Zhang, J., Chi, X.: MVMoE: multi-task vehicle routing solver with mixture-of-experts. In: Proceedings of the 41st International Conference on Machine Learning, pp. 61804–61824. PMLR (2024)

CGD: Modifying the Loss Landscape by Gradient Regularization

Shikhar Saxena$^{(\boxtimes)}$, Tejas Bodas , and Arti Yardi

International Institute of Information Technology (IIIT) Hyderabad,
Hyderabad, India
`shikhar.saxena@research.iiit.ac.in`, `{tejas.bodas,arti.yardi}@iiit.ac.in`

Abstract. Line-search methods are commonly used to solve optimization problems. The simplest line search method is steepest descent where one always moves in the direction of the negative gradient. Newton's method on the other hand is a second-order method that uses the curvature information in the Hessian to pick the descent direction. In this work, we propose a new line-search method called Constrained Gradient Descent (CGD) that implicitly changes the landscape of the objective function for efficient optimization. CGD is formulated as a solution to the constrained version of the original problem where the constraint is on a function of the gradient. We optimize the corresponding Lagrangian function thereby favourably changing the landscape of the objective function. This results in a line search procedure where the Lagrangian penalty acts as a control over the descent direction and can therefore be used to iterate over points that have smaller gradient values, compared to iterates of vanilla steepest descent. We establish global linear convergence rates for CGD and provide numerical experiments on synthetic test functions to illustrate the performance of CGD. We also provide two practical variants of CGD, CGD-FD which is a Hessian free variant and CGD-QN, a quasi-Newton variant and demonstrate their effectiveness.

Keywords: Numerical Optimization · Gradient Regularization

1 Introduction

Line-search methods such as gradient descent and Newton's method have been extremely popular in solving unconstrained optimization problems [17]. However, vanilla version of such algorithms typically produce unsatisfactory results when applied to large scale optimization problems such as those involving deep neural networks. In such problems, the objective/loss function has a very non-linear landscape that is embedded with several local maxima and minima making them difficult to optimize [12]. To tackle this problem, one of the approaches used is to modify the objective/loss function in such a way that the new function is easier to optimize. This is typically done by adding a regularization term that will make the loss landscape more smooth, and which would in turn reduce the chances of the optimizer reaching sub-optimal minima, thereby improving generalizability [1,5,7,9,22,23].

Y. Zhang et al. (Eds.): LION 2025, LNCS 15745, pp. 267–283, 2026.
https://doi.org/10.1007/978-3-032-09192-5_18

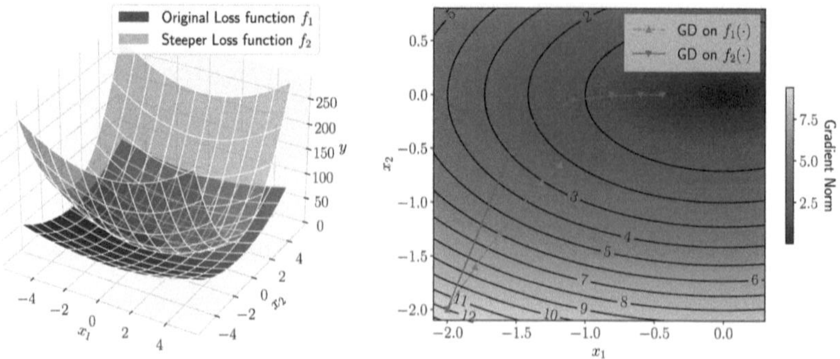

Fig. 1. *(Left)* Loss function $f_1(\mathbf{x}) = x_1^2 + 2x_2^2$ and a steeper loss function $f_2(\mathbf{x})$. *(Right)* 5 steps of GD on functions $f_1(\mathbf{x})$ and $f_2(\mathbf{x})$ with a fixed step-size $\alpha = 0.05$. The contours and the gradient norm heatmap are over the function $f_1(\mathbf{x})$.

In this work, we focus on this idea of modifying the objective function for effective numerical optimization and propose an algorithmic paradigm to achieve it. To illustrate the key idea, consider minimizing a loss function $f_1(\cdot)$ over $\mathbb{R}^2$ as shown in Fig. 1 *(Left)*. Additionally consider a steeper loss function $f_2(\cdot)$ such that for every $\mathbf{x} \in \mathbb{R}^2$ we have $f_1(\mathbf{x}) \leq f_2(\mathbf{x})$. Note that the functions are such that their minima coincides. Now perform gradient descent (GD) with a fixed step-size on both these functions. Their respective trajectories are plotted in Fig. 1 *(Right)* on the contours of f_1. It is clear from the figure that the iterates of GD on f_2 are much closer to the minima as compared to iterates of GD on f_1. We make this observation a focal point of this work and investigate if one can achieve the latter iterates (iterates from GD on f_2) on the former (f_1).

Towards this, we propose Constrained gradient descent (CGD), a variant of the GD algorithm which achieves this by constraining the gradient norm to be appropriately small. Instead of solving this constrained optimization problem, we consider the Lagrangian of this function and perform GD on it. The Lagrangian parameter λ controls the penalty on gradient norm and thereby controls the steepness of the modified function. It is on this modified function that we seek to apply GD to possibly attain iterates that are much closer to the local minima as compared to GD iterates. From Fig. 1 *(Right)* we additionally see that compared to GD, the CGD iterates (which is nothing but GD iterates on f_2) have a lower gradient norm. Since the path to minima is over points with lower gradient norm, this we believe is an attractive feature and even amounts to better generalization properties for high dimensional functions such as loss landscapes of neural networks. The idea of flat minima and their attractiveness goes long back to Hochreiter and Schmidhuber [8]: An optimal minimum is considered "flat" if the test error changes less in its neighbourhood. Keskar et al. [11] and Chaudhari et al. [4] observe better generalization results for neural networks at flat minima. Gradient regularization has only been recently explored, that too from a numerical perspective for deep neural networks (DNNs) where it has been shown that fixed learning rates result in implicit gradient regularization [1,9,19,22].

A key aim of this work is to understand gradient regularization from the perspective of a numerical optimization algorithm and identify properties and features that may not have been obvious earlier. We believe CGD and its variants discussed in this work have the ability to find flatter minima, something which is useful while training neural networks. We summarize our contributions below:

- We propose a new line-search procedure called Constrained Gradient Descent (CGD) that performs gradient descent on a gradient regularized loss function. We also provide global linear convergence for CGD in Theorem 1.
- While CGD requires the Hessian information, we also propose a first-order variant of CGD using finite-difference approximation of the Hessian called *CGD-FD*. We define appropriate stopping criteria in CGD-FD for settings where gradient computation can be expensive.
- We conduct experiments over synthetic test functions to compare the performance of CGD and its variants compared to standard line-search procedures.
- Our work also provides new insights to existing gradient regularization based methods. In fact, we re-interpret and identify pitfalls in the Explicit Gradient Regularization (EGR) Method [1] using our formulation.

The remainder of the paper is organized as follows. In the next section, we recall some preliminaries on line-search methods. We then discuss the CGD algorithm and propose its variants. We then illustrate the performance of our algorithm on several test functions and conclude with a discussion on future directions.

2   Notation and Preliminaries

The set of real numbers and non-negative real numbers is denoted by $\mathbb{R}$ and $\mathbb{R}^+$ respectively. We consider a vector $\mathbf{x} \in \mathbb{R}^n$ as a column vector given by $\mathbf{x} = \begin{bmatrix} x_1 & x_2 & \dots & x_n \end{bmatrix}^T$. We use a boldface letter to denote a vector and lowercase letters (with subscripts) to denote its components. L^p-norm of a vector $\mathbf{x}$ is defined as $\|\mathbf{x}\|_p = (\sum_i |x_i|^p)^{1/p}$. Setting $p = 2$ gives us L^2-norm: $\|\mathbf{x}\| \triangleq \|\mathbf{x}\|_2 = \sqrt{\sum_i x_i^2} = \sqrt{\mathbf{x}^T \mathbf{x}}$. $\lambda_{\max}(\cdot)$ and $\lambda_{\min}(\cdot)$ denote the maximum and minimum eigenvalue of a matrix respectively. Norm of a matrix is assumed to be the spectral norm *i.e.*, $\|A\| = \sqrt{\lambda_{\max}(A^T A)}$. I_n denotes the identity matrix of size $n \times n$. All zero vector of length n is denoted by $\mathbf{0}_n$.

2.1   Line-Search Methods

Let $f(\mathbf{x})$ be a twice differential function with domain $\mathcal{D} \subseteq \mathbb{R}^n$ and codomain $\mathbb{R}$, i.e., $f : \mathcal{D} \to \mathbb{R}$. Let $\nabla f(\mathbf{x}_k)$ and $H(\mathbf{x}_k)$ denote the gradient and Hessian of the function $f(\mathbf{x})$ evaluated at point $\mathbf{x}_k \in \mathcal{D}$. For the sake of simplicity, we will also use the notation ∇f_k and H_k to denote $\nabla f(\mathbf{x}_k)$ and $H(\mathbf{x}_k)$ respectively. For the given function $f(\mathbf{x})$, we consider the problem of finding its minimizer, i.e., we wish to find $\mathbf{x}^* \in \mathcal{D}$ such that

$$\mathbf{x}^* = \arg\min_{\mathbf{x} \in \mathcal{D}} f(\mathbf{x}). \tag{1}$$

We focus on the situation where $\mathbf{x}^*$ is obtained using an iterative line-search procedure (for details refer [13, Ch. 3]). The steepest descent (*gradient descent*) method, Newton's method, and quasi-Newton's method are some examples of line-search methods. In an iterative algorithm, the key idea is to begin with an initial guess $\mathbf{x}_0$ for the minimizer and generate a sequence of vectors $\mathbf{x}_0, \mathbf{x}_1, \ldots$ until convergence (or until desired level of accuracy is achieved). In such algorithms, $\mathbf{x}_{k+1}$ is obtained using $\mathbf{x}_k$ using a pre-defined update rule.

For line-search methods, the update rule can be written in a general form as, $\mathbf{x}_{k+1} = \mathbf{x}_k + \alpha \mathbf{p}_k$ where $\alpha \in \mathbb{R}^+$ is the step-size (or learning rate) and $\mathbf{p}_k \in \mathbb{R}^n$ corresponds to the direction in the kth iteration. For $\mathbf{p}_k$ to be a *descent* direction, the following condition must hold:

$$\nabla f_k^T \mathbf{p}_k < 0. \tag{2}$$

2.2 Quasi-Newton Methods

Quasi-Newton methods (QN methods) are line-search methods that maintain an approximation of the inverse of the Hessian to emulate Newton's direction. The iterates here are of the form,

$$\mathbf{x}_{k+1} = \mathbf{x}_k - \alpha G_k \nabla f_k \tag{3}$$

where G_k is a positive definite matrix that is updated at every step to approximate the inverse Hessian.

We consider DFP and BFGS algorithms that update G_k using symmetric rank-two updates at each descent step [3,6]. The update equations for DFP and BFGS are given as follows (for more details refer [13, Ch. 6]):

$$\textbf{(DFP)}\ G_{k+1} = G_k + \frac{\mathbf{s}_k \mathbf{s}_k^T}{\mathbf{y}_k^T \mathbf{s}_k} - \frac{G_k \mathbf{y}_k \mathbf{y}_k^T G_k}{\mathbf{y}_k^T G_k \mathbf{y}_k}$$

$$\textbf{(BFGS)}\ G_{k+1} = \left(I_n - \frac{\mathbf{s}_k \mathbf{y}_k^T}{\mathbf{y}_k^T \mathbf{s}_k} \right) G_k \left(I_n - \frac{\mathbf{y}_k \mathbf{s}_k^T}{\mathbf{y}_k^T \mathbf{s}_k} \right) + \frac{\mathbf{s}_k \mathbf{s}_k^T}{\mathbf{y}_k^T \mathbf{s}_k}$$

where $\mathbf{s}_k = \mathbf{x}_{k+1} - \mathbf{x}_k$ and $\mathbf{y}_k = \nabla f_{k+1} - \nabla f_k$.

We can derive the corresponding Hessian approximation $\tilde{G}_k$ from the update steps for G_k by applying the Sherman-Morrison-Woodbury formula [13, App. A]. These are given as follows:

$$\textbf{(DFP)}\ \tilde{G}_{k+1} = \left(I_n - \frac{\mathbf{y}_k \mathbf{s}_k^T}{\mathbf{y}_k^T \mathbf{s}_k} \right) \tilde{G}_k \left(I_n - \frac{\mathbf{s}_k \mathbf{y}_k^T}{\mathbf{y}_k^T \mathbf{s}_k} \right) + \frac{\mathbf{y}_k \mathbf{y}_k^T}{\mathbf{y}_k^T \mathbf{s}_k} \tag{4}$$

$$\textbf{(BFGS)}\ \tilde{G}_{k+1} = \tilde{G}_k + \frac{\mathbf{y}_k \mathbf{y}_k^T}{\mathbf{y}_k^T \mathbf{s}_k} - \frac{\tilde{G}_k \mathbf{s}_k \mathbf{s}_k^T \tilde{G}_k}{\mathbf{s}_k^T \tilde{G}_k \mathbf{s}_k} \tag{5}$$

3 Constrained Gradient Descent

In this section we propose *Constrained Gradient Descent* (CGD), an iterative line-search method to find a minimizer $\mathbf{x}^*$ of the given function $f(\mathbf{x})$ (see Eq. 1). The key idea in our approach is to focus on the set of $\mathbf{x} \in \mathcal{D}$ such that $\nabla f(\mathbf{x})$ is close to zero. For this consider a general constrained optimization problem:

$$\mathbf{x}^* = \underset{\substack{\mathbf{x} \in \mathcal{D} \\ h(\nabla f(\mathbf{x})) \leq \epsilon}}{\arg\min} \ f(\mathbf{x}) \tag{6}$$

where the constraint $h(\cdot)$ is defined on the gradient and ϵ is a small positive real number. The unconstrained optimization problem corresponding to Eq. 6 is obtained by penalizing the constraint with a Lagrange multiplier $\lambda > 0$:

$$\mathbf{x}^* = \underset{\mathbf{x} \in \mathcal{D}}{\arg\min} \left[f(\mathbf{x}) + \lambda \ h\left(\nabla f(\mathbf{x})\right) \right] \tag{7}$$

Note that ϵ doesn't affect the optimization and hence ignored from the objective. Also, observe that such a formulation provides us with a *modified* loss function to optimize over. A suitable constraint $h(\cdot)$ will make the objective steeper while also keeping the stationary points intact. We call this *penalization* or using a *gradient penalty* since the objective is penalized at points based on its gradient value. For CGD, we consider $h(\cdot)$ to be the square of L^2-norm of the gradient.

$$\mathbf{x}^* = \underset{\mathbf{x} \in \mathcal{D}}{\arg\min} \, g(\mathbf{x}) \triangleq f(\mathbf{x}) + \lambda\|\nabla f(\mathbf{x})\|^2 = f(\mathbf{x}) + \lambda \left(\nabla f(\mathbf{x})^T \nabla f(\mathbf{x})\right) \tag{8}$$

Steepest descent iterates over $g(\cdot)$ in Eq. 8 are given as

$$\begin{aligned}
\mathbf{x}_{k+1} &= \mathbf{x}_k - \alpha \nabla g_k \\
&= \mathbf{x}_k - \alpha \left(\nabla f_k + 2\lambda H_k \nabla f_k\right) \\
&= \mathbf{x}_k - \alpha B_k \nabla f_k
\end{aligned} \tag{9}$$

where $B_k \triangleq I_n + 2\lambda H_k$. Thus, $\mathbf{p}_k = -B_k \nabla f_k$ which is the direction taken at the kth iteration by CGD. Revisiting the example in Fig. 1, the function $f_1(\mathbf{x}) = x_1^2 + 2x_2^2$ was penalized with the square of L^2-norm of gradient as given in Eq. 8. Thus, the modified loss function $f_2(\mathbf{x}) = x_1^2 + 2x_2^2 + \lambda \left((2x_1)^2 + (4x_2)^2\right) = (1 + 4\lambda)x_1^2 + 2(1 + 8\lambda)x_2^2$ where λ was chosen to be 0.4.

We now provide a lemma that investigates if the penalized objective function has stationary points which are different from the original function and if so characterizes them.

Lemma 1. *Let $\mathcal{S}_{\hat{\mathbf{x}}}$ and $\mathcal{S}_{\mathbf{x}^*}$ be the stationary points of $f(\mathbf{x})$ and $g(\mathbf{x})$ as defined in Eq. 8. Then, $\mathcal{S}_{\hat{\mathbf{x}}} \subseteq \mathcal{S}_{\mathbf{x}^*}$. Furthermore, for any $\mathbf{x}^* \in \mathcal{S}_{\mathbf{x}^*}$ one of the following is true: (a) $\mathbf{x}^* \in \mathcal{S}_{\hat{\mathbf{x}}}$ or (b) $\nabla f(\mathbf{x}^*)$ is an eigenvector of $H(\mathbf{x}^*)$ with the eigenvalue $-\frac{1}{2\lambda}$.*

Proof. From Eq. 9, for any $\hat{\mathbf{x}} \in \mathcal{S}_{\hat{\mathbf{x}}}$, $\nabla f(\hat{\mathbf{x}}) = \mathbf{0}_n \Rightarrow \nabla g(\hat{\mathbf{x}}) = \mathbf{0}_n$. So, $\mathcal{S}_{\hat{\mathbf{x}}} \subseteq \mathcal{S}_{\mathbf{x}^*}$ holds trivially. Now for any $\mathbf{x}^* \in \mathcal{S}_{\mathbf{x}^*}$, $\nabla g(\mathbf{x}^*) = \mathbf{0}_n \Rightarrow (I_n + 2\lambda H(\mathbf{x}^*))\nabla f(\mathbf{x}^*) = \mathbf{0}_n$. If $\nabla f(\mathbf{x}^*) \neq \mathbf{0}_n$, we have $H(\mathbf{x}^*)\nabla f(\mathbf{x}^*) = -\frac{1}{2\lambda}\nabla f(\mathbf{x}^*)$ which proves the result. $\qquad\square$

The above lemma illustrates that additional stationary points could possibly be introduced and some of these points could also be local minima. The lemma also characterizes conditions under which this is true and therefore such points can easily be detected. A perturbation from the current λ in that case results in the iterate to descend further, possibly moving towards a better minimum.

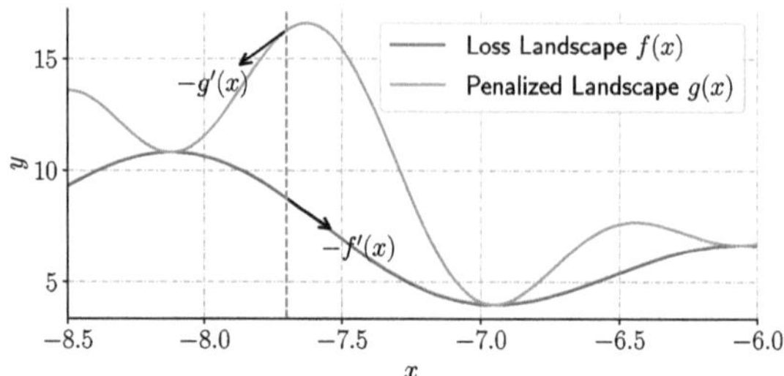

Fig. 2. Penalization introducing new stationary points at $x \approx -6.5$ and $x \approx -7.6$, and changing the local maximum at $x \approx -8.1$ to a local minimum.

Further note that within the current scheme, the nature of stationary points of $f(\mathbf{x})$ might change in $g(\mathbf{x})$. Particularly, the local maxima and saddle points of $f(\mathbf{x})$ might become local minima in $g(\mathbf{x})$ depending on λ. This happens because penalization does not change the function value at the stationary point while spiking up the function values at points in the neighbourhood.

As a result, descent on the penalized function might actually cause ascent over the original loss function. For example in Fig. 2, observe that at $x = -7.7$ (marked with the dashed line), steepest descent along $-\nabla g(\mathbf{x})$ direction would actually cause an ascent over the original loss function.

To fix this behaviour, we ensure that we only move along the direction $-\nabla g(\mathbf{x})$ when it is a descent direction. Therefore, we only move along $-\nabla g(\mathbf{x})$, if $\nabla f(\mathbf{x})^T \nabla g(\mathbf{x}) > 0$ (from Eq. 2). Otherwise, we set $\lambda = 0$ at this point and move along the $-\nabla f(\mathbf{x})$ direction. Note that this simple check also helps us to avoid stopping at artificially introduced stationary points. We summarize CGD in Algorithm 1.

3.1   Global Linear Convergence of CGD

We begin with the following definition followed by a theorem on convergence guarantees for CGD.

Algorithm 1. Constrained Gradient Descent (CGD)

Input: Objective function $f : \mathcal{D} \to \mathbb{R}$, initial point $\mathbf{x}_0$, max iterations T, step-size α, regularization coefficient λ.

Output: Final point $\mathbf{x}_T$.

1: **for** iteration $k = 0, \ldots, T-1$ **do**
2:     $\mathbf{p}_k \leftarrow - \left(I_n + 2\lambda_k H_k\right) \nabla f_k$
3:     **if** $\nabla f_k^T \mathbf{p}_k < 0$ **then**   ▷ *Check if $\mathbf{p}_k$ is a Descent Direction*
4:        $\mathbf{x}_{k+1} \leftarrow \mathbf{x}_k + \alpha \mathbf{p}_k$
5:     **else**
6:        $\mathbf{x}_{k+1} \leftarrow \mathbf{x}_k - \alpha \nabla f_k$
7: **return** $\mathbf{x}_T$

Definition 1 (PL Inequality [15]). *A function $f : \mathcal{D} \to \mathbb{R}$ satisfies the PL inequality if for some $\mu > 0$,*

$$\frac{1}{2}\|\nabla f(x)\|^2 \geq \mu(f(x) - f^*), \ \forall x \in \mathcal{D}. \tag{10}$$

where f^* is the optimal function value.

Theorem 1. *Let f be a convex, L-smooth function on $\mathcal{D} \subseteq \mathbb{R}^n$. Let $\{\mathbf{x}_k\}_{k \geq 0}$ be the sequence generated by the CGD method as described in Eq. 9. Then for a constant step-size $\alpha \in \left(0, \frac{2}{L(1+2\lambda L)^2}\right)$, the following holds true [2, Theorem 4.25]:*

1. *The sequence $\{f_k\}_{k \geq 0}$ is nonincreasing. In addition, for any $k \geq 0$, $f_{k+1} < f_k$ unless $\nabla f_k = \mathbf{0}$.*
2. *$\nabla f_k \to 0$ as $k \to \infty$.*

Furthermore, if f satisfies the PL inequality (Eq. 10) then the CGD method with a step-size of $\frac{1}{L(1+2\lambda L)^2}$, has a global linear convergence rate,

$$f_k - f^* \leq \left(1 - \frac{\mu}{L(1+2\lambda L)^2}\right)^k (f_0 - f^*).$$

Proof. Since f is convex and L-smooth we have,

- $\mathbf{0} \preceq H \preceq LI \Leftrightarrow 0 \leq \mathbf{v}^T H \mathbf{v} \leq L \|\mathbf{v}\|^2 \Leftrightarrow 0 \leq \|H\| \leq L$
- $f(\mathbf{y}) \leq f(\mathbf{x}) + \nabla f(\mathbf{x})^T (\mathbf{y} - \mathbf{x}) + \frac{L}{2} \|\mathbf{y} - \mathbf{x}\|^2$

Substituting $\mathbf{x}_{k+1}$ and $\mathbf{x}_k$ in place of $\mathbf{y}$ and $\mathbf{x}$. And $\mathbf{x}_{k+1} - \mathbf{x}_k = -\alpha \nabla g_k$,

$$f_{k+1} \leq f_k - \alpha \nabla f_k^T \nabla g_k + \frac{L}{2}\alpha^2 \|\nabla g_k\|^2 \tag{11}$$

Bounds on $\|\nabla g_k\|^2$ and $\nabla f_k^T \nabla g_k$ are found as follows:

$$\|\nabla g_k\|^2 = \|\nabla f_k + 2\lambda H_k \nabla f_k\|^2$$
$$= \|\nabla f_k\|^2 + 4\lambda \nabla f_k^T H_k \nabla f_k + 4\lambda^2 \|H_k \nabla f_k\|^2$$
$$\leq \|\nabla f_k\|^2 + 4\lambda \nabla f_k^T H_k \nabla f_k + 4\lambda^2 \|H_k\|^2 \|\nabla f_k\|^2$$
$$\leq \|\nabla f_k\|^2 \left(1 + 4\lambda L + 4\lambda^2 L^2\right) = (1 + 2\lambda L)^2 \|\nabla f_k\|^2$$

and $\nabla f_k^T \nabla g_k = \|\nabla f_k\|^2 + 2\lambda \nabla f_k^T H_k \nabla f_k \geq \|\nabla f_k\|^2$. Substituting in Eq. 11,

$$f_{k+1} - f_k \leq -\alpha \|\nabla f_k\|^2 + \frac{L}{2}\alpha^2(1 + 2\lambda L)^2 \|\nabla f_k\|^2$$

$$= -\alpha \left(1 - \frac{L(1 + 2\lambda L)^2 \alpha}{2}\right) \|\nabla f_k\|^2 \tag{12}$$

Thus, we have, $f_k - f_{k+1} \geq M\|\nabla f_k\|^2 \geq 0$ where $M = \alpha \left(1 - \frac{L(1+2\lambda L)^2\alpha}{2}\right) > 0$ for constant step-size $\alpha \in \left(0, \frac{2}{L(1+2\lambda L)^2}\right)$. So, the equality $f_{k+1} = f_k$ only holds when $\nabla f_k = \mathbf{0}$. Furthermore, since the sequence $\{f_k\}_{k\geq 0}$ is non-increasing and bounded below, it converges. So, $\nabla f_k \to 0$ as $k \to \infty$. This proves the first two parts. For the third part, consider the μ-PL assumption; from Eqs. 10 and 12 at step-size $\alpha = \frac{1}{L(1+2\lambda L)^2}$, we have

$$f_{k+1} - f_k \leq -\frac{1}{2L(1 + 2\lambda L)^2} \|\nabla f_k\|^2 \leq -\frac{\mu}{L(1 + 2\lambda L)^2}(f_k - f^*)$$

On subtracting f^* from both sides,

$$\therefore f_{k+1} - f^* \leq \left(1 - \frac{\mu}{L(1 + 2\lambda L)^2}\right)(f_k - f^*)$$

Applying this inequality recursively gives us the result. $\square$

Note that the PL inequality assumption is not necessary for functions that are strongly convex or strictly convex functions (over compact sets) since the inequality is already satisfied. Karimi et al. [10] further shows that this convergence result can be extended to functions of the form $h(A\mathbf{x})$ where h is a strongly (or strictly) convex function composed with a linear function $e.g.$, least-squares problem and logistic regression.

CGD on Strongly Quadratic Function. Next we investigate the rate of convergence of CGD on a strongly quadratic function as our objective. Such an analysis helps in identifying how penalization affects the *condition number*, which essentially controls the geometry of descent. For a matrix A, the condition number is defined as $\kappa(A) \triangleq \frac{\lambda_{\max}(A)}{\lambda_{\min}(A)}$. A higher condition number or *ill conditioning* implies a more zig-zagging behaviour and thus, slower convergence.

Let $f(\mathbf{x}) = \frac{1}{2}\mathbf{x}^T Q\mathbf{x} - \mathbf{b}^T\mathbf{x}$ where $Q \succ 0$. Let $l \triangleq \lambda_{\min}(Q)$ and $L \triangleq \lambda_{\max}(Q)$. From Eq. 9 we know $\mathbf{x}_{k+1} = \mathbf{x}_k - \alpha B_k \nabla f_k$ where $B_k = I + 2\lambda Q$ and $\nabla f(\mathbf{x}_k) =$

$Q\mathbf{x}_k - \mathbf{b}$. The error at each iterate is then given as:

$$\|\mathbf{e}_{k+1}\| = \|\mathbf{x}_{k+1} - \mathbf{x}^*\| = \|\mathbf{x}_k - \alpha B_k \nabla f_k - \mathbf{x}^*\| \qquad (13)$$
$$= \|\mathbf{x}_k - \alpha B_k (Q\mathbf{x}_k - \mathbf{b}) - \mathbf{x}^*\|$$
$$= \|(I - \alpha B_k Q)\,\mathbf{x}_k + \alpha B_k \mathbf{b} - \mathbf{x}^*\|$$
$$= \|(I - \alpha B_k Q)\,\mathbf{x}_k + \alpha B_k \mathbf{b} - \mathbf{x}^* + (\alpha B_k Q\mathbf{x}^* - \alpha B_k Q\mathbf{x}^*)\|$$
$$= \|(I - \alpha B_k Q)\,(\mathbf{x}_k - \mathbf{x}^*) + \alpha B_k (\mathbf{b} - Q\mathbf{x}^*)\| \qquad \text{(on rearrangement)}$$
$$= \|(I - \alpha B_k Q)\,\mathbf{e}_k\| \qquad \text{(Gradient at } \mathbf{x}^* = \mathbf{0})$$
$$\therefore \|\mathbf{e}_{k+1}\| \le \|I - \alpha B_k Q\|\,\|\mathbf{e}_k\| \qquad (14)$$

Since I and Q commute, we have $(1 + 2\lambda l)I \preccurlyeq B_k \preccurlyeq (1 + 2\lambda L)I$. Also since Q and B_k are both symmetric positive-definite matrices and commute, we can say the following about $B_k Q$,

$$\lambda_{\min}(B_k)\,\lambda_{\min}(Q)I \preccurlyeq B_k Q \preccurlyeq \lambda_{\max}(B_k)\,\lambda_{\max}(Q)I$$
$$\Leftrightarrow \lambda_{\min}(B_k)\,\lambda_{\min}(Q) \le \|B_k Q\| \le \lambda_{\max}(B_k)\,\lambda_{\max}(Q)$$

$$\therefore \|I - \alpha B_k Q\| \le \max\left\{|1 - \alpha\lambda_{\min}(B_k)\lambda_{\min}(Q)|, |1 - \alpha\lambda_{\max}(B_k)\lambda_{\max}(Q)|\right\}$$
$$= \max\left\{|1 - \alpha(1 + 2\lambda l)l|, |1 - \alpha(1 + 2\lambda L)L|\right\}$$

So in Eq. 14,

$$\|\mathbf{x}_{k+1} - \mathbf{x}^*\| \le \max\left\{|1 - \alpha(1 + 2\lambda l)l|, |1 - \alpha(1 + 2\lambda L)L|\right\}\|\mathbf{x}_k - \mathbf{x}^*\|$$
$$\le \left(\max\left\{|1 - \alpha(1 + 2\lambda l)l|, |1 - \alpha(1 + 2\lambda L)L|\right\}\right)^k \|\mathbf{x}_0 - \mathbf{x}^*\|$$

Hence for CGD to converge,

$$\max\left\{|1 - \alpha(1 + 2\lambda l)l|, |1 - \alpha(1 + 2\lambda L)L|\right\} < 1,\ \forall k$$

A suitable α that satisfies this condition is chosen to be

$$\alpha = \frac{2}{\lambda_{\max}(B_k)\lambda_{\max}(Q) + \lambda_{\min}(B_k)\lambda_{\min}(Q)} = \frac{2}{(1 + 2\lambda L)L + (1 + 2\lambda l)l}$$

Both terms within the $\max(\cdot)$ achieve the same value at this α.

$$\therefore \|\mathbf{x}_{k+1} - \mathbf{x}^*\| \le \left(\frac{(1 + 2\lambda L)L - (1 + 2\lambda l)l}{(1 + 2\lambda L)L + (1 + 2\lambda l)l}\right)^k \|\mathbf{x}_0 - \mathbf{x}^*\|$$
$$= \left(\frac{\lambda_{\max}(B_k)\lambda_{\max}(Q) - \lambda_{\min}(B_k)\lambda_{\min}(Q)}{\lambda_{\max}(B_k)\lambda_{\max}(Q) + \lambda_{\min}(B_k)\lambda_{\min}(Q)}\right)^k \|\mathbf{x}_0 - \mathbf{x}^*\|$$
$$= \left(\frac{\kappa - 1}{\kappa + 1}\right)^k \|\mathbf{x}_0 - \mathbf{x}^*\|$$

where $\kappa = \kappa(B_k)\kappa(Q)$.

Remark 1. Note that GD has a similar rate with the condition number there being $\kappa(Q)$ instead. And since $\kappa(B_k) > 0$, $\kappa > \kappa(Q)$. Thus, the condition number worsens while descending over the penalized landscape in comparison to the original objective. This indicates that our method would perform poorly for functions that have a high condition number and thus, take longer to converge.

3.2 CGD-FD: Finite Difference Approximation of the Hessian

Note that CGD is a second-order optimization algorithm as it requires the Hessian information to compute each iterate. We improve over this complexity by restricting CGD to be a first-order line search method wherein the Hessian is approximated using a finite difference [14]. Using Taylor series, we know that

$$\nabla f(\mathbf{x}_k + \Delta \mathbf{x}) = \nabla f_k + H_k \Delta \mathbf{x} + \mathcal{O}(\|\Delta \mathbf{x}\|^2).$$

Now let $\Delta \mathbf{x} = r\mathbf{v}$ where r is arbitrarily small. Then, we can rewrite the above expression as:

$$H\mathbf{v} = \frac{\nabla f(\mathbf{x}_k + r\mathbf{v}) - \nabla f_k}{r} + \mathcal{O}(\|r\|).$$

Substituting $\mathbf{v} = \nabla f_k$ and using this approximation in Eq. 9 gives us

$$\nabla g(\mathbf{x}_k) \approx \nabla f_k + 2\lambda \frac{\nabla f(\mathbf{x}_k + r\nabla f_k) - \nabla f_k}{r}$$
$$= (1 - \nu)\nabla f_k + \nu \nabla f(\mathbf{x}_k + r\nabla f_k) \tag{15}$$

where $\nu = 2\lambda/r$. We call this version *CGD with Finite Differences* (CGD-FD).

Note that this requires two gradient calls in every iteration, which might be prohibitive. Towards this it is natural to consider a stopping criterion to revert back to using steepest direction whenever the improvement through CGD-FD iterates is low. This is done to avoid the extra gradient evaluations after the optimizer has moved to a sufficiently optimal point in the domain. Such criteria are particularly helpful in loss functions where gradient evaluations are costly and therefore, budgeted. This also gives us the freedom to play around with which direction to choose (steepest or CGD-FD) and quantify when to switch to the other. For instance, we initially only move in the directions suggested by CGD-FD and switch to steepest once we make a good-enough drop in the function value from the initial point.

In our experiments, we only use CGD-FD for the first b iterations (out of the total budget T) to strive for a good-drop in function values initially. While doing so, we also ensure that we only move in the direction $\mathbf{p}_k$ (CGD-FD, Eq. 15) if it is a descent direction. Otherwise, we stop using CGD-FD and use the steepest direction henceforth. We summarize our CGD-FD method in Algorithm 2.

3.3 CGD-QN: Quasi-Newton Variants of CGD

We consider quasi-Newton variants of CGD (CGD-QN) where a positive-definite approximation of the Hessian is maintained. The update step for CGD-QN is given as: $\mathbf{x}_{k+1} = \mathbf{x}_k - \alpha \left(I_n + 2\lambda_k \tilde{G}_k \right) \nabla f_k$ where $\tilde{G}_k$ is the Hessian approximation at step k, which is updated after every step. Particularly Eqs. 4 and 5 correspond to the updates of DFP and BFGS variants of CGD respectively. We summarize CGD-QN method in Algorithm 3.

Algorithm 2. Constrained Gradient Descent using Finite Differences (CGD-FD)

Input: Objective function $f : \mathcal{D} \to \mathbb{R}$, initial point $\mathbf{x}_0$, max iterations T, step-size α, regularization coefficients λ, stopping threshold b.

Output: Final point $\mathbf{x}_T$.

  1: $\nu \leftarrow 2\lambda/r$, USE_CGD $\leftarrow$ TRUE
  2: Gradient Evaluations $c \leftarrow 0$ $\triangleright$ *Can be interpreted as cost*
  3: **for** $k = 0, \ldots, T - 1$ **do**
  4:      **if** $c = T$ **then return** $\mathbf{x}_k$ $\triangleright$ *if budget has been exhausted*
  5:      **if** USE_CGD **then**
  6:          $\mathbf{p}_k \leftarrow -(1 - \nu)\nabla f_k - \nu \nabla f(\mathbf{x}_k + r\nabla f_k)$
  7:          $c \leftarrow c + 2$
  8:          **if** $\nabla f_k^T \mathbf{p}_k < 0$ **then** $\triangleright$ *Check if* $\mathbf{p}_k$ *is a Descent Direction*
  9:             $\mathbf{x}_{k+1} \leftarrow \mathbf{x}_k + \alpha \mathbf{p}_k$
10:          **else**
11:             $\mathbf{x}_{k+1} \leftarrow \mathbf{x}_k - \alpha \nabla f_k$
12:             USE_CGD $\leftarrow$ FALSE
13:      **else**
14:          $\mathbf{x}_{k+1} \leftarrow \mathbf{x}_k - \alpha \nabla f_k$
15:          $c \leftarrow c + 1$
16:      **if** $k \geq b$ **then**
17:          USE_CGD $\leftarrow$ FALSE
18: **return** $\mathbf{x}_T$

Algorithm 3. Constrained Gradient Descent: Quasi Newton (CGD-QN)

Input: Objective function $f : \mathcal{D} \to \mathbb{R}$, initial point $\mathbf{x}_0$, max iterations T, step-size α, regularization coefficient λ.

Output: Final point $\mathbf{x}_T$.

  1: $\tilde{G}_0 \leftarrow I_n$
  2: **for** iteration $k = 0, \ldots, T - 1$ **do**
  3:      $\mathbf{p}_k \leftarrow -\left(I_n + 2\lambda_k \tilde{G}_k\right)\nabla f_k$
  4:      **if** $\nabla f_k^T \mathbf{p}_k < 0$ **then** $\triangleright$ *Check if* $\mathbf{p}_k$ *is a Descent Direction*
  5:          $\mathbf{x}_{k+1} \leftarrow \mathbf{x}_k + \alpha \mathbf{p}_k$
  6:      **else**
  7:          $\mathbf{x}_{k+1} \leftarrow \mathbf{x}_k - \alpha \nabla f_k$
  8:      Update $\tilde{G}$ according to the QN method used. For CGD-DFP and CGD-BFGS follow the updates in Equations 4 and 5 respectively.
  9: **return** $\mathbf{x}_T$

---

4   Numerical Experiments

We test CGD-FD and CGD-QN on synthetic test functions from *Virtual Library of Simulation Experiments: Test Functions and Datasets*[1].

For our experiments we chose total budget $T = 40$ and the stopping threshold b as $T/4$. For the hyperparameter λ, we empirically tested different schedules

---

[1] http://www.sfu.ca/~ssurjano.

and strategies for different kinds of functions. We observed that using constant λ works well with convex functions. For some non-convex functions, an increasing schedule is preferred. We denote this as $L(a, b)$ which is an increasing linear schedule of T values going from a to b. For the step-size α, a constant-value was found to be suitable in all our experiments.

To measure the initial drop in function value, we compare the *Improvement* for the first step of CGD-FD vs the first step of steepest descent. Improvement (in %) is given as $\frac{f(\mathbf{x}_0) - f(\mathbf{x}_1)}{f(\mathbf{x}_0)} * 100$.

Table 1. Initial Improvements (in %) over test functions (Dimensions=n) for choices of α and λ with a budget $T = 40$ and stopping threshold $b = T/4$.

Test Function	n	λ	α	GD	CGD-FD
Quadratic function	10	0.4	0.01	18.89	**97.91**
Rotated hyper-ellipsoid function	5	0.5	0.01	15.94	**82.76**
Levy function	2	L(0.01, 0.1)	0.05	23.73	**63.21**
Branin function	2	0.07	0.01	37.53	**87.07**
Griewank function	2	40.0	0.01	0.01	**0.08**
Matyas function	2	10.0	0.01	1.83	**34.40**

Table 1 summarizes our findings and Fig. 3 visualizes function values vs gradient evaluations for the chosen test functions for experiments on CGD-FD. Observe that in the first step of optimizing Quadratic function (Fig. 3a), CGD-FD has a 97.91% decrease in value compared to that of GD which only had an initial improvement of about 18.89%. Similar pattern can also be seen for other functions. This gives us the insight that CGD-FD penalizes the original function well enough to concentrate most of the decrease in value within the first few steps of the trajectory itself. Also, notice that in Griewank function (Fig. 3e) even though the initial step improvement is low (= 0.08%), we still see a much rapid decrease in value within the first 10 steps compared to that of the GD trajectory. Thus, using CGD-FD with appropriate penalization can provide for great boosts in function value well within the starting iterates of the optimization procedure.

Experiments for CGD-QN. We ran CGD-BFGS and CGD-DFP methods against BFGS and DFP for $T = 40$. The function values vs iterations are visualized in Fig. 4. We observe that for most suitable choices of λ, CGD-DFP and CGD-BFGS exhibit similar behaviour and hence, their trajectories coincide. Another thing to observe is that in CGD-QN methods, most improvements aren't concentrated in the initial steps since Hessian approximation is still not very great here. Instead the improvements are more gradual and appear over further steps. For example, in Zakharov function (Fig. 4a), we can observe that

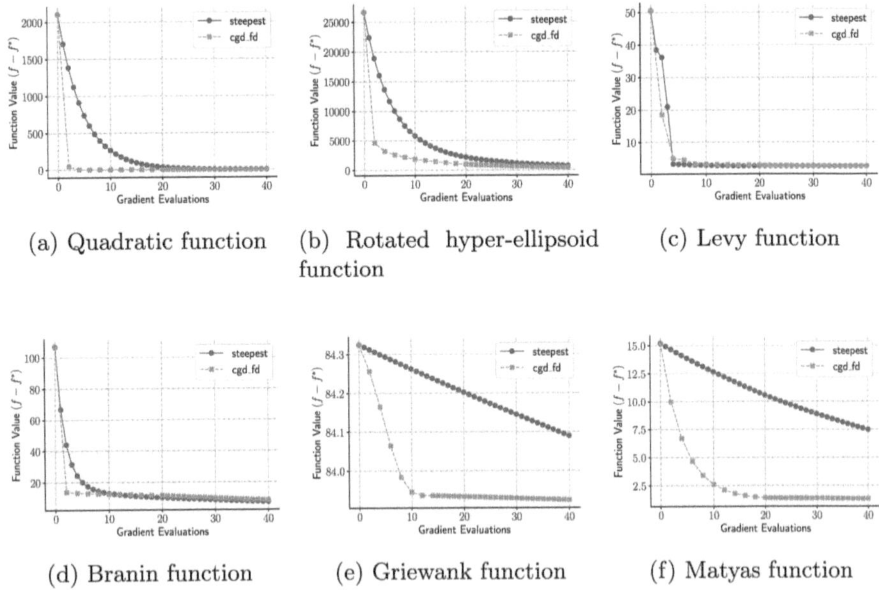

(a) Quadratic function   (b) Rotated hyper-ellipsoid   (c) Levy function
                          function

(d) Branin function   (e) Griewank function   (f) Matyas function

Fig. 3. Experiments for CGD-FD vs GD: Function value $f(\cdot) - f^*$ vs Gradient Evaluations. *Note*: The x-axis of each plot is not iterations but number of gradients evaluated.

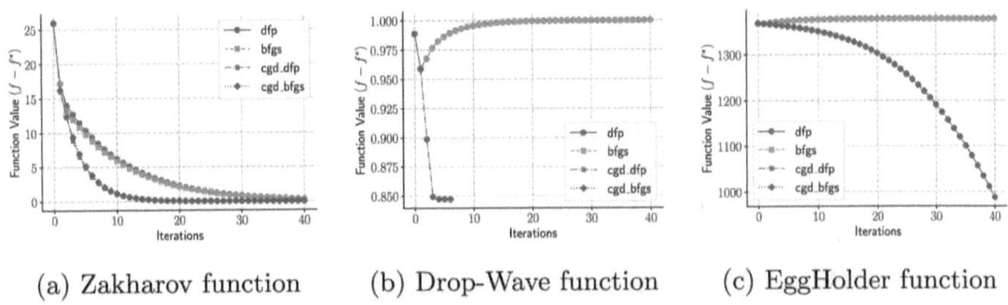

(a) Zakharov function   (b) Drop-Wave function   (c) EggHolder function

Fig. 4. Experiments for CGD-QN vs QN methods: Function value $f(\cdot) - f^*$ vs iterations.

all methods follow the same decrease in function value for the first 1–2 steps and then the CGD-QN methods start to drop more in value. In EggHolder function (Fig. 4c) this effect is most pronounced where the improvements are very gradual but providing with better results than their vanilla counterparts.

5   Reinterpretation of Explicit Gradient Regularization (EGR)

Barrett and Dherin [1] proposed **Explicit Gradient Regularization** (EGR) where the original loss function is regularized with the square of L^2-norm of the gradient. This regularized objective is then optimized with the intention that the model's parameters will converge to a *flat*-minima and thus, be more generalizable. However, an understanding of why this happens is missing.

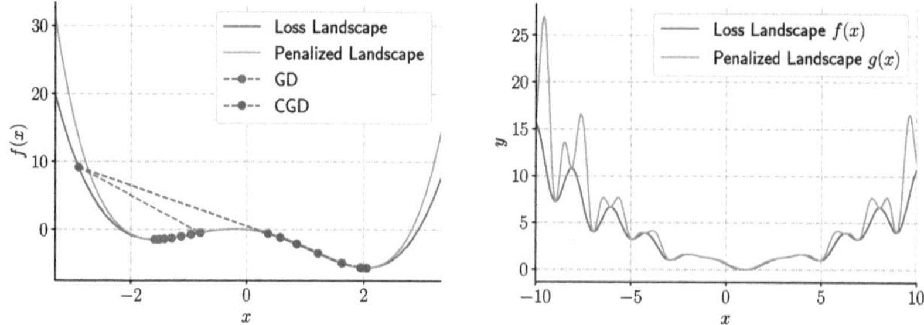

Fig. 5. *(Left)* CGD is able to move to a better local minimum by moving along the negative gradients over the modified loss function. *(Right)* Local maxima of the original loss function turning into local minima of the penalized loss function.

We explain EGR through the example visualized in Fig. 5 *(Left)*. Based on our formulation explained in Sect. 3, we can interpret the gradient-regularized function as a steeper version of the original loss function wherein the minima remain same. Due to the gradient penalty, a steep minimum (in the original loss function) would turn steeper while the increase in steepness wouldn't be this high at a flatter minimum. Thus, at a constant step-size, the optimizer will likely overshoot over the sharper minimum due to the extremely high gradient value, while still being able to converge to the less-steeper and preferred flat minimum point. In Fig. 5 *(Left)* we see how GD gets stuck at a suboptimal point of the function while CGD (GD over the penalized loss function) is able to avoid this point and converges to a better (more optimal) minimum point.

EGR has the drawbacks of having fictitious minima identified in Lemma 1. Since, the loss landscape of neural networks is highly uneven [12], it's likely that artificial stationary points are introduced and the optimizer might converge at local-maxima points since their nature changes with regularization. For example, in Fig. 5 *(Right)* we observe artificial stationary points being introduced between local-minima and local-maxima of the original loss function while some local-maxima also turn into local-minima of the penalized loss function. Therefore, one needs to ensure that there are suitable fixes for these scenarios for better performance. Specifically, the direction along of the penalized loss function, $-\nabla g(\cdot)$ should be a descent direction and that we shouldn't stop at points where $\nabla g(\cdot) = \mathbf{0}$ but $\nabla f(\cdot) \neq \mathbf{0}$ as discussed in Lemma 1.

Another downfall with EGR is that it requires hessian evaluation for each mini-batch while training. This might become a costly operation for bigger models and hence some approximation of the hessian should be used. Another thing we observe from the experimental results for CGD-FD, is that the improvement through optimizing the regularized function is mostly only during the initial steps. Hence, one should only use EGR for some initial steps to reach a good-enough starting point while not exhausting the budget for gradient evaluations.

6   Conclusions and Future Work

In this work, we considered a new line search method that penalizes the norm of the gradient and provides iterates that have lower gradient norm compared to vanilla gradient descent. We identify properties of this algorithm, provide variants that do not require the Hessian and illustrate connections to the widely popular explicit gradient regularization literature.

There are several future directions arising from this work. We would like to investigate the role of different penalty functions $h(\cdot)$ in greater detail. Another future direction is using a combination of negative and positive $\{\lambda_k\}$ values in our iterates. Particularly, negative $\{\lambda_k\}$ values might help in descending faster at points where the Hessian is negative definite. We would also like to explore applications of our method in more diverse settings like reinforcement learning (RL) and Bayesian optimization (BO). In particular, for RL, we can develop penalized variants of existing Policy Gradient Methods [18,20,21], which can be evaluated to assess the effectiveness of gradient regularization in RL settings. Similarly, for BO tasks, we can construct acquisition functions involving a gradient penalty, in a manner similar to [16], which might improve performance. We hope this work sparks further discussion on gradient regularization helping in neural network generalization.

Acknowledgments. We thank IHub-Data, IIIT Hyderabad for the research fellowship supporting this work. This work was also supported by the SERB MATRICS project (Grant No. MTR/2023/000042) from the Science and Engineering Research Board (SERB).

Disclosure of Interests. The authors have no competing interests to declare that are relevant to the content of this article.

References

1. Barrett, D.G.T., Dherin, B.: Implicit gradient regularization (2022). https://arxiv.org/abs/2009.11162
2. Beck, A.: Introduction to nonlinear optimization: theory, algorithms, and applications with MATLAB. Soc. Ind. Appl. Math. (2014). https://doi.org/10.1137/1.9781611973655
3. Broyden, C.G.: Quasi-newton methods and their application to function minimisation. Math. Comput. **21**(99), 368–381 (1967). https://doi.org/10.1090/S0025-5718-1967-0224273-2
4. Chaudhari, P., et al.: Entropy-SGD: biasing gradient descent into wide valleys. J. Stat. Mech: Theory Exp. **2019**(12), 124018 (2019). https://doi.org/10.1088/1742-5468/ab39d9

5. Du, J., Zhou, D., Feng, J., Tan, V., Zhou, J.T.: Sharpness-aware training for free. In: Advances in Neural Information Processing Systems, pp. 23439–23451 (2022). https://proceedings.neurips.cc/paper_files/paper/2022/file/948b1c9d660d7286dd767cd07dabd487-Paper-Conference.pdf
6. Fletcher, R.: A new approach to variable metric algorithms. Comput. J. **13**(3), 317–322 (1970). https://doi.org/10.1093/comjnl/13.3.317
7. Foret, P., Kleiner, A., Mobahi, H., Neyshabur, B.: Sharpness-aware minimization for efficiently improving generalization (2021). https://arxiv.org/abs/2010.01412
8. Hochreiter, S., Schmidhuber, J.: Flat minima. Neural Comput. **9**(1), 1–42 (1997). https://doi.org/10.1162/neco.1997.9.1.1
9. Karakida, R., Takase, T., Hayase, T., Osawa, K.: Understanding gradient regularization in deep learning: efficient finite-difference computation and implicit bias. In: Proceedings of the 40th International Conference on Machine Learning, pp. 15809–15827 (2023). https://proceedings.mlr.press/v202/karakida23a.html
10. Karimi, H., Nutini, J., Schmidt, M.: Linear convergence of gradient and proximal-gradient methods under the polyak-łojasiewicz condition. In: Machine Learning and Knowledge Discovery in Databases, pp. 795–811 (2016). https://doi.org/10.1007/978-3-319-46128-1_50
11. Keskar, N.S., Mudigere, D., Nocedal, J., Smelyanskiy, M., Tang, P.T.P.: On large-batch training for deep learning: generalization gap and sharp minima (2017). https://arxiv.org/abs/1609.04836
12. Li, H., Xu, Z., Taylor, G., Studer, C., Goldstein, T.: Visualizing the loss landscape of neural nets. In: Advances in Neural Information Processing Systems (2018). https://proceedings.neurips.cc/paper_files/paper/2018/file/a41b3bb3e6b050b6c9067c67f663b915-Paper.pdf
13. Nocedal, J., Wright, S.J.: Numerical Optimization, 2nd edn. Springer, New York (2006). https://doi.org/10.1007/978-0-387-40065-5
14. Pearlmutter, B.A.: Fast exact multiplication by the hessian. Neural Comput. **6**(1), 147–160 (1994). https://doi.org/10.1162/neco.1994.6.1.147
15. Polyak, B.: Gradient methods for the minimisation of functionals. USSR Comput. Math. Math. Phys. **3**(4), 864–878 (1963). https://doi.org/10.1016/0041-5553(63)90382-3
16. Prakash, U., Chollera, A., Khatwani, K., K. J., P., Bodas, T.: Practical first-order bayesian optimization algorithms. In: Proceedings of the 7th Joint International Conference on Data Science & Management of Data (11th ACM IKDD CODS and 29th COMAD), pp. 173–181 (2024). https://doi.org/10.1145/3632410.3632418
17. Ruder, S.: An overview of gradient descent optimization algorithms (2017). https://arxiv.org/abs/1609.04747
18. Schulman, J., Levine, S., Abbeel, P., Jordan, M., Moritz, P.: Trust region policy optimization. In: Proceedings of the 32nd International Conference on Machine Learning, pp. 1889–1897 (2015). https://proceedings.mlr.press/v37/schulman15.html
19. Smith, S.L., Dherin, B., Barrett, D.G.T., De, S.: On the origin of implicit regularization in stochastic gradient descent (2021). https://arxiv.org/abs/2101.12176
20. Sutton, R.S., McAllester, D., Singh, S., Mansour, Y.: Policy gradient methods for reinforcement learning with function approximation. In: Advances in Neural Information Processing Systems (1999). https://proceedings.neurips.cc/paper_files/paper/1999/file/464d828b85b0bed98e80ade0a5c43b0f-Paper.pdf
21. Williams, R.J.: Simple statistical gradient-following algorithms for connectionist reinforcement learning. Mach. Learn. **8**(3–4), 229–256 (1992). https://doi.org/10.1007/bf00992696

22. Zhao, Y., Zhang, H., Hu, X.: Penalizing gradient norm for efficiently improving generalization in deep learning. In: Proceedings of the 39th International Conference on Machine Learning, pp. 26982–26992 (2022). https://proceedings.mlr.press/v162/zhao22i.html
23. Zhuang, J., et al.: Surrogate gap minimization improves sharpness-aware training (2022). https://arxiv.org/abs/2203.08065

Data Sampling-Driven Adaptive Modification of Bus Routes Under Time-Varying Road Conditions

Zihao Guo$^{(\boxtimes)}$ (iD), Andrea Araldo (iD), and Mounim A. El Yacoubi (iD)

Samovar, Télécom SudParis, Institut Polytechinique de Paris,
Évry-Courcouronnes, France
{zihao.guo,andrea.araldo,mounim.el_yacoubi}@telecom-sudparis.eu

Abstract. In urban areas, fluctuating road speeds due to traffic congestion and accidents significantly impact bus operations and stop connectivity. Current approaches cannot maintain public transport (PT) network stability during adaptation to changing road conditions, undermining both operations and passenger experience. This paper proposes a data sampling-based adjustment strategy to adapt the time-varying road conditions. The innovation lies in utilising limited network modifications to enhance the existing static PT network instead of considering reconstruction from scratch or minor adjustments (such as stop-skipping), aiming to minimise both passenger travel time degradation and the operational duration of each transit line. Our proposed multi-objective optimization model leverages historical traffic data samples and integrates route variation quantification with penalty mechanisms to enable real-time adaptive routing decisions. The case studies utilising Mandl's network illustrate that our methodology can propose effective strategies for time-varying roads with any coefficient of variation. Experimental findings with high-variance samples indicate that our methodology decreases passenger travel time in roughly 80% of various scenarios compared to conventional static routes, providing a more efficient solution for public transport systems.

Keywords: Adaptive Network Optimization · Time-Varying Road Conditions · Data Sampling-driven

1 Introduction

Fluctuating road speeds in urban areas, caused by traffic congestion, accidents, and varying demand patterns, significantly impact bus operations within public transport (PT) network performance. These variations can disrupt bus schedules and connectivity between stops, presenting growing challenges as urban populations expand [1].

When road conditions alter, current bus routes—including both alignments and stops—must be modified to sustain service. This issue is termed the bus route adjustment problem, and the modified routes are identified as adaptive bus routes. Prior research typically suggests a singular adaptive pathway for each original route, whether through reconstructing PT networks from scratch or implementing minor adjustments (e.g., stop-skipping), that remains constant throughout the planning horizon [5,17]. However, the dynamic evolution of road conditions complicates practical implementation [30]. These approaches fail to minimize network modifications, thereby compromising operational manageability and potentially causing user disorientation.

In response to these limitations, this paper presents a novelty data sampling-based adjustment strategy designed to reduce extra passenger travel time resulting from fluctuating road conditions. Thus, initially, static bus routes can be dynamically adjusted as required, enhancing adaptability to constantly changing traffic conditions. When designing adaptive bus service networks, the main goal is usually to minimise the overall passenger cost, focusing solely on service-level factors while disregarding bus operating expenses. Nevertheless, designing adaptive networks without considering operational costs can be unfeasible or infrequently implemented [13]. Thus, the issue addressed here aims to minimise both overall passenger expenses and bus operational time. Furthermore, to comprehensively account for passenger expenses, our research considers passenger assignment as an endogenous variable by integrating in-vehicle travel costs, transfer costs, and waiting costs.

To the best of our knowledge, this paper is the first study to incorporate the minimisation of route modifications into the objective of adaptive bus network optimisation, thereby adapting the PT network to time-varying road conditions. Specifically, we propose a multi-objective optimisation model employing sampling methodology to reduce passenger travel time, minimise route alterations, and concurrently decrease the duration of public transport operations.

Based on extensive experiments with high-variance samples (coefficient of variation, $cv=1$), compared to conventional static PT routes, our proposed approach demonstrates passenger travel time reductions in approximately 80% of scenarios for each different test sample.

2   Related Work

2.1   Transit Network Design Problem

The Transit Network Design Problem (TNDP) is a critical strategic decision-making challenge within public transit. Traditional research generally delineates objective functions from two viewpoints: a user-centric perspective aimed at minimising the average travel time for passengers (comprising in-vehicle and transfer durations) and an operator-centric perspective concentrated on costs, typically represented by total route length. This fundamental version of the TNDP is often employed as a benchmark for evaluating various solutions. However, it has been established as NP-hard [19].

As research has advanced, the user dimension has been broadened in many ways. In addition to overall travel duration [14,16], recent studies also examine network coverage [4], route efficiency [22], and demand fulfillment [3]. Meanwhile, the operator dimension persists in prioritising the optimisation of overall route length and operational expenses [2,8]. Nonetheless, these traditional methodologies demonstrate limitations when confronted with dynamic urban settings and ever-evolving commuter requirements [30].

To balance diverse objectives, researchers have introduced multi-objective optimization models. Szeto and Jiang [23] established a bilevel transit network design framework, wherein the upper-level model aims to minimise passenger transfers, while the lower-level model focuses on transit assignment. Tian et al. [25] proposed an innovative trilevel programming model that not only considers congestion in public transit routes but also achieves more comprehensive network design through three levels of optimization: the upper-level minimizes both passenger and operating costs, the middle level formulates passenger routing strategies, and the lower-level describes the equilibrium state under congested public lines. Based on these approaches, Cervantes-Sanmiguel et al. [7] investigated the trade-off between cutting travel time and decreasing fares in further detail. Nonetheless, as noted by [24], existing route planning methodologies exhibit significant deficiencies, especially their incapacity to adjust to real-time variations in traffic patterns.

2.2 Bus Route Adjustment Problem

Various approaches can be employed in the bus route adjustment problem to generate adaptive bus routes, including rerouting, short-turning, skip-stopping, and branching [6,18,29]. Previous research on bus route localised adjustments has often aimed at two main objectives: accommodating demand fluctuations and enhancing integration with subway networks.

Guan et al. [11] focused on customizing bus routes for passengers with multiple trip requests. They proposed a bilevel planning model based on a spatiotemporal state network, aiming to optimize customized bus routes while considering trip characteristics, time windows, capacity constraints, and mixed loads. However, their approach creates entirely new routes for each demand pattern. Zhang et al. [31] investigated long-distance bus routes located near subway stations, seeking to adapt them to urban rail networks. To achieve this, the authors developed a bilevel planning model to optimize both bus route adjustments and vehicle headways and then solved it using a tailored genetic algorithm. While effective for subway integration, this method only addresses localized adjustments near stations rather than system-wide adaptation to varying road conditions. Under circumstances where passenger flow control is implemented in a subway system, Zhou et al. [33] adjusted existing bus routes to evacuate passengers stranded at regulated subway stations. Their study integrated passenger flow control with bus route adjustments in an optimization framework, aiming to relieve passenger congestion and reduce travel times. Wang et al. [28] proposed a two-stage method to mitigate disruptions to bus routes. First, they identified affected routes based on newly developed indicators and generated a set of candidate adaptive routes using skip-stop and detour strategies. Subsequently, an optimization model was formulated to determine the combination of these adaptive routes that maximizes the number of passengers served. While practical for local disruptions, their skip-stop and detour strategies are insufficient for handling system-wide performance degradation under highly variable conditions. Zheng et al. [32] proposed a methodology for generating adaptive bus routes that incorporate diversion, short-run, and cancellation strategies. Nonetheless, their localised adjustments depend on prior awareness of disruptions, rendering them less suitable for fluctuating road conditions where decisions must be made regarding the necessity of systematic network alterations to implement adaptive routes. Additionally, route cancellations can severely impact network connectivity.

Existing methods fall into two extremes: complete network redesign, which is computationally expensive and disrupts stability, or minor local adjustments (e.g., skip-stopping, detouring) with limited adaptability. More critically, current research rarely quantifies the operational costs of route modifications, yet frequent changes increase complexity and confuse passengers. Thus, balancing network stability with adaptation to time-varying conditions remains an unresolved challenge.

2.3   Our Contribution

As previously reviewed, to minimize travel time and the length of public transport's operating time, a large number of studies have delved into the transit network design and bus route adjusting problem by integrating route adjustment and passenger assignment optimization. *However*, to the best of our knowledge, to adapt to time-varying traffic conditions, no prior research has done such an integration while *minimizing route modifications*. Specifically, this study proposes a multi-objective optimization model that samples historical traffic data as inputs and incorporates route variation quantification and penalty mechanisms to make real-time decisions on adaptive routing, generating effective strategies that adapt to road fluctuations.

3   Methodology

In this section, we first present the graph model of the public transport (PT) network and describe our problem statement. We then formulate a multi-objective optimization problem to obtain the optimal PT routes that adapt to travel time variations while maintaining network stability.

3.1   Graph Model of PT Network

We model the original static PT network before travel time variation as graph $\mathcal{G}_{\mathrm{PT}}^{\mathrm{orig}} = (\mathcal{V}_{\mathrm{PT}}^{\mathrm{orig}}, \mathcal{E}_{\mathrm{PT}}^{\mathrm{orig}})$, where $\mathcal{V}_{\mathrm{PT}}^{\mathrm{orig}}$ represents the set of bus stops and $\mathcal{E}_{\mathrm{PT}}^{\mathrm{orig}}$ represents the set of links between consecutive bus stops. To clarify, in this paper, the term public transport exclusively denotes buses. $\mathcal{G}_{\mathrm{PT}}^{\mathrm{orig}}$ composed of multiple

PT lines. Each line ℓ has a headway h^ℓ, defined as the distance between two consecutive vehicles expressed in time [27]. The headway can be calculated using the formula $h^\ell = \frac{t^\ell}{N^\ell}$, where t^ℓ represents the round trip time along the entire line, and N^ℓ denotes the number of vehicles operating on that line (Eq. 4.4 of [27]). To avoid introducing nonlinearity into our model, we treat N^ℓ as a parameter rather than a decision variable in this work. A line consists of a sequence of stops connected by edges. For simplicity, we ignore the dwell time at stops. Assuming random passenger arrivals, the average waiting time for passengers' first boarding on line ℓ is $\frac{h^\ell}{2}$ [21]. We define t_{ij}, the time taken to travel between any two successive stops i and j. For interchange at stop i from line ℓ to line ℓ', we consider an average waiting time $\frac{h^{\ell'}}{2}$ at stop i.

3.2   Time-Varying Road Network and Travel Time Sampling

Let $\mathcal{CV}$ be the set of coefficient of variation values. For a substrate network $\mathcal{G}_{\text{sub}} = (\mathcal{V}_{\text{sub}}, \mathcal{E}_{\text{sub}})$, representing the underlying physical road infrastructure upon which transit services operate. For simplicity, we consider $\mathcal{V}_{\text{sub}} = \mathcal{V}_{\text{PT}}^{\text{orig}}$, representing the set of bus stops, and $\mathcal{E}_{\text{sub}} = \mathcal{E}_{\text{PT}}^{\text{orig}}$, representing the road segments connecting these stops. We consider travel times with varying coefficients of variation $cv \in \mathcal{CV}$. For each cv, we generate a set of samples Ω_{cv}, where each sample $\omega \in \Omega_{cv}$ corresponds to a set of travel times $\{t_{i,j}^{cv,\omega} | \forall (i,j) \in \mathcal{E}_{\text{sub}}\}$ across the network. Given the periodic nature of traffic patterns [10], these samples can be viewed as historical observations that represent potential future travel times under the same CV. In this work, we use log-normal distribution to generate these samples, which is described in Sect. 4.2. For future work, we'll use actual historical data as Ω_{cv}.

3.3   Problem Statement

Our research addresses the challenge of optimizing conventional static PT networks under these varying road conditions. We aim to minimize both increased trip durations and total route lengths while maintaining network stability through limited line modifications. Figure 1 illustrates this concept with a time-varying PT network comprising 16 bus stops. Based on the coefficient of variation $cv \in \mathcal{CV}$ for each time t, we sample a $\omega \in \Omega_{cv}$ to derive travel times $\{t_{i,j}^{cv,\omega} | \forall (i,j) \in \mathcal{E}_{sub}\}$ that reflect the prevailing road conditions, which informs our adaptation of bus routes.

Our optimization problem makes the following three decisions: (a) How to dynamically modify PT lines' topology to adapt to actual road conditions when they severely deviate from the expected one? (b) How to determine which routes should be made available for users to travel among OD pairs? and (c) How to optimally locate the terminals of each line?

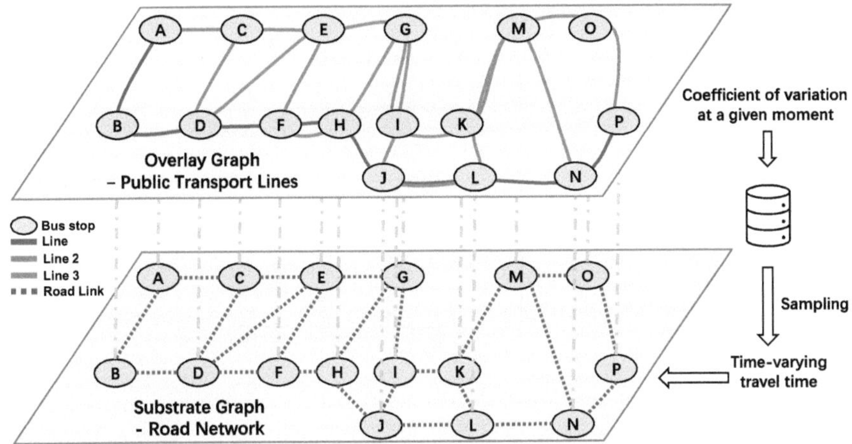

Fig. 1. Illustration of Bus Routes and Corresponding Road Network under Time-Varying Travel Times.

3.4 Multi-Objective Optimization Model

For the origin-destination (OD) pair $k \in \mathcal{K}$, following sampling, we acquire a particular realisation of the travel times for each link $t_{i,j}$ within the road network $\mathcal{G}_{\text{sub}}$. Let t_k^o and t_k denote the travel times (comprising transfer, waiting, and in-vehicle time) attained with the original static PT network $\mathcal{G}_{\text{PT}}^{\text{orig}}$ and the adaptive PT network $\mathcal{G}_{\text{PT}}^{\text{adpt}}$, respectively. Our objective is to optimize the adaptive PT network to reduce both passenger travel time degradation and the operational duration of each transit line via constrained network alterations.

Objective Functions. We formulate our public transport network adaptation problem as a three-objective optimization. The first objective Z_1 minimizes passenger travel time deterioration, where q_k is the demand for OD pair k, t_k and t_k^o are the travel times under adaptive and original networks respectively. Using $\max(t_k - t_k^o, 0)$ ensures only deteriorations are penalized. The second objective Z_2 minimizes operational costs through total vehicle-hours, where t_{ij} is the link travel time and x_{ij}^ℓ indicates whether link (i, j) is used by line ℓ. The third objective Z_3 minimizes network modifications, calculated as the sum of absolute differences $|x_{ij}^\ell - x_{ij}^{\ell o}|$ between adaptive and original network configurations, where $x_{ij}^{\ell o}$ denotes the original link usage. This preserves operational stability and reduces user disorientation. For a detailed explanation of the notation and variables used, please refer to Table 1.

$$
\min_{\mathcal{G}_{\mathrm{PT}}^{\mathrm{adpt}}}
\begin{pmatrix} Z_1 \\ Z_2 \\ Z_3 \end{pmatrix}
=
\begin{pmatrix}
\underbrace{\sum_{k \in \mathcal{K}} q_k \cdot \max(t_k - t_k^o, 0)}_{\text{passenger travel time deterioration}} \\
\underbrace{\sum_{\ell \in \mathcal{L}} \sum_{i,j \in \mathcal{N}} t_{ij} x_{ij}^\ell}_{\text{total operating time}} \\
\underbrace{\sum_{\ell \in \mathcal{L}} \sum_{i,j \in \mathcal{N}} |x_{ij}^\ell - x_{ij}^{\ell\,o}|}_{\text{penalty for route modification}}
\end{pmatrix}
\tag{1}
$$

Subject to:

$$
ft_k^\ell \geq \frac{t^\ell}{2N^\ell} - M\left(1 - \sum_{j \in \mathcal{N}} y_{o_k,j}^{k\ell}\right) \qquad \forall k \in \mathcal{K}, \ell \in \mathcal{L} \tag{2}
$$

$$
ft_k^\ell \leq M \cdot \sum_{j \in \mathcal{N}} y_{o_k,j}^{k\ell} \qquad \forall k \in \mathcal{K}, \ell \in \mathcal{L} \tag{3}
$$

$$
tt_{ki}^\ell \geq \frac{t^\ell}{2N^\ell} - M(1 - tb_{ki}^\ell) \qquad \forall k \in \mathcal{K}, i \in \mathcal{N}, \ell \in \mathcal{L} \tag{4}
$$

$$
tt_{ki}^\ell \leq M \cdot tb_{ki}^\ell \qquad \forall k \in \mathcal{K}, i \in \mathcal{N}, \ell \in \mathcal{L} \tag{5}
$$

Line Constraints

$$
\sum_{i \in \mathcal{N}} a_i^\ell M \geq \sum_{(i,j) \in \mathcal{E}} x_{ij}^\ell, \ \sum_{i \in \mathcal{N}} b_i^\ell M \geq \sum_{(i,j) \in \mathcal{E}} x_{ij}^\ell \qquad \forall \ell \in \mathcal{L} \tag{6}
$$

$$
\sum_{i \in \mathcal{N}} a_i^\ell \leq 1, \ \sum_{i \in \mathcal{N}} b_i^\ell \leq 1 \qquad \forall \ell \in \mathcal{L} \tag{7}
$$

$$
1 \leq u_i^\ell \leq |N| \qquad \forall i \in \mathcal{N}, \ell \in \mathcal{L} \tag{8}
$$

$$
u_i^\ell - u_j^\ell + 1 \leq (|N| - 1)(1 - x_{ij}^\ell) \qquad \forall (i,j) \in \mathcal{E}, \ell \in \mathcal{L} \tag{9}
$$

$$
\sum_{j:(j,i) \in \mathcal{E}} x_{ji}^\ell + a_i^\ell = \sum_{j:(i,j) \in \mathcal{E}} x_{ij}^\ell + b_i^\ell \qquad \forall i \in \mathcal{N}, \ell \in \mathcal{L} \tag{10}
$$

$$
\sum_{l \in \mathcal{L}} x_{ij}^\ell = \sum_{l \in \mathcal{L}} x_{ji}^\ell \qquad \forall (i,j) \in \mathcal{E} \tag{11}
$$

Table 1. Notation for Adaptive PT Network Design Problem

Symbol	Description
Sets:	
$\mathcal{N}$	Set of nodes, indexed by $i, j \in \mathcal{N}$
$\mathcal{L}$	Set of transit lines, indexed by $\ell \in \mathcal{L}$
$\mathcal{K}$	Set of origin-destination pairs, indexed by $k \in \mathcal{K}$
$\mathcal{E}$	Set of feasible edges in the network, indexed by $(i,j) \in \mathcal{E}$
Parameters:	
$q_k, ; k \in \mathcal{K}$	Demands for OD pair k (passengers)
$t_{ij}{}^o, ; (i,j) \in \mathcal{E}$	Travel time on link (i,j) before changes (minutes)
$t_{ij}, ; (i,j) \in \mathcal{E}$	Travel time on link (i,j) (minutes)
$t_k{}^o, ; k \in \mathcal{K}$	Original travel time for OD pair k before changes (minutes)
α, β	Weights for objective functions
$N^\ell, ; \ell \in \mathcal{L}$	Number of vehicles on line ℓ (vehicles per hour)
M	Large constant
$o_k, d_k; k \in \mathcal{K}$	Origin and destination node of OD pair k
$x_{ij}^\ell{}^o, ; i,j \in \mathcal{N}, \ell \in \mathcal{L}$	Original network configuration (1 if link (i,j) is used by line ℓ)
Decision Variables:	
Binary Variables:	
$x_{ij}^\ell, ; i,j \in \mathcal{N}, \ell \in \mathcal{L}$	1 if link (i,j) is used by line ℓ in new network
$y_{ij}^{kl}, ; i,j \in \mathcal{N}, k \in \mathcal{K}, \ell \in \mathcal{L}$	1 if OD pair k uses link (i,j) on line ℓ
$a_i^\ell, b_i^\ell, ; i \in \mathcal{N}, \ell \in \mathcal{L}$	1 if node i is the start/end terminal of line ℓ
$tb_{ki}^\ell, ; k \in \mathcal{K}, i \in \mathcal{N}, \ell \in \mathcal{L}$	1 if OD pair k transfers by boarding line ℓ at node i
$dx_{ij}^{l+}, ; i,j \in \mathcal{N}, \ell \in \mathcal{L}$	1 if link (i,j) is added to line ℓ in new network
$dx_{ij}^{l-}, ; i,j \in \mathcal{N}, \ell \in \mathcal{L}$	1 if link (i,j) is removed from line ℓ in new network
Continuous Variables:	
$u_i^\ell, ; i \in \mathcal{N}, \ell \in \mathcal{L}$	Auxiliary variable for subtour elimination (MTZ)
$s_i^k, ; i \in \mathcal{N}, k \in \mathcal{K}$	Order of node i in the path of OD pair k (0 if not visited)
$tt_{ki}^\ell, ; k \in \mathcal{K}, i \in \mathcal{N}, \ell \in \mathcal{L}$	Transfer waiting time for OD pair k at node i for line ℓ
$ft_k^\ell, ; k \in \mathcal{K}, \ell \in \mathcal{L}$	Initial waiting time for OD pair k on line ℓ
$t^\ell, ; \ell \in \mathcal{L}$	Total travel time of line ℓ
$it_k, ; k \in \mathcal{K}$	Increased travel time for OD pair k

Passenger Flow Constraints

$$y_{ij}^{k\ell} \leq x_{ij}^{\ell} \qquad\qquad \forall (i,j) \in \mathcal{E}, k \in \mathcal{K}, \ell \in \mathcal{L} \qquad (12)$$

$$\sum_{j:(o_k,j)\in\mathcal{E}} \sum_{\ell\in\mathcal{L}} y_{o_k,j}^{k\ell} = 1 \qquad\qquad \forall k \in \mathcal{K} \qquad (13)$$

$$\sum_{j:(j,d_k)\in\mathcal{E}} \sum_{\ell\in\mathcal{L}} y_{j,d_k}^{k\ell} = 1 \qquad\qquad \forall k \in \mathcal{K} \qquad (14)$$

$$\sum_{j:(j,i)\in\mathcal{E}} \sum_{\ell\in\mathcal{L}} y_{ji}^{k\ell} = \sum_{j:(i,j)\in\mathcal{E}} \sum_{\ell\in\mathcal{L}} y_{ij}^{k\ell} \qquad \forall k \in \mathcal{K}, i \in \mathcal{N} \setminus \{o_k, d_k\} \qquad (15)$$

$$s_{o_k}^{k} = 1 \qquad\qquad \forall k \in \mathcal{K} \qquad (16)$$

$$s_j^k \geq s_i^k + 1 - M(1 - \sum_{\ell\in\mathcal{L}} y_{ij}^{k\ell}) \qquad \forall k \in \mathcal{K}, (i,j) \in \mathcal{E} \qquad (17)$$

Transfer Constraints

$$tb_{ki}^{\ell} \leq \sum_{j:(j,i)\in\mathcal{E}} \sum_{\ell'\neq\ell} y_{ji}^{k\ell'} \qquad\qquad \forall k \in \mathcal{K}, i \in \mathcal{N}, \ell \in \mathcal{L} \quad (18)$$

$$tb_{ki}^{\ell} \leq \sum_{j:(i,j)\in\mathcal{E}} y_{ij}^{k\ell} \qquad\qquad \forall k \in \mathcal{K}, i \in \mathcal{N}, \ell \in \mathcal{L} \quad (19)$$

$$tb_{ki}^{\ell} \geq \sum_{j:(j,i)\in\mathcal{E}} \sum_{\ell'\neq\ell} y_{ji}^{k\ell'} + \sum_{j:(i,j)\in\mathcal{E}} y_{ij}^{k\ell} - 1 \qquad \forall k \in \mathcal{K}, i \in \mathcal{N}, \ell \in \mathcal{L} \quad (20)$$

$$tb_{ko_k}^{\ell} = 0, tb_{kd_k}^{\ell} = 0 \qquad\qquad \forall k \in \mathcal{K}, \ell \in \mathcal{L}, i = o_k \text{ or } d_k \quad (21)$$

$$\sum_{j:(i,j)\in\mathcal{E}} x_{ij}^{\ell} \geq tb_{ki}^{\ell} \qquad\qquad \forall k \in \mathcal{K}, i \in \mathcal{N}, \ell \in \mathcal{L} \quad (22)$$

$$\sum_{j:(j,i)\in\mathcal{E}} y_{ji}^{k\ell} = \sum_{j:(i,j)\in\mathcal{E}} y_{ij}^{k\ell} + \sum_{\ell'\neq\ell} tb_{ki}^{\ell'} \qquad \forall k \in \mathcal{K}, i \in \mathcal{N} \setminus \{o_k, d_k\}, \ell \in \mathcal{L} \quad (23)$$

Constraints (2)-(5) define passenger waiting time and transfer time at stops. Line constraints (6)-(11) ensure the feasibility of the transit network structure. Constraints (6)-(7) specify unique start and end points for each line, constraints (8)-(9) eliminate potential subtours, constraint (10) maintains line continuity, and constraint (11) ensures bidirectional connectivity at the network level by maintaining an equal number of transit lines between each node pair in both directions. Passenger flow constraints (12)-(17) govern passenger movement within the network. Constraint (12) restricts passenger flows to established transit lines, constraints (13)-(14) ensure all origin-destination demands are satisfied, and constraints (15)-(17) guarantee the validity of passenger paths by preventing cycles and ensuring logical progression from origin to destination. Transfer constraints (18)-(23) handle passenger transfers within the network. Constraints (18)-(20) define the conditions for transfers to occur, constraint (21) prohibits transfers at origin and destination nodes, constraint (22) ensures the presence of necessary lines at transfer points, and constraint (23) maintains flow balance considering transfers between lines.

3.5   Linearization

The objective function (1) contains nonlinear components, and in order to obtain an exact solution using the solver, we need to linearize it using the following constraints.

$$it_k \geq \sum_{(i,j)\in\mathcal{E}} \sum_{\ell\in\mathcal{L}} y_{ij}^{k\ell} t_{ij} + \sum_{i\in\mathcal{N}} \sum_{\ell\in\mathcal{L}} tt_{ki}^{\ell} + \sum_{\ell\in\mathcal{L}} ft_k^{\ell} - t_k^o \qquad \forall k \in \mathcal{K} \quad (24)$$

$$it_k \geq 0 \qquad \forall k \in \mathcal{K} \quad (25)$$

$$x_{ij}^{\ell} + dx_{ij}^{l-} = x_{ij}^{lo} + dx_{ij}^{l+} \qquad \forall(i,j) \in \mathcal{E}, \ell \in \mathcal{L} \quad (26)$$

$$dx_{ij}^{l-} + dx_{ij}^{l+} \leq 1 \qquad \forall(i,j) \in \mathcal{E}, \ell \in \mathcal{L} \quad (27)$$

Constraints (24)-(25) calculate and control changes in passenger travel time after network optimization. Constraints (26)-(27) manage line changes during network optimization.

After linearization, the optimization problems become as follows:

$$\text{Minimize} \quad Z = \beta \underbrace{\sum_{k\in\mathcal{K}} d_k \cdot it_k}_{\text{passenger inconvenience}} + (1-\beta) \underbrace{\sum_{\ell\in\mathcal{L}} \sum_{i,j\in\mathcal{N}} x_{ij}^{\ell} t_{ij}}_{\text{operating cost}}$$

$$+ \alpha \underbrace{\sum_{\ell\in\mathcal{L}} \sum_{i,j\in\mathcal{N}} (dx_{ij}^{l+} + dx_{ij}^{l-})}_{\text{network modification}} \qquad (28)$$

s.t.

　　constraints (2)–(27)

4   Numerical Results

The performance of the proposed methodology is evaluated through numerical experiments. A case study on Mandl's Swiss network (see Sect. 4.1) is presented, along with a travel time variance sample pool generated using a log-normal distribution (see Sect. 4.2). The corresponding mixed-integer programming model is implemented in Python and solved using the Gurobi 11.0.3 solver on an Intel Core i5 PC operating at 4.6 GHz with 16.0 GB of RAM. A maximum CPU time of 1200 s is set for all computations.

4.1   Mandl's Swiss Network

Mandl's Swiss network (see Fig. 2) is a small network consisting of 15 nodes and 21 links, initially used by Mandl [20]. The network includes a total of 15,570 trips, and the demand is symmetric. This dataset is one of the few publicly accessible resources for the route planning problem and has emerged as the predominant benchmark instance. The predominant route length restriction confines each

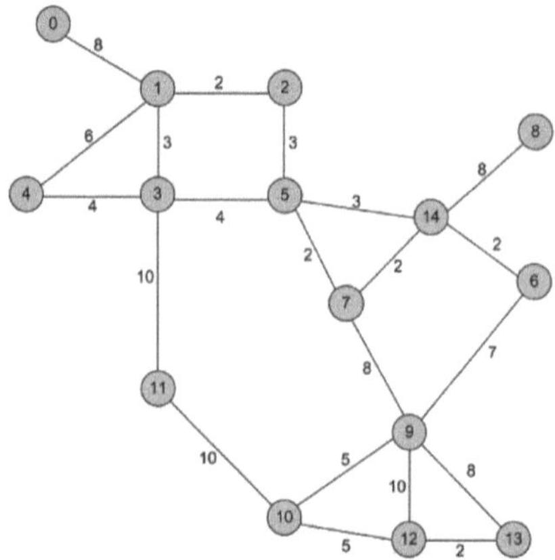

Fig. 2. Mandl's network [20].

route to a maximum of 8 nodes. In this paper, we minimise human bias by permitting an unlimited number of nodes per route, thereby enabling the model to determine the optimal routes autonomously. These "unlimited" routes remain constrained by the condition that each node may be visited no more than once per route.

We adopted the PT network reported by Vermeir et al. [26] that performed best in minimizing total passenger travel time. Under their assumption that passengers always choose the shortest path, we obtained the passenger travel times T_k^{ver} under their PT network. Nonetheless, as our goal encompasses not only the optimisation of passenger travel time but also the operational duration of lines, and considering the inherent complexity of the Transit Network Design Problem with differing assumptions across various studies, direct comparisons prove to be difficult [9]. We used these passenger travel times T_k^{ver} with the penalty parameter $\alpha = 0$ in our objective function (28) to generate our own transport network that optimizes both passenger travel times and operating times, which serves as our original existing static PT network $\mathcal{G}_{\text{PT}}^{\text{orig}}$.

4.2   Travel Time Variability Modeling

In this work, we account for the variability in travel times by modeling them with a log-normal distribution, which has been shown to capture the skewed nature of travel time data effectively [12]. For a log-normal distribution with parameters μ and σ, the mean and variance are given by $\mathrm{E}[X] = \exp\left(\mu + \frac{\sigma^2}{2}\right)$ and $\mathrm{Var}(X) = \left[\exp(\sigma^2) - 1\right] \exp\left(2\mu + \sigma^2\right)$, respectively. Therefore, the coefficient of variation (CV), defined as the ratio of the standard deviation to the mean, is

$$\mathrm{CV} = \frac{\sqrt{\mathrm{Var}(X)}}{\mathrm{E}[X]} = \frac{\sqrt{\left[\exp(\sigma^2) - 1\right]\exp\left(2\mu + \sigma^2\right)}}{\exp\left(\mu + \frac{\sigma^2}{2}\right)} = \sqrt{\exp(\sigma^2) - 1}.$$

Based on this property, for each link (i,j), we can compute the median as $\exp(\mu_{i,j})$ and the coefficient of variation as $cv_{i,j} = \sqrt{\exp(\sigma_{i,j}^2) - 1}$. Considering the Mandl's network (Fig. 2), we assume the nominal travel time on link (i,j) in the original dataset as the median, which we denote as $m_{i,j}$. Since $m_{i,j} = \exp(\mu_{i,j})$, we have $\mu_{i,j} = \ln(m_{i,j})$. Given a desired coefficient of variation $cv \in \mathcal{CV}$, we can derive $\sigma_{i,j} = \sqrt{\ln(cv^2 + 1)}$. The travel time $t_{i,j}^{cv,\omega}$ for each sample $\omega \in \Omega_{cv}$ is then generated as a realization of log-normal$(\mu_{i,j}, \sigma_{i,j}^2)$.

4.3   Performance Comparison Under Different Scenarios

We now show the improvement in network performance, achieved via our method that adapts the PT network to current road conditions. To demonstrate our method's adaptability to travel time variations, for each $cv \in \mathcal{CV}$, we consider a subset of samples $\Omega'_{cv} = \{1, ..., 5\} \subset \Omega_{cv}$. For each sampled travel time realization $\{t_{i,j}^{cv,\omega} | \forall (i,j) \in \mathcal{E}\}$ where $\omega \in \Omega'_{cv}$. Let $z \in \mathcal{Z}$ denote the trips, each corresponding to a certain OD pair $k \in \mathcal{K}$. We denote by $T_z^{o,cv,\omega}$ and $T_z^{*,cv,\omega}$ the travel time of that trip when performed on the original PT network and the optimized one, respectively. We evaluate the network performance improvement by:

$$\frac{T_z^{o,cv,\omega} - T_z^{*,cv,\omega}}{T_z^{o,cv,\omega}} \times 100\% \tag{29}$$

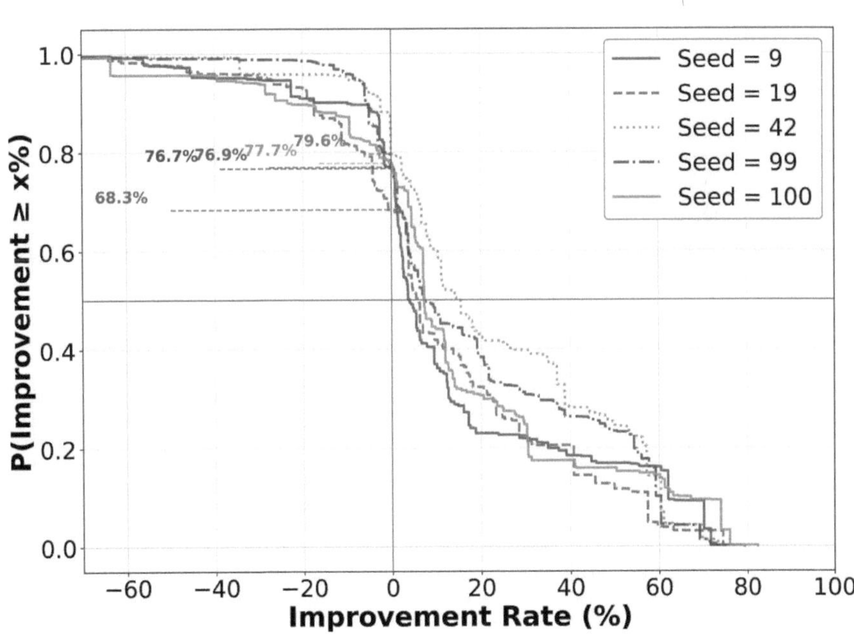

Fig. 3. Inverse Cumulative Distribution of Improvement Rates for Different Seeds (based on $cv = 1$).

Figure 3 shows the inverse cumulative distribution of improvement rates for different seeds with $cv = 1$. For any point (x, y) on the curve, y represents the probability that a randomly selected trip will experience an improvement rate of at least $x\%$. The horizontal dashed lines highlight the probability of achieving a positive improvement for each seed value. The results reveal that our method yields positive enhancements in approximately 80% of trips across different instances. Furthermore, the consistency of the distribution curves across various seeds indicates that our method identifies the optimal value across all sample cases.

To ensure brevity, we perform the following analysis utilising a singular representative sample ($\omega = 1 \in \Omega'_{cv}$) for each $cv \in \mathcal{CV}$. The results presented herein are derived with $\alpha = 10$ and $\beta = 0.5$. In Fig. 4, we represent, for different values of $cv \in \mathcal{CV}$, the distribution of travel times of the original PT network $\{T_z^{o,cv,\omega} | z \in \mathcal{Z}\}$ versus the ones on the optimized adaptive PT network $\{T_z^{*,cv,\omega} | z \in \mathcal{Z}\}$. We observe the improvement of travel times is consistent, and it is particularly strong when variability is high ($cv = 1$); indeed, in this case, adaptivity to actual road conditions is particularly beneficial. We also spec-

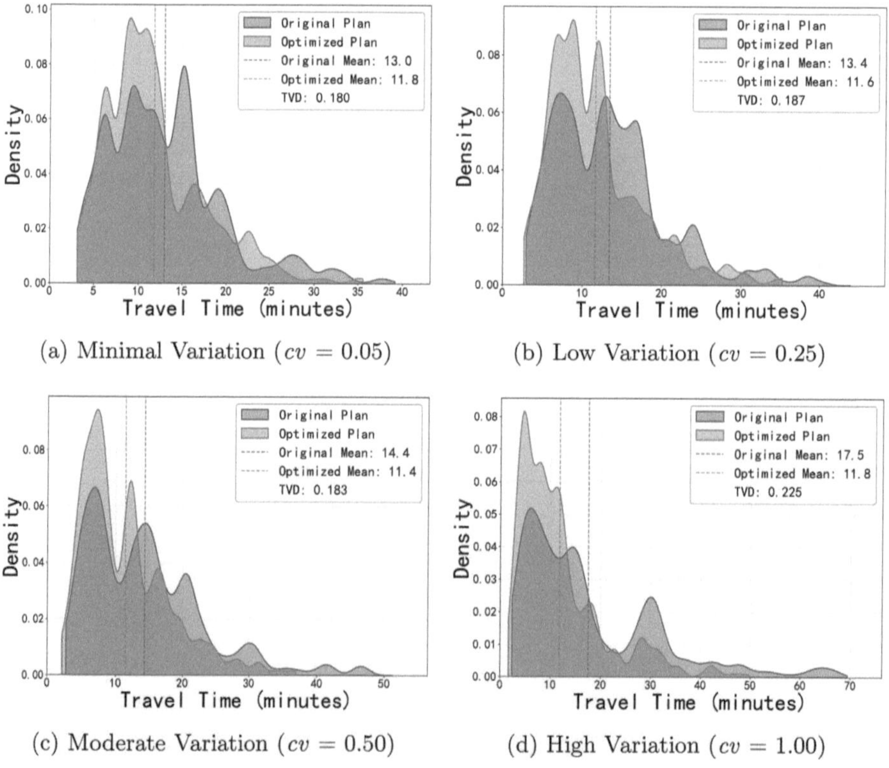

(a) Minimal Variation ($cv = 0.05$)   (b) Low Variation ($cv = 0.25$)

(c) Moderate Variation ($cv = 0.50$)   (d) High Variation ($cv = 1.00$)

Fig. 4. Trip Time Distribution Analysis: Comparison of Original Optimal Static and Adaptive Optimal PT Networks.

Table 2. Comparison of Route Adjustments Under Different cv

Coefficient of variation (cv)	Line	Path Sequence	Overlap (%)	Time differ. (mins)
Original	1	11→10→8→6→3→2→4→5	–	–
	2	2→4→6→15→9	–	–
	3	5→4→12→11→13→14→10→7→15→8→6→3→2→1	–	–
0.05	1	11→10→8→6→3→2→1	66.7	+1.7
	2	10→7→15→8→6→4→5	33.3	+3.9
	3	10→14→13→11→12→4→2→3→6→15→9	66.7	**−10.2**
0.25	1	1→2→3→6→8→15→9	36.4	+4.6
	2	11→13→14→10→8→6→4	20.0	+11.8
	3	5→2→4→12→11→10→7→15→6	64.3	**−24.0**
0.50	1	1→2→3→6→8→15→9	36.4	**−6.4**
	2	11→10→8→6→4→5	22.2	+8.6
	3	5→2→4→12→11→13→14→10→8→15→7	78.6	**−31.1**
1.00	1	9→15→6→3→2→4→12→11→13→14→10	46.2	+18.4
	2	11→10→8→6→3→2→5	20.0	+5.7
	3	1→2→4→6→8→15→7→10→13→14	71.4	**−71.6**

ify in the figure the Total Variation Distance (TVD) between the distribution of $\{T_z^{o,cv,\omega}|z \in \mathcal{Z}\}$ and $\{T_z^{*,cv,\omega}|z \in \mathcal{Z}\}$, which as expected increases with cv.

To quantify the similarity between the old and new routes. We adopt the Jaccard index [15], defined as $R = \frac{|A \cap B|}{|A \cup B|}$, where A and B denote the node sets of the original and new routes, respectively. Table 2 illustrates that our method adeptly reconciles route preservation with performance enhancement: it achieves reductions in operating time while ensuring substantial overlap with the original routes. *It's worth noting that* the most substantial time savings are realised in longer routes, as these routes present greater opportunities for optimisation through strategic adjustments. Line 3, the longest route in the network, exhibits a significant time reduction of 71.6 mins under $cv = 1$, while preserving a 71.4% overlap with its original trajectory.

5   Conclusion

This paper presents a method to modify the existing static bus lines to adapt to time-varying road conditions. Its novelty is that, instead of considering rebuilding from scratch or localized adjustments, we aim to optimize the original public transport (PT) network to reduce both passenger travel time degradation and the operational duration of each transit line via constrained network alterations.

We formulate the adaptive PT design problem as a mixed-integer program. The model samples historical traffic data as input parameters and incorporates route variation quantification and penalisation processes to develop resilient strategies that adjust to variations in road speed. The case studies utilising Mandl's network demonstrate that our methodology can suggest effective strategies for time-varying roads with any coefficient of variation.

Disclosure of Interests. The authors declared no potential conflicts of interest with respect to the research, authorship, and/or publication of this article.

References

1. Agatz, N., Erera, A., Savelsbergh, M., Wang, X.: Optimization for dynamic ride-sharing: a review. Eur. J. Oper. Res. **223**(2), 295–303 (2012)
2. Ahern, Z., Paz, A., Corry, P.: Approximate multi-objective optimization for integrated bus route design and service frequency setting. Transp. Res. Part B Methodol. **155**, 1–25 (2022)
3. Barahimi, A.H., Eydi, A., Aghaie, A.: Multi-modal urban transit network design considering reliability: multi-objective bi-level optimization. Reliab. Eng. Syst. Saf. **216**, 107922 (2021)
4. Cadarso, L., Codina, E., Escudero, L.F., Marín, A.: Rapid transit network design: considering recovery robustness and risk aversion measures. Transp. Res. Procedia **22**, 255–264 (2017)
5. Calabrò, G., Araldo, A., Oh, S., Seshadri, R., Inturri, G., Ben-Akiva, M.: Adaptive transit design: optimizing fixed and demand responsive multi-modal transportation via continuous approximation. Transp. Res. Part A Policy Pract. **171**, 103643 (2023)
6. Cao, Z., Ceder, A.A.: Autonomous shuttle bus service timetabling and vehicle scheduling using skip-stop tactic. Transp. Res. Part C Emerg. Technol. **102**, 370–395 (2019)
7. Cervantes-Sanmiguel, K.I., Chavez-Hernandez, M.V., Ibarra-Rojas, O.J.: Analyzing the trade-off between minimizing travel times and reducing monetary costs for users in the transit network design. Transp. Res. Part B Methodol. **173**, 142–161 (2023)
8. Chen, J., Liu, Z., Wang, S., Chen, X.: Continuum approximation modeling of transit network design considering local route service and short-turn strategy. Transp. Res. Part E Logist. Transp. Rev. **119**, 165–188 (2018)
9. Durán-Micco, J., Vansteenwegen, P.: A survey on the transit network design and frequency setting problem. Public Transp, **14**(1), 155–190 (2022)
10. Fan, J., Weng, W., Chen, Q., Wu, H., Wu, J.: PDG2SEQ: periodic dynamic graph to sequence model for traffic flow prediction. Neural Netw. **183**, 106941 (2025)
11. Guan, Y., Xiang, W., Wang, Y., Yan, X., Zhao, Y.: Bi-level optimization for customized bus routing serving passengers with multiple-trips based on state-space-time network. Phys. A **614**, 128517 (2023)
12. Guessous, Y., Aron, M., Bhouri, N., Cohen, S.: Estimating travel time distribution under different traffic conditions. Transp. Res. Procedia **3**, 339–348 (2014)
13. Guihaire, V., Hao, J.K.: Transit network design and scheduling: a global review. Transp. Res. Part A Policy Pract. **42**(10), 1251–1273 (2008)
14. Heyken Soares, P., Mumford, C.L., Amponsah, K., Mao, Y.: An adaptive scaled network for public transport route optimisation. Public Transp. **11**(2), 379–412 (2019). https://doi.org/10.1007/s12469-019-00208-x

15. Jaccard, P.: Étude comparative de la distribution florale dans une portion des alpes et des jura. Bull. Soc. Vaudoise Sci. Nat. **37**, 547–579 (1901)
16. John, M.P., Mumford, C.L., Lewis, R.: An improved multi-objective algorithm for the urban transit routing problem. In: European Conference on Evolutionary Computation in Combinatorial Optimization, pp. 49–60. Springer (2014)
17. Leffler, D., Burghout, W., Cats, O., Jenelius, E.: An adaptive route choice model for integrated fixed and flexible transit systems. Transportmetrica B Transp. Dyn. **12**(1), 2303047 (2024)
18. Ma, H., Yang, M., Li, X.: Integrated optimization of customized bus routes and timetables with consideration of holding control. Comput. Ind. Eng. **175**, 108886 (2023)
19. Magnanti, T.L., Wong, R.T.: Network design and transportation planning: models and algorithms. Transp. Sci. **18**(1), 1–55 (1984)
20. Mandl, C.E.: Evaluation and optimization of urban public transportation networks. Eur. J. Oper. Res. **5**(6), 396–404 (1980)
21. Pinto, H.K., Hyland, M.F., Mahmassani, H.S., Verbas, I.Ö.: Joint design of multi-modal transit networks and shared autonomous mobility fleets. Transp. Res. Part C Emerg. Technol. **113**, 2–20 (2020)
22. Suman, H.K., Bolia, N.B.: Improvement in direct bus services through route planning. Transp. Policy **81**, 263–274 (2019)
23. Szeto, W.Y., Jiang, Y.: Transit route and frequency design: bi-level modeling and hybrid artificial bee colony algorithm approach. Transp. Res. Part B Methodol. **67**, 235–263 (2014)
24. Tan, W., Peng, X., Huang, J., Wang, Y., Qiu, J., Liu, X.: Optimizing transit network departure frequency considering congestion effects. J. Adv. Transp. **2024**(1), 8131887 (2024)
25. Tian, Q., Wang, D.Z., Lin, Y.H.: Service operation design in a transit network with congested common lines. Transp. Res. Part B Methodol. **144**, 81–102 (2021)
26. Vermeir, E., Engelen, W., Philips, J., Vansteenwegen, P.: An exact solution approach for the bus line planning problem with integrated passenger routing. J. Adv. Transp. **2021**(1), 6684795 (2021)
27. Vuchic, V.R., Clarke, R., Molinero, A.: Timed transfer system planning, design and operation. United States Department of Transportation Urban Mass Transportation Administration (731) (1981)
28. Wang, Y., Chen, J., Yu, X., An, Q., Wu, X.: Optimization of bus rerouting to alleviate the impact of rail transit construction. J. Adv. Transp. **2022**(1), 5773362 (2022)
29. Wu, J., Kulcsár, B., Qu, X., et al.: A modular, adaptive, and autonomous transit system (MAATS): an in-motion transfer strategy and performance evaluation in urban grid transit networks. Transp. Res. Part A Policy Pract. **151**, 81–98 (2021)
30. Xiao, M., Chen, L., Feng, H., Peng, Z., Long, Q.: Smart city public transportation route planning based on multi-objective optimization: a review. In: Archives of Computational Methods in Engineering, pp. 1–25 (2024)
31. Zhang, S.J., Jia, S.P., Bai, Y., Mao, B.H., Ma, C.R., Zhang, T.: Optimal adjustment schemes on the long through-type bus lines considering the urban rail transit network. Discret. Dyn. Nat. Soc. **2018**(1), 8029137 (2018)

32. Zheng, H., Sun, H., Wu, J., Kang, L.: Alternative service network design for bus systems responding to time-varying road disruptions. Transp. Res. Part B Methodol. **188**, 103042 (2024)
33. Zhou, W., Hu, P., Huang, Y., Deng, L.: Integrated optimization of bus route adjustment with passenger flow control for urban rail transit. IEEE Access **9**, 63073–63093 (2021)

Author Index

Y. Zhang et al. (Eds.): LION 2025, LNCS 15745, pp. 301–302, 2026.
https://doi.org/10.1007/978-3-032-09192-5

MIX
Papier aus verantwortungsvollen Quellen
Paper from responsible sources
FSC® C105338

If you have any concerns about our products,
you can contact us on
ProductSafety@springernature.com

In case Publisher is established outside the EU,
the EU authorized representative is:
Springer Nature Customer Service Center GmbH
Europaplatz 3, 69115 Heidelberg, Germany

Printed by Libri Plureos GmbH
in Hamburg, Germany